METHODS IN MOLECULAR BIOLOGY™

Series Editor
John M. Walker
School of Life Sciences
University of Hertfordshire
Hatfield, Hertfordshire, AL10 9AB, UK

For other titles published in this series, go to
www.springer.com/series/7651

Neuroproteomics

Methods and Protocols

Edited by

Andrew K. Ottens* **and Kevin K.W. Wang**[†]

**Department of Anatomy and Neurobiology,*
Virginia Commonwealth University, Richmond, VA, USA
[†]McKnight Brain Institute of the University of Florida, Gainesville, FL, USA Banyan Biomakers, Inc.,
Alachua, FL, USA

Humana Press

Editors
Andrew K. Ottens
Department of Anatomy and Neurobiology
Virginia Commonwealth University
Richmond, VA
USA

Kevin K.W. Wang
McKnight Brain Institute of the University of Florida,
Gainesville, FL, USA
Banyan Biomakers, Inc.,
Alachua, FL, USA

ISBN 978-1-61779-754-5
ISBN 978-1-59745-562-6 (eBook)
ISSN 1064-3745
e-ISSN 1940-6029
DOI 10.1007/978-1-59745-562-6
Springer Dordrecht Heidelberg London New York

Softcover reprint of the hardcover 1st edition 2009

Printed on acid-free paper

Springer is part of Springer Science+Business Media (www.springer.com)

Preface

Neuroproteomics: Methods and Protocols presents experimental details for applying proteomics to the study of the central nervous system (CNS) and its dysfunction through trauma and disease. The target audience includes clinical or basic scientists who look to apply proteomics to the neurosciences. Often researchers hear of proteomics without an adequate explanation of the methodology and inherent limitations. This volume conveys where proteomic methodology is in its application to CNS research and what results can be expected. We also address clinical translation of neuroproteomics, specifically in the area of biomarker research.

The inception of neuroproteomics capitalized on rapid progress in large-molecule mass spectrometry over the last decade. Two seminal advances have spurred research – development of reliable polypeptide ionization processes and bioinformatics to rapidly process tandem mass spectra for peptide identification and quantification. What has followed is the exponential application of mass spectrometry to proteome characterization across biological and biomedical disciplines. Arguably, the most elaborate proteomic implementation is in studying the CNS, the most enigmatic and complex animal system. Neuroscience is characterized by grandiose questions – what is consciousness, how does thought or memory work. Neuroproteomics researchers, however, have primarily involved themselves dysfunction, based on a pressing need (and invariably funding), in answering questions on CNS dysfunction, based on a pressing need (and invariably funding), and because such questions hold more accessible answers. Dysfunction is readily contrasted against normal function and presumably produces a lasting differential protein signature.

Neuroproteomics: Methods and Protocols provides an account of tools used by researchers to address questions in neuroscience. The contributors are some of the first to investigate CNS dysfunction with proteomics, including experts in neurological and analytical sciences. The volume is organized into four methodological parts. **Part I** is focused on CNS animal models used for neuroproteomics research – from neurotrauma caused by ischemic stroke, spinal cord injury or traumatic brain injury to neurodegeneration caused by drug abuse or adult onset disease. Animal models are an essential tool in neuroproteomics research, as human CNS tissue is difficult to acquire and degrades rapidly postmortem. Models that accurately reflect the clinical condition provide a means to assess disease-associated molecular dynamics and evaluate new diagnostics and treatments under controlled conditions with reduced biological variability. **Part II** includes methods for separating the neuroproteome and analyzing select discrete components. While the technology has markedly improved as of late, complete neuroproteome characterization is still well beyond our ability. Rather, researchers must focus on subsets, whether organized through selective sampling of subcellular components, or analysis of specific neuroproteome features, such as protein-binding partners or a target posttranslational modification. **Part III** examines large-scale approaches for CNS proteome characterization and quantification. In particular, this section includes a detailed chapter on bioinformatics, an essential, yet poorly defined, component of any neuroproteomic platform involved in processing complex datasets. Mass spectrometry vendors are developing quantification software to include statistical analysis and integration with annotation and pathway packages. **Part IV** imparts methods that evaluate biofluids and translate neuroproteomic results into clinical platforms. Given limited access to CNS tissue, biofluids have long been targeted as a biomarker source, though with notable limitations. The move from animal to clinical protocols also amplifies issues with biological variability, validating results, and handling regulatory constraints, which must be considered carefully.

The *Methods in Molecular Biology* series aims to provide the reader with "how-to" information from experts in the field. The contributors to *Neuroproteomics: Methods and Protocols* provide minimal background on their science, which is aptly covered elsewhere. Instead, chapters provide an inside account of protocols and tips for performing the actual experiments used by the authors. A few chapters are included to cover perspective accounts on nonexperimental protocols relevant to neuroproteomics research. The included methods are by no means exhaustive; rather, our aim is to cover a full range of topics starting at sample generation, on through sample processing, fractionation, analysis, data interpretation, and translation of results. Simply, the volume presents a distilled account of current neuroproteomic methodology, with the intent to endow the reader with expert insight such that they can critically assess what can be accomplished and how to perform and evaluate neuroproteomic experiments in their own research.

Richmond, VA ***Andrew K. Ottens***

Contents

Part III Neuroproteomic Methodology and Bioinformatics

Part IV Biofluid Analysis and Clinical Translation

Contributors

DAVID R. BORCHELT • *Department of Neuroscience, Santa Fe Health Alzheimer's Disease Research Center, McKnight Brain Institute of the University of Florida, Gainesville, FL, USA*

ANDRÁS BÜKI • *Department of Neurosurgery, Pécs University, Pécs, Hungary*

D. ALLAN BUTTERFIELD • *Department of Chemistry, Sanders-Brown Center on Aging, and Center of Membrane Sciences, University of Kentucky, Lexington, KY, USA*

ANSHU CHEN • *Departments of Physical Medicine and Rehabilitation, Anatomy and Neurobiology and Spinal Cord and Brain Injury Research Center, University of Kentucky Medical Center, Lexington, KT, USA*

JOSHUA J. COON • *Departments of Chemistry and Biomolecular Chemistry, University of Wisconsin, Madison, WI, USA*

DIANE CRIPPS • *The Greenebaum Cancer Center, University of Maryland, Baltimore, MD, USA*

ENDRE CZEITER • *Department of Neurosurgery, Pécs University, Pécs, Hungary*

JITENDRA R. DAVE • *Department of Applied Neurobiology, Division of Psychiatry and Neuroscience, Walter Reed Army Institute of Research, Silver Spring, MD, USA*

DOMINIC M. DESIDERIO • *Charles B. Stout Neuroscience Mass Spectrometry Laboratory, Departments of Neurology and Molecular Sciences and the Cancer Institute, University of Tennessee Health Science Center, Memphis, TN, USA*

MICHAEL FOUNTOULAKIS • *Roche Center for Medical Genomics, F. Hoffmann-La Roche AG, Basel, Switzerland*

BRIAN F. FULLER • *Department of Anatomy & Neurobiology and Biochemistry, Virginia Commonwealth University, Richmond, VA, USA*

MARK S. GOLD • *Departments of Psychiatry, Neuroscience and Community Health and Family Medicine, McKnight Brain Institute of the University of Florida, Gainesville, FL, USA*

DAVID M. GOOD • *Department of Chemistry, University of Wisconsin, Madison, WI, USA*

RONALD L. HAYES • *Center of Innovative Research, Clinical Department, Banyan Biomarkers Inc., Alachua, FL, USA*

SCOTT E. HEMBY • *Departments of Physiology & Pharmacology and Psychiatry, Wake Forest University School of Medicine, Winston-Salem, NC, USA*

JUN HU • *Center for Advanced Proteomics Research and Department of Biochemistry and Molecular Biology, University of Medicine and Dentistry of New Jersey – New Jersey Medical School Cancer Center, Newark, NJ, USA*

HYE JIN HWANG • *Department of Pathology, University of Washington School of Medicine, Seattle, WA, USA*

Firas H. Kobeissy • *Center for Neuroproteomics and Biomarkers Research, Department of Psychiatry, McKnight Brain Institute of the University of Florida, Gainesville, FL, USA*

Erzsebet Kovesdi • *Department of Neurosurgery, Pécs University, Pécs, Hungary*

Kurt Krapfenbauer • *Novartis Institutes for Biomedical Research, Vienna, Austria*

Hong Li • *Center for Advanced Proteomics Research and Department of Biochemistry and Molecular Biology, University of Medicine and Dentistry of New Jersey – New Jersey Medical School Cancer Center, Newark, NJ, USA*

Lujian Liao • *Proteomic Mass Spectrometry Lab, Department of Chemical Physiology, The Scripps Research Institute, La Jolla, CA, USA*

Tong Liu • *Center for Advanced Proteomics Research and Department of Biochemistry and Molecular Biology, University of Medicine and Dentistry of New Jersey – New Jersey Medical School Cancer Center, Newark, NJ, USA*

Bingwen Lu • *Proteomic Mass Spectrometry Lab, Department of Chemical Physiology, The Scripps Research Institute, La Jolla, CA, USA*

X.-C. May Lu • *Department of Applied Neurobiology, Division of Psychiatry and Neuroscience, Walter Reed Army Institute of Research, Silver Spring, MD, USA*

Juan A. Martínez • *Department of Psychiatry, McKnight Brain Institute of the University of Florida, Gainesville, FL, USA*

Daniel B. McClatchy • *Proteomic Mass Spectrometry Lab, Department of Chemical Physiology, The Scripps Research Institute, La Jolla, CA, USA*

Uwe Muller • *Clinical Department, Banyan Biomarkers Inc., Alachua, FL, USA*

Monika W. Oli • *Research and Development Department, Banyan Biomarkers Inc., Alachua, FL, USA*

Andrew K. Ottens • *Departments of Anatomy & Neurobiology and Biochemistry, Virginia Commonwealth University, Richmond, VA, USA*

Jozsef Pal • *Department of Neurosurgery, Pécs University, Pécs, Hungary*

Sung Kyu Park • *Proteomic Mass Spectrometry Lab, Department of Chemical Physiology, The Scripps Research Institute, La Jolla, CA, USA*

Thomas Quinn • *Department of Pathology, University of Washington School of Medicine, Seattle, WA, USA*

Gillian Robinson • *Clinical Department, Banyan Biomarkers Inc., Alachua, FL, USA*

Shankar Sadasivan • *Center for Neuroproteomics and Biomarkers Research, Department of Psychiatry, McKnight Brain Institute of the University of Florida, Gainesville, FL, USA*

Joe E. Springer • *Departments of Physical Medicine and Rehabilitation, Anatomy and Neurobiology and Spinal Cord and Brain Injury Research Center, University of Kentucky Medical Center, Lexington, KT, USA*

Rukhsana Sultana • *Department of Chemistry, Sanders-Brown Center on Aging, University of Kentucky, Lexington, KY, USA*

Nilesh Tannu • *Departments of Physiology & Pharmacology and Psychiatry, Wake Forest University School of Medicine, Winston-Salem, NC, USA*

ANDREW T.N. TEBBENKAMP • *Santa Fe Health Alzheimer's Disease Research Center, Department of Neuroscience, McKnight Brain Institute of the University of Florida, Gainesville, FL, USA*

STEFANI N. THOMAS • *Department of Radiation Oncology and the Greenebaum Cancer Center, University of Maryland, Baltimore, MD, USA*

FLORIAN TRIBL • *Medizinisches Proteom-Center, Ruhr-Universität Bochum, Bochum, Germany*

FRANK C. TORTELLA • *Department of Applied Neurobiology, Division of Psychiatry and Neuroscience, Walter Reed Army Institute of Research, Silver Spring, MD, USA*

KEVIN K.W. WANG • *Center for Neuroproteomics and Biomarkers Research, Department of Psychiatry, McKnight Brain Institute of the University of Florida, Gainesville, FL and Center for Innovative Research, Banyan Biomarkers Inc., Alachua, FL, USA*

ANTHONY J. WILLIAMS • *Department of Applied Neurobiology, Division of Psychiatry and Neuroscience, Walter Reed Army Institute of Research, Silver Spring, MD, USA*

TAO XU • *Proteomic Mass Spectrometry Lab, Department of Chemical Physiology, The Scripps Research Institute, La Jolla, CA, USA*

AUSTIN J. YANG • *Department of Anatomy and Neurobiology and the Greebebaum Cancer Center, University of Maryland, Baltimore, MD, USA*

CHANGPING YAO • *Department of Applied Neurobiology, Division of Psychiatry and Neuroscience, Walter Reed Army Institute of Research, Silver Spring, MD, USA*

JOHN R. YATES, III • *Proteomic Mass Spectrometry Lab, Department of Chemical Physiology, The Scripps Research Institute, La Jolla, CA, USA*

XIANQUAN ZHAN • *Charles B. Stout Neuroscience Mass Spectrometry Laboratory, Department of Neurology, University of Tennessee Health Science Center, Memphis, TN, USA*

JING ZHANG • *Department of Pathology, University of Washington School of Medicine, Seattle, WA, USA*

ZHIQUN ZHANG • *Center for Neuroproteomics and Biomarkers Research, Department of Psychiatry, McKnight Brain Institute of the University of Florida, Gainesville, FL, USA*

Chapter 1

The Methodology of Neuroproteomics

Andrew K. Ottens

Summary

The human central nervous system (CNS) is the most complex organ in nature, composed of ten trillion cells forming complex neural networks using a quadrillion synaptic connections. Proteins, their modifications, and their interactions are integral to CNS function. The emerging field of neuroproteomics provides us with a wide-scope view of posttranslation protein dynamics within the CNS to better our understanding of its function, and more often, its dysfunction consequent to neurodegenerative disorders. This chapter reviews methodology employed in the neurosciences to study the neuroproteome in health and disease. The chapter layout parallels this volume's four parts. Part I focuses on modeling human neuropathology in animals as surrogate, accessible, and controllable platforms in our research. Part II discusses methodology used to focus analysis onto a subneuroproteome. Part III reviews analytical and bioinformatic technologies applied in neuroproteomics. Part IV discusses clinical neuroproteomics, from processing of human biofluids to translation in biomarkers research. Neuroproteomics continues to mature as a discipline, confronting the extreme complexity of the CNS proteome and its dynamics, and providing insight into the molecular mechanisms underlying how our nervous system works and how it is compromised by injury and disease.

Key words: Neuroproteomics, Proteomics, Brain, Neurodegenerative, Disease, Injury, Chromatography, Mass spectrometry

1. Introduction

The term neuroproteomics began use in 2003 *(1)*, and continues to gain acceptance *(2)* as a subdiscipline of proteomics. A product of the -omics revolution, proteomics is the study of the protein profile of an organism, its proteome, and the dynamics of that proteome in time and space. Such research began in the 1970s with the advent of multidimensional protein separations by gel electrophoresis; however, conceptually, the discipline started in

Andrew K. Ottens and Kevin K.W. Wang (eds.), *Neuroproteomics,* Methods in Molecular Biology, vol. 566
DOI 10.1007/978-1-59745-562-6_1, © Humana Press, a part of Springer Science+Business Media, LLC 2009

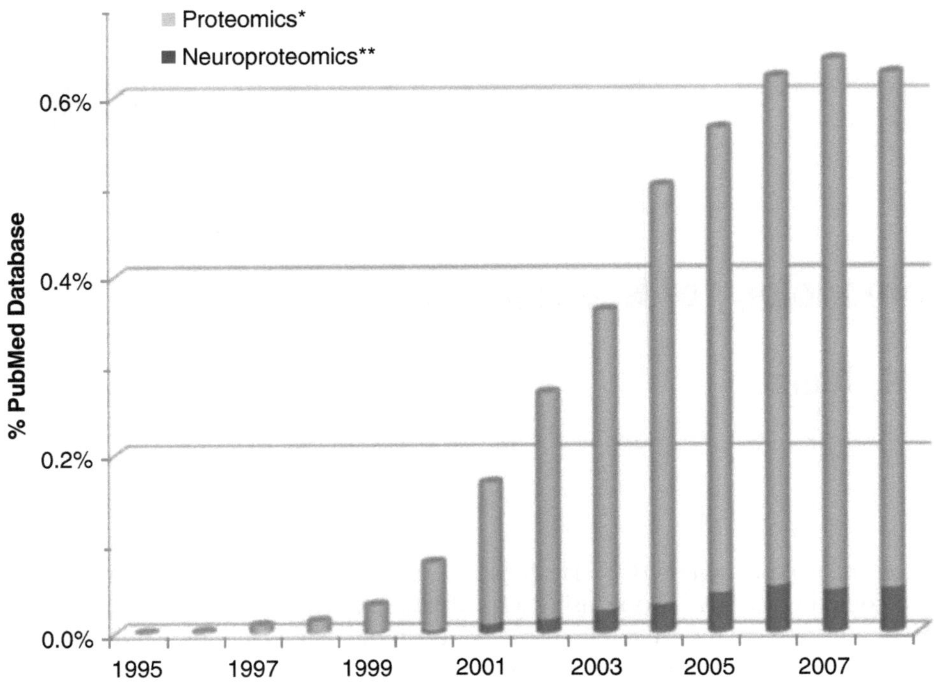

Fig. 1. Growth of neuroproteomics. Number of proteomics publications and those focused on neuroproteomics normalized to the total number of PubMed entries per calendar year. A sharp increase in proteomics research occurred following the development of MS search algorithms in the late 1990s, which has leveled off just above 0.6% in the last few years. Publication of neuroproteomics research has remained modest but growing, up from 4% to 8% of total proteomic-related entries in PubMed over the last 10 years. *Asterisk*, PubMed search for "proteomics or proteome" per annum. *Double asterisk*, Pubmed search for "proteomics or proteome" and "neuronal or CNS or neuroscience or brain or spinal or neuron or nerve or neurodegeneration" per annum.

the mid 1990s *(3)*. The first neuroproteomic studies followed in 1999, contributing a modest 4% of total proteomics research (**Fig. 1**). Since, neuroproteomics has steadily grown, with over 360 publications in 2007. In particular, neuroproteomic methods are now applied to a wide variety of diseases and disorders of the central nervous system (CNS).

Neuroproteomics is challenged by the dramatic complexity and heterogeneity of the CNS. Brain anatomy involves a complex integration of structures and subnuclei with specialized functions. At the cellular level, the human brain consists of 10^{12} neurons supported by ten times as many glia *(4, 5)*. Several thousand different types of neuronal and glial cells have been distinguished histologically within the brain's many nuclei *(6, 7)*. Function, however, is ultimately attained through the complex neural networks interconnecting brain structures with 10^{15} synaptic connections *(4)*. Disruption of synaptic networks, whether through aberrant pruning or neuronal death, is now thought to underlie numerous neurodegenerative conditions, from Alzheimer's disease (AD) to traumatic brain injury (TBI).

CNS complexity continues down at the molecular level (**Fig. 2**). Between 12,000 and 14,000 genes are believed to be expressed throughout life in the human CNS *(8)*. All told, more than half of the human genome is involved in brain development, function, and maintenance *(7)*. Alternate splicing, alternative cleavage, and polyadenylation events are estimated to result in a transcriptome tenfold larger than its corresponding genome *(9)*, with 92–94% of human genes producing two or more protein isoforms *(10)*. Yet, we are just beginning to realize the complexity of the neuroproteome. Posttranslational modifications (PTMs) influence the conformation, cellular location, interaction, and function of proteins *(11)*. Most proteins are estimated to have between 2 and 20 PTMs *(7)*. Several hundred PTM processes are known, with phosphorylation, acetylation, glycosylation, methylation, and ubiquitination being common examples. Indeed, between 30% and 50% of all proteins are proposed to be phosphorylated *(12)*, with an estimated 100,000 phosphorylated residues in the mammalian proteome *(13)*. PTMs such as phosphorylation and acetylation serve as reversible motifs in protein signaling pathways, which are regulated through single site or combinatorial modification patterns *(14)*. PTMs can act as binary (on or off) or analog signaling cues (modulated by the degree of site modification) *(11)*. Glycosylation, for example, is a PTM used to

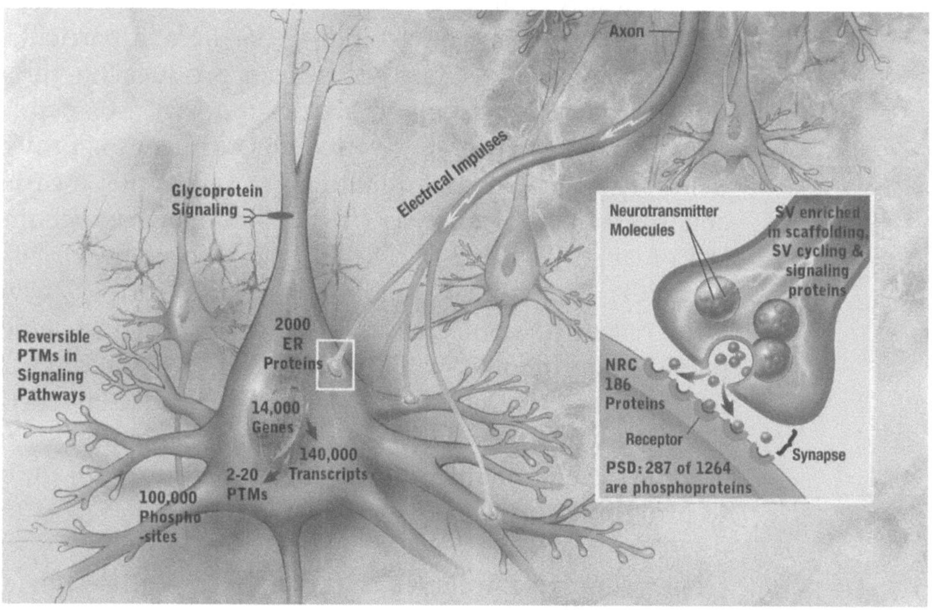

Fig. 2. Cartoon depiction of the complex neuroproteome. An estimated 14,000 genes are expressed in the human CNS, which encode ten times as many transcripts. Proteins are posttranslationally modified between 2 and 20 times. Subcellular structures, such as the endoplasmic reticulum (ER) or synaptosomes [includes synaptic vesicles (SV) and the postsynaptic density (*PSD*)] have their own protein compliment. Background image provided by the Alzheimer's Disease Education and Referral Center, a service of the National Institute on Aging.

regulate cellular interactions via as many as 2,700 known sugar structures (*7*). Often, glycosylation is used on membrane proteins to relay cell status, such as stress and disease (*15*). All told, the mammalian CNS may produce >10^6 protein products, posing an immense challenge for neuroproteomic methodology.

2. Methods

Neuroproteomics aims to characterize the structure, function, interaction, and dynamics of proteins modulated by PTM signaling. To address the complex nature of such research, protein separations and identification methodology have significantly improved over the last 10 years. Indeed, mass spectrometry (MS) played a primary role in the inception of neuroproteomics. The advent of polypeptide ionization techniques in the mid 1990s and peptide identification bioinformatics at the turn of the century has dawned a new era in which sophisticated, hybrid, mass analyzers are now developed for proteomic analysis. Yet, MS remains limited in its resolving capability and sensitivity, and must be complimented by multidimensional molecular separations.

Given the technical limitations, neuroproteomic experiments should be limited in scope to some combination of an anatomical region, a cell type, a subcellular structure, or a specific protein group (e.g., binding partners, a particular PTM). Even analysis of a subproteome can produce on the order of 10^5 MS data measurements, associated with 10^4 peptides from 10^3 proteins. Fortunately, vetted algorithms now provide confident, reproducible, and quantifiable polypeptide measurements, and the statistical analysis of the MS data has become standard practice *(8, 16, 17)*. The Human Proteome Organization's (HUPO) Proteomics Standards Initiative (http://www.psidev.info) recently established a set of data reporting standards for publishing proteomics results *(18)*. The Proteomics Data Collection project (ProDaC, http://www.fp6-prodac.eu) further envisions standardized data generation and submission processes, and the development of a centralized repository for global access to proteomics data *(19)*.

2.1. Modeling Neurodegenerative Disease

Generating a sample set is the first critical design aspect of any neuroproteomics study. The variables involved include source and type of biological material, defining experimental groups, group size, amount of and the means to collect samples, and so on. Fundamentally, a sample set must be honed to discriminately address the hypothesis being tested. While this statement is true of research in general, extra care must be taken in neuroproteomics due to the extreme complexity of the nervous system. In our own neurotrauma research, we carefully assess an appropriate animal model and anatomical region in order to test select secondary injury processes. For example, a fluid percussion brain injury model is relevant to the study of diffuse axonal injury (DAI); however, the study design must focus down on a region where DAI would occur, for example, part of the white matter where the imparted stress of the mechanical force would focus to cause DAI (*see* Chapter 3). At the extreme, laser capture microdissection has been used to narrow neuroproteomic analysis to selected brain cells ***(8, 20, 21)***. Though thousands of such cells were required to generate sufficient material, such resolution will no doubt improve the output of neuroproteomics research.

Selection of a sample source in itself, whether cell culture, animal, or clinical, involves a number of important considerations. Cell cultures may be useful to answer certain fundamental questions; however, neuroproteomic studies often require tissue samples to examine the molecular effects caused by CNS physiological disruption under neurodegenerative conditions. Not surprisingly, human brain and spinal cord tissues are difficult to attain. Postmortem tissue is most available, though it is susceptible to degradation ***(22)***. A common solution is to employ animal models of the human condition of interest. Many neurodegenerative diseases have been simulated in one or more animal models; however, researchers must be mindful of model limitations, and that results must eventually be validated in humans. This is particularly true in neuroscience, since the human brain's architecture is considerably different from other animals. For example, the ratio of white to gray matter is much greater in the human relative to the rodent brain. Yet, animal models remain an important basic research tool for the study of CNS dysfunction.

Cerebral ischemia is a good example to demonstrate the utility of animal models in neuroproteomics (*see* Chapter 2). Ischemia is the main pathophysiological factor in the 80% of strokes caused by blockage of a blood vessel [see review ***(23)***]. The loss of blood flow to part of the brain will rapidly influence cellular processes and cause dyshomeostasis ***(24)***. If blood flow is not restored quickly, ischemia is promptly followed by secondary insult processes, including neuroinflammation and cell death ***(25–27)***. Neuroproteome dynamics correlate with the resulting pathology and outcome ***(28, 29)***. However, removal of brain tissue from stroke patients is not ethical, so it is otherwise impossible to study neuroproteome dynamics following ischemia without a model system. Animal models, such as occlusion of the middle cerebral artery in rodents ***(30)***, provide a direct biomechanical correlate to ischemia in humans. Further, the resulting effect on the neuroproteome is more accurate in the rodent brain than is possible with cell cultures that lack the architecture and vasculature of tissue. Animal models, thus, prove useful for studying injury mechanisms and treatment ***(31–39)***.

Ischemia is also a contributing component to traumatic brain and spinal cord injuries. In both, vascular effects compound already compromised CNS tissue. Ischemia can also occur as a secondary injury process; dying tissue near the lesion site may result in loss of collateral blood flow to adjoining tissue. Indeed, CNS trauma is a complex assemblage of primary and secondary insults; thus, multiple models are necessary to consider different aspects of the injury (*see* Chapters 2 and 3). Models vary by the mode of mechanical force applied to the CNS tissue ***(40–43)***. Forces can be applied as a direct, rapid impact deformation of the tissue, as a slower crush without the transfer of energy to distal areas, as a rapid acceleration injury, or as a penetrating, laceration type injury. The varying modalities mimic the wide variety of injury types and energies found in the clinical condition ***(44, 45)***. Ultimately, the hypothesis being tested must be carefully matched with a model to ensure correct representation in the mechanisms studied with neuroproteomics [see reviews ***(46–48)***].

Substance abuse-induced brain dynamics is another prominent area of neuroproteomics research [see reviews ***(49, 50)***]. Numerous animal models have been developed to simulate substance use and abuse (*see* Chapters 5 and 15). More sophisticated animal models involve self-administration paradigms (*see* Chapter 5), where the drug acts as a positive reinforcement as in the human correlate. Importantly, the brain pathways involved in the reinforcement are similar across mammals. A wide variety of drugs have been studied for their effect on the neuroproteome: alcohol ***(51)***, cocaine ***(52)***, morphine ***(53)***, methamphetamine ***(54, 55, 56)***, amphetamine ***(57)***, and nicotine ***(58, 59)***. The effects of these substances range from modest to extreme changes in brain anatomy ***(60)***. Neuroproteome dynamics in limbic circuitry are of particular interest, with a focus on dopamingeric and serotonergic systems in brain reward centers ***(61)***.

In the above-mentioned models, mechanical or neurotoxic agents are used to simulate a human condition. Adult onset neurodegenerative diseases, however, are harder to simulate in animals, which are not prone to correlative disorders. Animal models are unable to reproduce all clinical features or the spatiotemporal distribution of the human disease; however, models are powerful tools for biochemical and neuroproteomic studies. The majority of neurodegenerative models involve transgenic animals. Transgenic models of AD have been around since the mid 1990s [see review ***(62)***]. Often, mutant forms of amyloid precursor protein (APP) are inserted to produce Aβ variants and plaque pathology relevant to AD *(63–66)*. Other transgenic models look at the pathogenic mutations of tau in relation to neurofibrillary tangle pathology, separately or together with APP mutations *(67–69)*. Tissue from transgenic models (*see* Chapters 6 and 13) is used to identify disease modifiers, molecular pathways, and susceptible genes involved in the disease, and may be useful in developing treatments *(70)*. Already, neuroproteomic studies involving transgenic models of AD have provided insight into molecular aspects of the disease *(71–74)*.

Other neurodegenerative diseases, such as Huntington's (HD) and Parkinson's (PD), are also modeled in animals. Models for HD include transgenic manipulation of CAG repeats in the Huntingtin gene, or the use of neurotoxins, such as quinolinic acid, to destroy corticostriatal projections [see review *(75)*]. Animal models of HD have also been applied in a few neuroproteomic studies to reveal protein expression and modification changes potentially relevant to disease pathology *(76, 77)*. Similarly, a few neuroproteomic studies have looked at animal models of PD [see review *(78)*]. Like with HD, both transgenic (e.g., α-synuclein mutants) and neurotoxin lesion (e.g., 6-hydroxydopamine) PD models exist. Using models, neuroproteomics has revealed novel insights into the relevance of ubiquitin–proteosome dysfunction *(79, 80)* and the presence of cytoskeletal dynamics *(81)* in PD-like pathology.

2.2. Subproteome Analysis in Neuroproteomics

Location, location, location is a chapter theme that bears repeating. As in real estate, sampling location in neuroproteomic studies is of immense importance, as aptly discussed in a recent review by Zabel et al. *(82)*. Neurodegenerative pathology, for example, localizes to specific brain areas delineated by the disorder – AD afflicts the hippocampus and temporal cortex while PD pathology appears focused in structures of the basal ganglia.

2.2.1. Subcellular Neuroproteomics

The pathobiology of a disorder extends down to the cellular and subcellular domains of the CNS tissue under study. We must be reductionists in order to increase the resolution with which we address our research in compliment with the detection capabilities of the methodology. To this end, a focus on subcellular structures has immense potential in neuroproteomics as recently reviewed by Tribl et al. *(83)*. For example, the substantia nigra pars compacta is severely degraded by PD. This brain region is distinct in function and in color due to the presence of neuromelanin granules (NM), which were found to bind α-synuclien and may be involved in the formation of insoluble aggregates *(78, 84, 85)*. NM isolation methodology allows for focused neuroproteomic studies on the function of NM in PD (*see* Chapter 7).

Existing methods for the isolation of other subcellular structures are also being adapted for neuroproteomics research. For example, Stevens et al. recently published an elegant neuroproteomic study of endoplasmic reticulum (ER) microsomes from mouse brain *(86)*. Of the two thousand proteins identified, nearly half were determined to be membrane proteins that are synthesized and transported from the ER to mitochondrial or plasma membranes. Importantly, such membrane proteins are rapidly altered in response to acute CNS treatment. Methods for direct proteomic study of membrane structures have also been reviewed recently *(87, 88)*, and have shown improved separation efficiency. Key to any subcellular isolation method is an evaluation of sample purity. Electron microscopy and immunochemical assays are routinely used to assess the integrity of subcellular samples (*see* Chapter 7).

The synapse is one of the most important membrane structures in the CNS. Circuits involved in perception, learning, memory, and other higher order functions employ chemical synapses - neurotransmitter-receptor signal junctions between neurons (**Fig. 2**). Chemical synapses are impacted by neurodegenerative disease, neurotoxicity and trauma, and are of particular interest in neuroproteomics, evidences by multiple recent reviews on the topic *(4, 83, 89–91)*. Axons terminate in the presynaptic membrane, where protein-rich synaptic vesicles (SV) are released into the synaptic cleft to be picked up on the receptor-rich postsynaptic density (PSD) found on dendritic spines or cell bodies. SV contain a high density of surface and integral membrane proteins for receptor docking, signaling, and vesicle cycling **(Fig. 3)**. The PSD is a three-layered structure organized with cytoskeletal proteins. Actin filaments define the outer two layers, with a tubulin-composed inner layer. Within and between each layer are several hundred other proteins, though about half the protein mass is comprised of 32–35 prominent proteins *(92, 93)*. These proteins form the PSD scaffold that supports an array of neurotransmitter and other receptors. Isolation begins with a discontinuous sucrose gradient where synaptosomes, the SV and PSD together, are resolved from other organelles. Detergents are then used to resolve the insoluble PSD from the SV, though the purity of each fraction is low. Given the specificity of SV and PSD proteins, immunoaffinity methods have been the most effective means for purifying SV *(94–96)* and the PSD *(93, 97)* for downstream neuroproteomic analysis.

2.2.2. Posttranslational Modifications and Interactions in the Neuroproteome

Identifying what proteins are present within the neuroproteome is difficult; yet, a greater challenge is in monitoring post-translational modification of the neuroproteome. PTMs control the function of proteins through affecting their conformation, cellular location, and interactions within the cell (**Fig. 4**). Indeed, some of the more prominent advances in neuroproteomic methodology have been in the isolation of PTMs. For example, methods were reported to study phosphorylation *(98, 99)* and O-linked *N*-acetylglucosamine (O-GlcNAc) *(100)* modified proteins in the PSD. These methods employed affinity chromatography, immobilized metal-ion

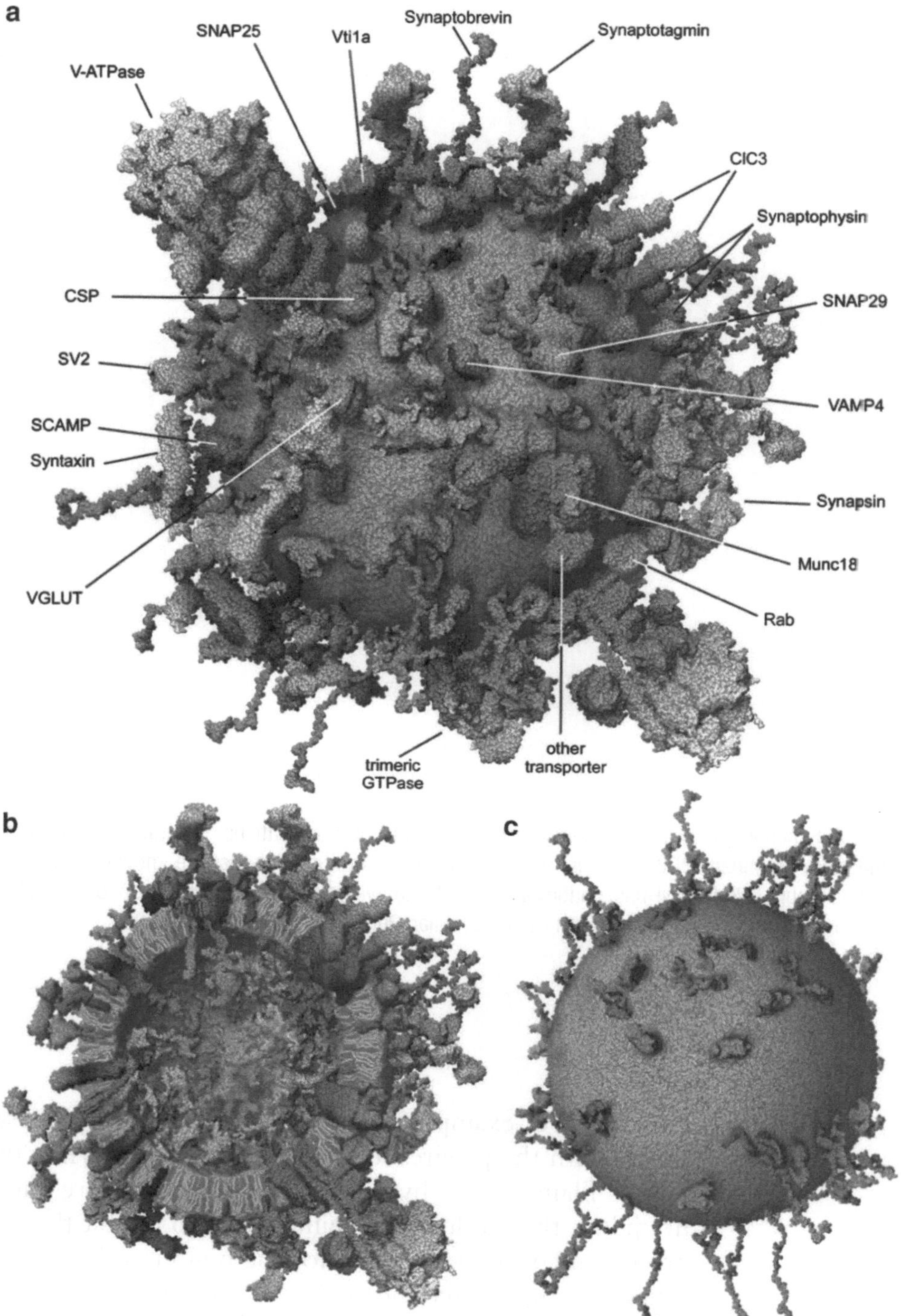

Fig. 3. Model of the synaptic vesicle neuroproteome. Synaptic vesicles exhibit a high density of surface and transmembrane proteins as illustrated by the outer surface (**a**) and transected (**b**) views. Synaptobrevin, the most abundant synaptic vesicle protein is found at a density upward of 70 molecules per vesicle (**c**). Reprinted from (*133*) with permission from Elsevier.

(IMAC), and lectin weak (LWAC), respectively, to isolate 723 unique phosphorylation sites on over a 1,000 proteins and 65 unique O-GlcNAc modification sites from the PSD.

An account of neuroproteome PTM dynamics is critical for understanding the molecular pathogenesis of neurodegenerative

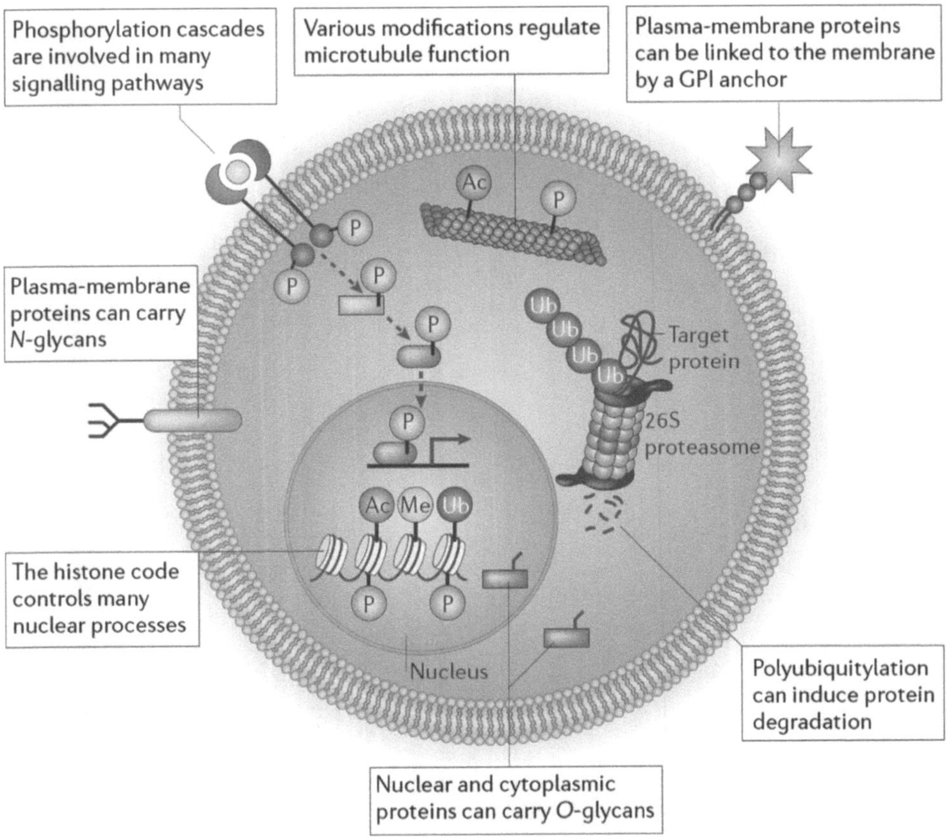

Fig. 4. Functional relevance of common posttranslational modifications. Illustrated are some of the more prominent PTMs and their cellular function from hundreds that are known to exist. *Ac* Acetyl group, *GPI* glycosyl-phosphatidylinositol, *Me* methyl group, *P* phosphoryl group, *Ub* ubiquitin. Reprinted from (*11*) with permission from Macmillan Publishers Ltd, copyright 2006.

disorders. For example, hyperphosphorylation of tau protein correlates with the progression of AD ***(101, 102)***. While the paired helical filaments of hyperphosphorylated tau are targeted by ubiquitin, the modified protein appears to inhibit the ubiquitin–proteasome system, allowing for the buildup of filaments ***(103)***. Relevant neuroproteomic methodology would begin with immunoprecipitation of hyperphosphorylated tau protein, followed by IMAC selection of phosphopeptides of tau and the characterization of the phosphorylated and ubiquitinated residues by advanced MS (*see* Chapter 8). In addition, the neuroproteome of AD brain is distinguished by increased oxidative modification ***(104)***. AD pathobiology includes increased peroxidation of lipids from oxidative stress, which results in the carbonylation and hydroxynonenal addition to cysteine, lysine, and histidine residues. A powerful approach to detect oxidatively modified proteins involves the comparative analysis of the AD neuroproteome separated by two-dimensional gel electrophoresis and two-dimensional immunoblot detection of the target PTM. With the appropriate antibody, proteins that are oxidized are discerned from the immunoblot and can be, subsequently, identified from the two-dimensional gel by MS (*see* Chapter 9).

Similar methodology can be applied for the detection of other oxidative PTMs, such as nitration of tyrosine residues *(105)*. Oxidative stress results in the production of reactive nitrogen species, which are able to oxidatively modify tryrosine, competing with phosphorylation signaling cues. The electron density of the residue is reduced, which alters protein conformation. Both two-dimensional immunoblots (described above) and immunoprecipitation have been used to isolate the nitrotyrosine subproteome for analysis by MS *(106)*. The two approaches proved complimentary for characterization of distinct nitrotyrosine modified proteins in human pituitary tumor tissue, and are applicable to neurodegenerative diseases such as AD and PD, where oxidative stress is prevalent (*see* Chapter 10).

All told, affinity coupling methods largely provide the selectively necessary for PTM analysis *(107)*. Another example is the use of lectin- or heparin-immobilized affinity chromatography to isolate glycosylated proteins (*see* Chapter 11) *(108, 109)*. Alternatively, affinity chromatography is a powerful tool for the determination of protein interaction partners *(110, 111)*, such as for the regulatory protein calmodulin (CaM). Calcium controls CaM binding, in particular with enzymes, which is advantageous for selective binding and release of protein partners to immobilized CaM (*see* Chapter 12) *(112)*. Ultimately, a wide variety of affinity purification methods are applicable tools to isolate a subsection of the neuroproteome.

2.3. Analytical Methodology and Bioinformatics

2.3.1. Protein Separations

Proteomic scale separations began in the 1970s with the advent of two-dimensional gel electrophoresis (2D-GE) *(113, 114)*. 2D-GE remains a powerful and highly utilized method for resolving a proteome (*see* Chapters 9 and 10). Modern inceptions of the method employ sophisticated normalization and detection strategies. For example, difference gel electrophoresis (DIGE) employs multiple fluorescent cyanine dies for within gel quantification and normalization across replicate gels (*see* Chapters 2 and 5) *(115, 116)*. 2D-GE combines separation by isoelectric point and protein size to resolve thousands of protein products. Shin et al. used a multiplexed 2D-GE platform to resolve 17,000 spots from mouse brain *(117)*. Interestingly, only 1,841 proteins were identified; the remaining spots were likely alternate isoforms and PTM products. In their methodology, Shin et al. also employed sample fractionation by liquid chromatography ahead of 2D-GE. Ion exchange (IEC) and hydrophobic interaction (HIC) chromatographies are among the more popular modes used for protein separations. Fractionated samples can subsequently be processed by other modes of protein and peptide separation for enhanced proteome coverage. IEC, in particular, has been popular for upfront processing of protein (*see* Chapter 13) and peptide samples in multidimensional strategies.

A common alternative to 2D-GE is shotgun proteomics, a two-dimensional liquid chromatography platform (2D-LC) for the separation of a digested proteome ***(118, 119)***. The conventional approach uses strong cation IEC to fractionate a complex peptide mixture directly onto reversed-phase liquid chromatography (RPLC) online with a mass spectrometer (*see* Chapters 14 and 17). The tandem configuration provided minimal sample loss and coverage of hundreds to thousands of proteins under sample-limited applications. 2D-LC has also expanded to include dual RPLC separations, differentiated by the pH of the mobile phase for each stage ***(120, 121)***. Shotgun proteomics can resolve roughly the same number of protein components as 2D-GE, but is complimentary in the subset of proteins revealed. For example, 2D-LC is better for analysis of high-mass and hydrophobic proteins than 2D-GE; however, 2D-LC inherently lacks information regarding the intact state of proteins due to the prerequisite enzymatic digestion step.

2.3.2. Mass Spectrometry

The field of proteomics took off with the advent of soft-ionization processes for peptide analysis by MS. Indeed, the importance of biological macromolecule ionization was recognized with the 2002 Nobel Prize in chemistry, awarded in part to John B. Fenn, for electrospray ionization (ESI), and Koichi Tanaka, for matrix-assisted laser desorption/ionization (MALDI). ESI transfers ionized biomolecules, such as polypeptides, from a liquid into the gas phase, which allows for online coupling with liquid chromatography (LC-MS). MALDI involves desorption of polypeptides from within a crystallized matrix into the gas phase. ESI-MS (*see* Chapters 8, 14, 15, 17, and 18) and MALDI-MS (*see* Chapter 10) provide complimentary peptide analysis platforms, the former integrated often with 2D-LC and the later with 2D-GE. However, both platforms remain limited in their ability to reproducibly analyze low-copy proteins mixed among the wide range of more abundant proteins found in biological samples. Protein concentration in blood products, for example, spans ten orders in magnitude, where the top few proteins comprise 90% of the protein mass; thus, depletion methods are now routinely used to assay lesser abundant proteins ***(122)***.

Neuroproteomics often involves comparative quantification between sample groups. A range of methods have been developed for quantifying 2D-GE spots or peptides identified by LC-MS [see review ***(123)***]. DIGE was mentioned earlier as a method for 2D-GE quantitative proteome analysis. Software, such as DeCyder from GE Healthcare, is employed to integrate spot density for each fluorophore and compare between gels to determine the mean quantity and variance of a spot across experimental groups. Labeling methodologies are also available for quantification by LC-MS. Some employ binary isotopic tagging strategies, such as isotope-coded affinity tag (ICAT) ***(124, 125)***, stable isotope-labeled amino acids in culture (SILAC) ***(126)***, or enzymatic labeling with ^{18}O ***(127)***. Alternative labeling technologies involve isobaric tagging of peptide samples that are distinguished in tandem mass spectra by a set of reporter ions with differing mass-to-charge values ***(128, 129)***. An advantage of isobaric over isotopic strategies is that more than two samples can be labeled and combined together in one LC-MS run to save time and reduce experimental variance. Commercial products are available from Applied Biosciences (iTRAQ; *see* Chapter 14) and ThermoFisher Scientific (TMT). A third option is label-free LC-MS quantification, which avoids known limitations associated with chemical labeling of peptides, for example the notable cost. Label-free methods, however, require separate analysis of each sample, involving greater instrument time and a potential decrease in precision. One implementation of label-free quantification involves multiple reaction monitoring (MRM), also known as selected reaction monitoring (SRM), where two or three peptide fragment ions are selectively quantified per peptide for a list of target proteins (*see* Chapter 8). Software is now available to automatically build MRM lists quantifying hundreds of target proteins per sample. Such technology will provide reliable assessment with repeated measures of a target subproteome, providing the statistical power and precision necessary for neuroproteomics research. Advanced quantitative methods will also herald a paradigm shift in neuroproteomics. While twofold or greater expression changes have appeal as diagnostic markers, most regulatory shifts in the neuroproteome are modest ***(11, 82)***. Quantitative differences will no longer be priorities by the magnitude of change, but by the biological relevance. Advanced bioinformatics is essential to allow greater characterization within and between sample groups, with the appropriate replication and statistical analysis. The reader is directed to Chapter 16 for an insightful review on available bioinformatics tools to process complex proteomic datasets.

2.4. Clinical Research and Translation

Clinical translation of neuroproteomics research has focused on the development of disease markers, whether with a clear biochemical association (biomarkers) or otherwise (surrogate markers). Indeed, the term clinical proteomics has been defined as protein biomarkers research ***(130)*** to include the discovery and validation of markers in preclinical and clinical paradigms; however, such research also extends into the development and evaluation of therapeutics. Markers provide a means to measure the effect of a therapy, its correlation with the target mechanism, and its safety *(8)*. Yet, despite the promise of neuroproteomics-derived markers, few examples exist. One clinical success is the measure of 14-3-3 protein in cerebrospinal fluid (CSF) found to be altered in patients with Creuzfeldt–Jakob disease *(131)*. Yet, the reality is such that no one marker is likely to be both sensitive and specific enough to diagnose or monitor a neurodegenerative disorder. Fortunately, neuroproteomics is versatile in identifying multiple proteins that are altered between conditions. A marker array could provide the combinatorial power to address the diagnostic needs.

Another confound in biomarker research is the poor transmission of protein components into biofluids such as CSF and blood products. For example, brain Aβ levels correlate with AD progression; however, Aβ levels in plasma are not reflective of brain levels *(130)*. Consequently, it is difficult to begin assessing marker levels in tissue with the expectation of detecting those markers in clinically accessible biofluids. Alternatively, neuroproteomics researchers may look in clinical biofluids such as CSF (*see* Chapter 17), blood or urine (*see* Chapter 18) for marker discovery. Human biofluids, however, can be difficult to acquire in sufficient numbers, particularly in the case of CSF. Further, the degree of biological variability is considerable within clinical sample sets (variations in patient genetics, history, diet, etc.), and the ability to collect and store samples in a routine manner is a challenge. For example, the CSF proteome varies depending on time of day for collection. Standard operating procedures are essential for every project and may eventually be established across the field *(132)*. Such considerations must be addressed ahead of the marker discovery process to enable effective translation into clinical validation platforms (*see* Chapters 19 and 20).

3. Concluding Remarks

This volume provides in-depth coverage of established methodology and protocols for animal and clinical neuroproteomics research. It details advanced protocols for PTM analysis and quantitative assays allowing the assessment of moderate, yet consistent changes in protein profiles. The field of neuroproteomics will continue to mature, with yearly advances in methodology and experimental protocols. Efforts across the world, such as the Human Proteome Organization (http://www.hupo.org), the BrainNet Europe (http://www.brainnet-europe.org) or the Brain Proteome Project (http://www.hbpp.org) continue to address challenges in neuroproteomics and establish guidelines for its successful use in addressing CNS questions on function and disease.

References

1. Kim, S. I., Voshol, H., van Oostrum, J., Hastings, T. G., Cascio, M., and Glucksman, M. J. (2004) Neuroproteomics: Expression profiling of the brain's proteomes in health and disease. *Neurochem. Res.* **29**, 1317–1331.
2. Butcher, J. (2007) Neuroproteomics comes of age. *Lancet Neurol.* **6**, 850–851.
3. Wasinger, V. C., Cordwell, S. J., Cerpa-Poljak, A., Yan, J. X., Gooley, A. A., Wilkins, M. R., Duncan, M. W., Harris, R., Williams, K. L., and Humphery-Smith, I. (1995) Progress with gene-product mapping of the mollicutes: *Mycoplasma genitalium*. *Electrophoresis* **16**, 1090–1094.
4. Pocklington, A. J., Armstrong, J. D., and Grant, S. G. (2006) Organization of brain complexity – synapse proteome form and function. *Brief Funct. Genomic Proteomic.* **5**, 66–73.
5. Temburni, M. K., and Jacob, M. H. (2001) New functions for glia in the brain. *Proc. Natl Acad. Sci. USA* **98**, 3631–3632.
6. Masland, R. H. (2004) Neuronal cell types. *Curr. Biol.* **14**, R497–R500.
7. Becker, M., Schindler, J., and Nothwang, H. G. (2006) Neuroproteomics – the tasks lying ahead. *Electrophoresis* **27**, 2819–2829.
8. Ekegren, T., Hanrieder, J., and Bergquist, J. (2008) Clinical perspectives of high-resolution mass spectrometry-based proteomics in neuroscience: Exemplified in amyotrophic lateral sclerosis biomarker discovery research. *J. Mass Spectrom.* **43**, 559–571.
9. Gustincich, S., Sandelin, A., Plessy, C., Katayama, S., Simone, R., Lazarevic, D., Hayashizaki, Y., and Carninci, P. (2006) The complexity of the mammalian transcriptome. *J. Physiol.* **575**, 321–332.
10. Wang, E. T., Sandberg, R., Luo, S., Khrebtukova, I., Zhang, L., Mayr, C., Kingsmore, S. F., Schroth, G. P., and Burge, C. B. (2008) Alternative isoform regulation in human tissue transcriptomes. *Nature* **456**, 470–476.
11. Jensen, O. N. (2006) Interpreting the protein language using proteomics. *Nat. Rev. Mol. Cell Biol.* **7**, 391–403.
12. Kalume, D. E., Molina, H., and Pandey, A. (2003) Tackling the phosphoproteome: Tools and strategies. *Curr. Opin. Chem. Biol.* **7**, 64–69.
13. Paradela, A., and Albar, J. P. (2008) Advances in the analysis of protein phosphorylation. *J. Proteome Res.* **7**, 1809–1818.
14. Yang, X. J. (2005) Multisite Protein Modification and Intramolecular Signaling. *Oncogene* **24**, 1653–1662.
15. Dube, D. H., and Bertozzi, C. R. (2005) Glycans in cancer and inflammation – potential for therapeutics and diagnostics. *Nat. Rev. Drug Discov.* **4**, 477–488.
16. Carpentier, S. C., Panis, B., Swennen, R., and Lammertyn, J. (2008) Finding the significant markers: Statistical analysis of proteomic data. *Methods Mol. Biol.* **428**, 327–347.
17. Isserlin, R., and Emili, A. (2008) Interpretation of large-scale quantitative shotgun proteomic profiles for biomarker discovery. *Curr. Opin. Mol. Ther.* **10**, 231–242.
18. Taylor, C. F., Paton, N. W., Lilley, K. S., Binz, P. A., Julian, R. K. Jr., Jones, A. R., Zhu, W., Apweiler, R., Aebersold, R., Deutsch, E. W., Dunn, M. J., Heck, A. J., Leitner, A., Macht, M., Mann, M., Martens, L., Neubert, T. A., Patterson, S. D., Ping, P., Seymour, S. L., Souda, P., Tsugita, A., Vandekerckhove, J., Vondriska, T. M., Whitelegge, J. P., Wilkins, M. R., Xenarios, I., Yates, J. R. 3rd, and Hermjakob, H. (2007) The minimum information about a proteomics experiment (MIAPE). *Nat. Biotechnol.* **25**, 887–893.
19. Eisenacher, M., Hardt, T., Martens, L., Hakkinen, J., Apweiler, R., Hamacher, M., Meyer, H. E., and Stephan, C. (2008) Proteomics data collection – 3rd ProDaC Workshop April 22nd 2008, Toledo, Spain. *Proteomics* **8**, 4163–4167.
20. Lu, Q., Murugesan, N., Macdonald, J. A., Wu, S. L., Pachter, J. S., and Hancock, W. S. (2008) Analysis of mouse brain microvascular endothelium using immuno-laser capture microdissection coupled to a hybrid linear ion trap with fourier transform-mass spectrometry proteomics platform. *Electrophoresis* **29**, 2689–2695.
21. Mustafa, D., Kros, J. M., and Luider, T. (2008) Combining laser capture microdissection and proteomics techniques. *Methods Mol. Biol.* **428**, 159–178.
22. Ferrer, I., Santpere, G., Arzberger, T., Bell, J., Blanco, R., Boluda, S., Budka, H., Carmona, M.,

Giaccone, G., Krebs, B., Limido, L., Parchi, P., Puig, B., Strammiello, R., Strobel, T., and Kretzschmar, H. (2007) Brain protein preservation largely depends on the postmortem storage temperature: Implications for study of proteins in human neurologic diseases and management of brain banks: A brainnet europe study. *J. Neuropathol. Exp. Neurol.* **66**, 35–46.

23. Van Elzen, R., Moens, L., and Dewilde, S. (2008) Expression profiling of the cerebral ischemic and hypoxic response. *Expert Rev. Proteomics* **5**, 263–282.
24. Lipton, P. (1999) Ischemic cell death in brain neurons. *Physiol. Rev.* **79**, 1431–1568.
25. Stanimirovic, D., and Satoh, K. (2000) Inflammatory mediators of cerebral endothelium: A role in ischemic brain inflammation. *Brain Pathol.* **10**, 113–126.
26. Won, S. J., Kim, D. Y., and Gwag, B. J. (2002) Cellular and molecular pathways of ischemic neuronal death. *J. Biochem. Mol. Biol.* **35**, 67–86.
27. Danton, G. H., and Dietrich, W. D. (2003) Inflammatory mechanisms after ischemia and stroke. *J. Neuropathol. Exp. Neurol.* **62**, 127–136.
28. Aurell, A., Rosengren, L. E., Karlsson, B., Olsson, J. E., Zbornikova, V., and Haglid, K. G. (1991) Determination of S-100 and glial fibrillary acidic protein concentrations in cerebrospinal fluid after brain infarction. *Stroke* **22**, 1254–1258.
29. Sotgiu, S., Zanda, B., Marchetti, B., Fois, M. L., Arru, G., Pes, G. M., Salaris, F. S., Arru, A., Pirisi, A., and Rosati, G. (2006) Inflammatory biomarkers in blood of patients with acute brain ischemia. *Eur. J. Neurol.* **13**, 505–513.
30. Longa, E. Z., Weinstein, P. R., Carlson, S., and Cummins, R. (1989) Reversible middle cerebral artery occlusion without craniectomy in rats. *Stroke* **20**, 84–91.
31. Bao, W. L., Williams, A. J., Faden, A. I., and Tortella, F. C. (2001) Selective mGluR5 receptor antagonist or agonist provides neuroprotection in a rat model of focal cerebral ischemia. *Brain Res.* **922**, 173–179.
32. Berti, R., Williams, A. J., Velarde, L. C., Moffett, J. R., Elliott, P. J., Adams, J., Yao, C., Dave, J. R., and Tortella, F. C. (2003) Effect of the proteasome inhibitor MLN519 on the expression of inflammatory molecules following middle cerebral artery occlusion and reperfusion in the rat. *Neurotox. Res.* **5**, 505–514.
33. Phillips, J. B., Williams, A. J., Adams, J., Elliott, P. J., and Tortella, F. C. (2000) Proteasome inhibitor PS519 reduces infarction and attenuates leukocyte infiltration in a rat model of focal cerebral ischemia. *Stroke* **31**, 1686–1693.
34. Williams, A. J., Dave, J. R., Phillips, J. B., Lin, Y., McCabe, R. T., and Tortella, F. C. (2000) Neuroprotective efficacy and therapeutic window of the high-affinity *N*-methyl-d-aspartate antagonist conantokin-G: In vitro (primary cerebellar neurons) and in vivo (rat model of transient focal brain ischemia) studies. *J. Pharmacol. Exp. Ther.* **294**, 378–386.
35. Williams, A. J., Hale, S. L., Moffett, J. R., Dave, J. R., Elliott, P. J., Adams, J., and Tortella, F. C. (2003) Delayed treatment with MLN519 reduces infarction and associated neurologic deficit caused by focal ischemic brain injury in rats via antiinflammatory mechanisms involving nuclear factor-kappaB activation, gliosis, and leukocyte infiltration. *J. Cereb. Blood Flow Metab.* **23**, 75–87.
36. Williams, A. J., Ling, G., Berti, R., Moffett, J. R., Yao, C., Lu, X. M., Dave, J. R., and Tortella, F. C. (2003) Treatment with the snail peptide CGX-1007 reduces DNA damage and alters gene expression of c-Fos and Bcl-2 following focal ischemic brain injury in rats. *Exp. Brain Res.* **153**, 16–26.
37. Williams, A. J., Ling, G., McCabe, R. T., and Tortella, F. C. (2002) Intrathecal CGX-1007 is neuroprotective in a rat model of focal cerebral ischemia. *Neuroreport* **13**, 821–824.
38. Williams, A. J., and Tortella, F. C. (2002) Neuroprotective effects of the sodium channel blocker RS100642 and attenuation of ischemia-induced brain seizures in the rat. *Brain Res.* **932**, 45–55.
39. Yao, C., Williams, A. J., Lu, X. C., Price, R. A., Cunningham, B. S., Berti, R., Tortella, F. C., and Dave, J. R. (2003) The sodium channel blocker RS100642 reverses down-regulation of the sodium channel alpha-subunit Na(v) 1.1 expression caused by transient ischemic brain injury in rats. *Neurotox. Res.* **5**, 245–253.
40. Cernak, I. (2005) Animal models of head trauma. *NeuroRx* **2**, 410–422.
41. Onifer, S. M., Rabchevsky, A. G., and Scheff, S. W. (2007) Rat models of traumatic spinal cord injury to assess motor recovery. *ILAR J.* **48**, 385–395.
42. Pearse, D. D., and Bunge, M. B. (2006) Designing cell- and gene-based regeneration strategies to repair the injured spinal cord. *J. Neurotrauma* **23**, 438–452.
43. Ravikumar, R., McEwen, M. L., and Springer, J. E. (2007) Post-treatment with the cyclosporin derivative, NIM811, reduced indices of cell death and increased the volume

of spared tissue in the acute period following spinal cord contusion. *J. Neurotrauma* **24**, 1618–1630.

44. Reilly, P. L. (2001) Brain injury: The pathophysiology of the first hours. 'Talk and Die revisited'. *J. Clin. Neurosci.* **8**, 398–403.
45. Stiefel, M. F., Tomita, Y., and Marmarou, A. (2005) Secondary ischemia impairing the restoration of ion homeostasis following traumatic brain injury. *J. Neurosurg.* **103**, 707–714.
46. Prieto, D. A., Ye, X., and Veenstra, T. D. (2008) Proteomic analysis of traumatic brain injury: The search for biomarkers. *Expert Rev. Proteomics* **5**, 283–291.
47. Ottens, A. K., Kobeissy, F. H., Fuller, B. F., Liu, M. C., Oli, M. W., Hayes, R. L., and Wang, K. K. (2007) Novel neuroproteomic approaches to studying traumatic brain injury. *Prog. Brain Res.* **161**, 401–418.
48. Ottens, A. K., Kobeissy, F. H., Golden, E. C., Zhang, Z., Haskins, W. E., Chen, S. S., Hayes, R. L., Wang, K. K., and Denslow, N. D. (2006) Neuroproteomics in neurotrauma. *Mass Spectrom. Rev.* **25**, 380–408.
49. Matsumoto, I., Alexander-Kaufman, K., Iwazaki, T., Kashem, M. A., and Matsuda-Matsumoto, H. (2007) CNS proteomes in alcohol and drug abuse and dependence. *Expert Rev. Proteomics* **4**, 539–552.
50. Kobeissy, F. H., Sadasivan, S., Liu, J., Gold, M. S., and Wang, K. K. (2008) Psychiatric research: Psychoproteomics, degradomics and systems biology. *Expert Rev. Proteomics* **5**, 293–314.
51 Matsumoto, H., and Matsumoto, I. (2008) Alcoholism: Protein expression profiles in a human hippocampal model. *Expert Rev. Proteomics* **5**, 321–331.
52. Hemby, S. E. (2006) Assessment of genome and proteome profiles in cocaine abuse. *Prog. Brain Res.* **158**, 173–195.
53 Bodzon-Kulakowska, A., Bierczynska-Krzysik, A., Drabik, A., Noga, M., Kraj, A., Suder, P., and Silberring, J. (2005) Morphinome–Proteome of the Nervous System After Morphine Treatment. *Amino Acids* **28**, 13–19.
54. Iwazaki, T., McGregor, I. S., and Matsumoto, I. (2008) Protein expression profile in the amygdala of rats with methamphetamine-induced behavioral sensitization. *Neurosci. Lett.* **435**, 113–119.
55. Iwazaki, T., McGregor, I. S., and Matsumoto, I. (2007) Protein expression profile in the striatum of rats with methamphetamine-induced behavioral sensitization. *Proteomics* **7**, 1131–1139.
56. Kobeissy, F. H., Warren, M. W., Ottens, A. K., Sadasivan, S., Zhang, Z., Gold, M. S., and Wang, K. K. (2008) Psychoproteomic analysis of rat cortex following acute methamphetamine exposure. *J. Proteome Res.* **7**, 1971–1983.
57. Freeman, W. M., Brebner, K., Amara, S. G., Reed, M. S., Pohl, J., and Phillips, A. G. (2005) Distinct proteomic profiles of amphetamine self-administration transitional states. *Pharmacogenomics J.* **5**, 203–214.
58. Hwang, Y. Y., and Li, M. D. (2006) Proteins differentially expressed in response to nicotine in five rat brain regions: Identification using a 2-DE/MS-based proteomics approach. *Proteomics* **6**, 3138–3153.
59. Yeom, M., Shim, I., Lee, H. J., and Hahm, D. H. (2005) Proteomic analysis of nicotine-associated protein expression in the striatum of repeated nicotine-treated rats. *Biochem. Biophys. Res. Commun.* **326**, 321–328.
60. Thompson, P. M., Hayashi, K. M., Simon, S. L., Geaga, J. A., Hong, M. S., Sui, Y., Lee, J. Y., Toga, A. W., Ling, W., and London, E. D. (2004) Structural abnormalities in the brains of human subjects who use methamphetamine. *J. Neurosci.* **24**, 6028–6036.
61. Pierce, R. C., and Kumaresan, V. (2006) The mesolimbic dopamine system: The final common pathway for the reinforcing effect of drugs of abuse? *Neurosci. Biobehav. Rev.* **30**, 215–238.
62. Gotz, J., Deters, N., Doldissen, A., Bokhari, L., Ke, Y., Wiesner, A., Schonrock, N., and Ittner, L. M. (2007) A decade of tau transgenic animal models and beyond. *Brain Pathol.* **17**, 91–103.
63. Games, D., Adams, D., Alessandrini, R., Barbour, R., Berthelette, P., Blackwell, C., Carr, T., Clemens, J., Donaldson, T., and Gillespie, F. (1995) Alzheimer-type neuropathology in transgenic mice overexpressing V717F beta-amyloid precursor protein. *Nature* **373**, 523–527.
64. Hsiao, K., Chapman, P., Nilsen, S., Eckman, C., Harigaya, Y., Younkin, S., Yang, F., and Cole, G. (1996) Correlative memory deficits, abeta elevation, and amyloid plaques in transgenic mice. *Science* **274**, 99–102.
65. Sturchler-Pierrat, C., Abramowski, D., Duke, M., Wiederhold, K. H., Mistl, C., Rothacher, S., Ledermann, B., Burki, K., Frey, P., Paganetti, P. A., Waridel, C., Calhoun, M. E., Jucker, M., Probst, A., Staufenbiel, M., and Sommer, B. (1997) Two amyloid precursor protein transgenic mouse models with alzheimer disease-like pathology. *Proc. Natl Acad. Sci. USA* **94**, 13287–13292.

66. Stalder, M., Phinney, A., Probst, A., Sommer, B., Staufenbiel, M., and Jucker, M. (1999) Association of microglia with amyloid plaques in brains of APP23 transgenic mice. *Am. J. Pathol.* **154**, 1673–1684.

67. Lewis, J., McGowan, E., Rockwood, J., Melrose, H., Nacharaju, P., Van Slegtenhorst, M., Gwinn-Hardy, K., Paul Murphy, M., Baker, M., Yu, X., Duff, K., Hardy, J., Corral, A., Lin, W. L., Yen, S. H., Dickson, D. W., Davies, P., and Hutton, M. (2000) Neurofibrillary tangles, amyotrophy and progressive motor disturbance in mice expressing mutant (P301L) tau protein. *Nat. Genet.* **25**, 402–405.

68. Lewis, J., Dickson, D. W., Lin, W. L., Chisholm, L., Corral, A., Jones, G., Yen, S. H., Sahara, N., Skipper, L., Yager, D., Eckman, C., Hardy, J., Hutton, M., and McGowan, E. (2001) Enhanced neurofibrillary degeneration in transgenic mice expressing mutant tau and APP. *Science* **293**, 1487–1491.

69. Gotz, J., Chen, F., van Dorpe, J., and Nitsch, R. M. (2001) Formation of neurofibrillary tangles in P3011 tau transgenic mice induced by abeta 42 fibrils. *Science* 293, 1491–1495.

70. Gotz, J., Streffer, J. R., David, D., Schild, A., Hoerndli, F., Pennanen, L., Kurosinski, P., and Chen, F. (2004) Transgenic animal models of Alzheimer's disease and related disorders: Histopathology, behavior and therapy. *Mol. Psychiatry* **9**, 664–683.

71. Abdul, H. M., Sultana, R., St. Clair, D. K., Markesbery, W. R., and Butterfield, D. A. (2008) Oxidative damage in brain from human mutant APP/PS-1 double knock-in mice as a function of age. Free Radic. *Biol. Med.* **45**, 1420–1425.

72. Owen, J. B., Opii, W. O., Ramassamy, C., Pierce, W. M., and Butterfield, D. A. (2008) Proteomic analysis of brain protein expression levels in NF-Kappabeta p50 (–/–) homozygous knockout mice. *Brain Res.* **1240**, 22–30.

73. Hartl, D., Rohe, M., Mao, L., Staufenbiel, M., Zabel, C., and Klose, J. (2008) Impairment of adolescent hippocampal plasticity in a mouse model for Alzheimer's disease precedes disease phenotype. *PLoS ONE* **3**, e2759.

74. Martin, B., Brenneman, R., Becker, K. G., Gucek, M., Cole, R. N., and Maudsley, S. (2008) ITRAQ analysis of complex proteome alterations in 3xTgAD Alzheimer's mice: Understanding the interface between physiology and disease. *PLoS ONE* **3**, e2750.

75. Vonsattel, J. P. (2008) Huntington disease models and human neuropathology: Similarities and differences. *Acta Neuropathol.* **115**, 55–69.

76. Perluigi, M., Poon, H. F., Maragos, W., Pierce, W. M., Klein, J. B., Calabrese, V., Cini, C., De Marco, C., and Butterfield, D. A. (2005) Proteomic analysis of protein expression and oxidative modification in r6/2 transgenic mice: A model of Huntington disease. *Mol. Cell. Proteomics* **4**, 1849–1861.

77. Chiang, M. C., Juo, C. G., Chang, H. H., Chen, H. M., Yi, E. C., and Chern, Y. (2007) Systematic uncovering of multiple pathways underlying the pathology of Huntington disease by an acid-cleavable isotope-coded affinity tag approach. *Mol. Cell. Proteomics* **6**, 781–797.

78. Fasano, M., and Lopiano, L. (2008) Alpha-synuclein and Parkinson's disease: A proteomic view. Expert Rev. *Proteomics* **5**, 239–248.

79. Betarbet, R., Canet-Aviles, R. M., Sherer, T. B., Mastroberardino, P. G., McLendon, C., Kim, J. H., Lund, S., Na, H. M., Taylor, G., Bence, N. F., Kopito, R., Seo, B. B., Yagi, T., Yagi, A., Klinefelter, G., Cookson, M. R., and Greenamyre, J. T. (2006) Intersecting pathways to neurodegeneration in Parkinson's disease: Effects of the pesticide rotenone on DJ-1, alpha-synuclein, and the ubiquitin–proteasome system. *Neurobiol.* Dis. **22**, 404–420.

80. Betarbet, R., Sherer, T. B., and Greenamyre, J. T. (2005) Ubiquitin-proteasome system and Parkinson's diseases. *Exp. Neurol.* **191**, S17–S27.

81. Ichibangase, T., Saimaru, H., Takamura, N., Kuwahara, T., Koyama, A., Iwatsubo, T., and Imai, K. (2008) Proteomics of caenorhabditis elegans over-expressing human alpha-synuclein analyzed by fluorogenic derivatization-liquid chromatography/tandem mass spectrometry: Identification of actin and several ribosomal proteins as negative markers at early Parkinson's disease stages. Biomed. *Chromatogr.* **22**, 232–234.

82. Zabel, C., Andreew, A., Mao, L., and Hartl, D. (2008) Protein expression overlap: More important than which proteins change in expression? Expert Rev. *Proteomics* **5**, 187–205.

83. Tribl, F., Meyer, H. E., and Marcus, K. (2008) Analysis of organelles within the nervous system: Impact on brain and organelle functions. Expert Rev. *Proteomics* **5**, 333–351.

84. Fasano, M., Giraudo, S., Coha, S., Bergamasco, B., and Lopiano, L. (2003) Residual substantia nigra neuromelanin in Parkinson's disease is cross-linked to alpha-synuclein. *Neurochem.* Int. **42**, 603–606.

85. Halliday, G. M., Ophof, A., Broe, M., Jensen, P. H., Kettle, E., Fedorow, H., Cartwright, M. I., Griffiths, F. M., Shepherd, C. E., and

Double, K. L. (2005) Alpha-synuclein redistributes to neuromelanin lipid in the substantia nigra early in Parkinson's disease. *Brain* **128**, 2654–2664.

86. Stevens, S. M. Jr., Duncan, R. S., Koulen, P., and Prokai, L. (2008) Proteomic analysis of mouse brain microsomes: Identification and bioinformatic characterization of endoplasmic reticulum proteins in the mammalian central nervous system. *J. Proteome Res.* 7, 1046–1054.
87. Grant, K. J., and Wu, C. C. (2007) Advances in neuromembrane proteomics: Efforts towards a comprehensive analysis of membrane proteins in the brain. Brief Funct. *Genomic Proteomic.* **6**, 59–69.
88. Macher, B. A., and Yen, T. Y. (2007) Proteins at membrane surfaces – a review of approaches. *Mol. Biosyst.* **3**, 705–713.
89. Emes, R. D., Pocklington, A. J., Anderson, C. N., Bayes, A., Collins, M. O., Vickers, C. A., Croning, M. D., Malik, B. R., Choudhary, J. S., Armstrong, J. D., and Grant, S. G. (2008) Evolutionary expansion and anatomical specialization of synapse proteome complexity. *Nat. Neurosci.* **11**, 799–806.
90. Li, K. W., and Jimenez, C. R. (2008) Synapse proteomics: Current status and quantitative applications. *Expert Rev. Proteomics* **5**, 353–360.
91. Bai, F., and Witzmann, F. A. (2007) Synaptosome proteomics. Subcell. *Biochem.* 43, 77–98.
92. Cheng, D., Hoogenraad, C. C., Rush, J., Ramm, E., Schlager, M. A., Duong, D. M., Xu, P., Wijayawardana, S. R., Hanfelt, J., Nakagawa, T., Sheng, M., and Peng, J. (2006) Relative and absolute quantification of postsynaptic density proteome isolated from rat forebrain and cerebellum. *Mol. Cell. Proteomics* **5**, 1158–1170.
93. Liu, S. H., Cheng, H. H., Huang, S. Y., Yiu, P. C., and Chang, Y. C. (2006) Studying the protein organization of the postsynaptic density by a novel solid phase- and chemical cross-linking-based technology. *Mol. Cell. Proteomics* **5**, 1019–1032.
94. Burre, J., Beckhaus, T., Schagger, H., Corvey, C., Hofmann, S., Karas, M., Zimmermann, H., and Volknandt, W. (2006) Analysis of the synaptic vesicle proteome using three gel-based protein separation techniques. *Proteomics* **6**, 6250–6262.
95. Burre, J., Zimmermann, H., and Volknandt, W. (2007) Immunoisolation and subfractionation of synaptic vesicle proteins. *Anal. Biochem.* **362**, 172–181.
96. Morciano, M., Burre, J., Corvey, C., Karas, M., Zimmermann, H., and Volknandt, W. (2005) Immunoisolation of two synaptic vesicle pools from synaptosomes: A proteomics analysis. *J. Neurochem.* **95**, 1732–1745.
97. Dosemeci, A., Tao-Cheng, J. H., Vinade, L., and Jaffe, H. (2006) Preparation of postsynaptic density fraction from hippocampal slices and proteomic analysis. *Biochem. Biophys. Res. Commun.* **339**, 687–694.
98. Trinidad, J. C., Specht, C. G., Thalhammer, A., Schoepfer, R., and Burlingame, A. L. (2006) Comprehensive identification of phosphorylation sites in postsynaptic density preparations. *Mol. Cell. Proteomics* **5**, 914–922.
99. Trinidad, J. C., Thalhammer, A., Specht, C. G., Lynn, A. J., Baker, P. R., Schoepfer, R., and Burlingame, A. L. (2008) Quantitative analysis of synaptic phosphorylation and protein expression. *Mol. Cell. Proteomics* 7, 684–696.
100. Vosseller, K., Trinidad, J. C., Chalkley, R. J., Specht, C. G., Thalhammer, A., Lynn, A. J., Snedecor, J. O., Guan, S., Medzihradszky, K. F., Maltby, D. A., Schoepfer, R., and Burlingame, A. L. (2006) O-linked *N*-acetylglucosamine proteomics of postsynaptic density preparations using lectin weak affinity chromatography and mass spectrometry. *Mol. Cell. Proteomics* **5**, 923–934.
101. Augustinack, J. C., Schneider, A., Mandelkow, E. M., and Hyman, B. T. (2002) Specific tau phosphorylation sites correlate with severity of neuronal cytopathology in Alzheimer's disease. *Acta Neuropathol.* **103**, 26–35.
102. Hampel, H., Burger, K., Pruessner, J. C., Zinkowski, R., DeBernardis, J., Kerkman, D., Leinsinger, G., Evans, A. C., Davies, P., Moller, H. J., and Teipel, S. J. (2005) Correlation of cerebrospinal fluid levels of tau protein phosphorylated at threonine 231 with rates of hippocampal atrophy in Alzheimer disease. *Arch. Neurol.* **62**, 770–773.
103. Cripps, D., Thomas, S. N., Jeng, Y., Yang, F., Davies, P., and Yang, A. J. (2006) Alzheimer disease-specific conformation of hyperphosphorylated paired helical filament-tau is polyubiquitinated through Lys-48, Lys-11, and Lys-6 ubiquitin conjugation. *J. Biol. Chem.* **281**, 10825–10838.
104. Butterfield, D. A., Reed, T., Newman, S. F., and Sultana, R. (2007) Roles of amyloid beta-peptide-associated oxidative stress and brain protein modifications in the pathogenesis of Alzheimer's disease and mild cognitive impairment. *Free Radic. Biol. Med.* **43**, 658–677.

105. Zhan, X., and Desiderio, D. M. (2004) The human pituitary nitroproteome: Detection of nitrotyrosyl-proteins with two-dimensional western blotting, and amino acid sequence determination with mass spectrometry. *Biochem. Biophys. Res. Commun.* **325**, 1180–1186.

106. Zhan, X., and Desiderio, D. M. (2006) Nitroproteins from a human pituitary adenoma tissue discovered with a nitrotyrosine affinity column and tandem mass spectrometry. *Anal. Biochem.* **354**, 279–289.

107. Azarkan, M., Huet, J., Baeyens-Volant, D., Looze, Y., and Vandenbussche, G. (2007) Affinity chromatography: A useful tool in proteomics studies. *J. Chromatogr. B. Analyt Technol. Biomed. Life. Sci.* **849**, 81–90.

108. Karlsson, K., Cairns, N., Lubec, G., and Fountoulakis, M. (1999) Enrichment of human brain proteins by heparin chromatography. *Electrophoresis* **20**, 2970–2976.

109. Xiong, S., Zhang, L., and He, Q. Y. (2008) Fractionation of proteins by heparin chromatography. *Methods Mol. Biol.* **424**, 213–221.

110. Abu-Farha, M., Elisma, F., and Figeys, D. (2008) Identification of protein-protein interactions by mass spectrometry coupled techniques. *Adv. Biochem. Eng. Biotechnol.* **110**, 67–80.

111. Berggard, T., Linse, S., and James, P. (2007) Methods for the detection and analysis of protein-protein interactions. *Proteomics* 7, 2833–2842.

112. Zhang, Z., Ottens, A. K., Golden, E. C., Hayes, R. L., and Wang, K. K. (2006) Using calmodulin-affinity capture to study the rat brain calmodulin binding proteome and its vulnerability to calpain and caspase proteolysis. *Calcium Binding Proteins* **1**, 125–134.

113. Klose, J. (1975) Protein mapping by combined isoelectric focusing and electrophoresis of mouse tissues. A novel approach to testing for induced point mutations in mammals. *Humangenetik* **26**, 231–243.

114. O'Farrell, P. H. (1975) High resolution two-dimensional electrophoresis of proteins. *J. Biol. Chem.* **250**, 4007–4021.

115. Alban, A., David, S. O., Bjorkesten, L., Andersson, C., Sloge, E., Lewis, S., and Currie, I. (2003) A novel experimental design for comparative two-dimensional gel analysis: Two-dimensional difference gel electrophoresis incorporating a pooled internal standard. *Proteomics* **3**, 36–44.

116. Tannu, N. S., and Hemby, S. E. (2006) Methods for proteomics in neuroscience. Prog. *Brain Res.* **158**, 41–82.

117. Shin, J. H., Krapfenbauer, K., and Lubec, G. (2006) Large-scale identification of cytosolic mouse brain proteins by chromatographic prefractionation. *Electrophoresis* **27**, 2799–2813.

118. McDonald, W. H., and Yates, J. R. 3rd. (2002) Shotgun proteomics and biomarker discovery. *Dis. Markers* **18**, 99–105.

119. Wolters, D. A., Washburn, M. P., and Yates, J. R. 3rd. (2001) An automated multidimensional protein identification technology for shotgun proteomics. *Anal. Chem.* **73**, 5683–5690.

120. Nakamura, T., Kuromitsu, J., and Oda, Y. (2008) Evaluation of comprehensive multidimensional separations using reversed-phase, reversed-phase liquid chromatography/mass spectrometry for shotgun proteomics. *J. Proteome Res.* 7, 1007–1011.

121. Gilar, M., Olivova, P., Daly, A. E., and Gebler, J. C. (2005) Two-dimensional separation of peptides using RP-RP-HPLC system with different pH in first and second separation dimensions. *J. Sep. Sci.* **28**, 1694–1703.

122. Righetti, P. G., Boschetti, E., Lomas, L., and Citterio, A. (2006) Protein equalizer technology: The quest for a "democratic proteome". *Proteomics* **6**, 3980–3992.

123. Panchaud, A., Affolter, M., Moreillon, P., and Kussmann, M. (2008) Experimental and computational approaches to quantitative proteomics: Status quo and outlook. *J. Proteomics* 71, 19–33.

124. Gygi, S. P., Rist, B., Gerber, S. A., Turecek, F., Gelb, M. H., and Aebersold, R. (1999) Quantitative analysis of complex protein mixtures using isotope-coded affinity tags. *Nat. Biotechnol.* **17**, 994–999.

125. Han, D. K., Eng, J., Zhou, H., and Aebersold, R. (2001) Quantitative profiling of differentiation-induced microsomal proteins using isotope-coded affinity tags and mass spectrometry. *Nat. Biotechnol.* **19**, 946–951.

126. Ong, S. E., Blagoev, B., Kratchmarova, I., Kristensen, D. B., Steen, H., Pandey, A., and Mann, M. (2002) Stable isotope labeling by amino acids in cell culture, SILAC, as a simple and accurate approach to expression proteomics. *Mol. Cell. Proteomics* **1**, 376–386.

127. Yao, X., Freas, A., Ramirez, J., Demirev, P. A., and Fenselau, C. (2001) Proteolytic 18O labeling for comparative proteomics: Model studies with two serotypes of adenovirus. *Anal. Chem.* **73**, 2836–2842.

128. Thompson, A., Schafer, J., Kuhn, K., Kienle, S., Schwarz, J., Schmidt, G., Neumann, T., Johnstone, R., Mohammed, A. K., and Hamon, C. (2003) Tandem mass tags: A novel

quantification strategy for comparative analysis of complex protein mixtures by MS/MS. *Anal. Chem.* **75**, 1895–1904.

129. Ross, P. L., Huang, Y. N., Marchese, J. N., Williamson, B., Parker, K., Hattan, S., Khainovski, N., Pillai, S., Dey, S., Daniels, S., Purkayastha, S., Juhasz, P., Martin, S., Bartlet-Jones, M., He, F., Jacobson, A., and Pappin, D. J. (2004) Multiplexed protein quantitation in saccharomyces cerevisiae using amine-reactive isobaric tagging reagents. *Mol. Cell. Proteomics* **3**, 1154–1169.
130. Zetterberg, H., Ruetschi, U., Portelius, E., Brinkmalm, G., Andreasson, U., Blennow, K., and Brinkmalm, A. (2008) Clinical proteomics in neurodegenerative disorders. *Acta Neurol. Scand.* **118**, 1–11.
131. Hsich, G., Kenney, K., Gibbs, C. J., Lee, K. H., and Harrington, M. G. (1996) The 14-3-3 brain protein in cerebrospinal fluid as a marker for transmissible spongiform encephalopathies. *N. Engl. J. Med.* **335**, 924–930.
132. Veenstra, T. D., and Marcus, K. (2008) Multidimensional advancement of neuroproteomics. *Expert Rev. Proteomics* **5**, 149–151.
133. Takamori, S., Holt, M., Stenius, K., Lemke, E. A., Gronborg, M., Riedel, D., Urlaub, H., Schenck, S., Brugger, B., Ringler, P., Muller, S. A., Rammner, B., Grater, F., Hub, J. S., De Groot, B. L., Mieskes, G., Moriyama, Y., Klingauf, J., Grubmuller, H., Heuser, J., Wieland, F., and Jahn, R. (2006) Molecular anatomy of a trafficking organelle. *Cell* **127**, 831–846.

Part I

Disease Models in Neuroproteomics

Chapter 2

Modeling Cerebral Ischemia in Neuroproteomics

Jitendra R. Dave, Anthony J. Williams, Changping Yao, X.-C. May Lu, and Frank C. Tortella

Summary

Protein changes induced by traumatic or ischemic brain injury can serve as diagnostic markers as well as therapeutic targets for neuroprotection. The focus of this chapter is to provide a representative overview of preclinical brain injury and proteomics analysis protocols for evaluation and discovery of novel biomarkers. Detailed surgical procedures have been provided for inducing MCAo and implantation of chronic indwelling cannulas for drug delivery. Sample collection and tissue processing techniques for collection of blood, CSF, and brain are also described including standard biochemical methodology for the proteomic analysis of these tissues.

The dynamics of proteomic analysis is a multistep process comprising sample preparation, separation, quantification, and identification of proteins. Our approach is to separate proteins first by two-dimensional gel electrophoresis according to charge and molecular mass. Proteins are then fragmented and analyzed using matrix-assisted laser desorption ionization time-of-flight mass spectrometry (MALDI-TOF MS). Identification of proteins can be achieved by comparing the mass-to-charge data to protein sequences in respective databases.

Key words: Ischemic brain injury, MCAo, Proteomics, 2D-gel electrophoresis, Mass spectroscopy, Models of cerebral ischemia, Brain injury, Biomarkers

1. Introduction

1.1. Definition of Cerebral Ischemia

Cerebral ischemia involves a loss of blood flow to the brain and the ensuing pathophysiological changes that eventually lead to brain cell death. The metabolic crisis provoked by an ischemic insult to the brain includes the inhibition of oxidative phosphorylation and eventual loss of cellular ionic homeostasis *(1)*. The ensuing maelstrom of cellular dysregulation incorporates a variety of secondary injury processes including activation of

Andrew K. Ottens and Kevin K.W. Wang (eds.), *Neuroproteomics,* Methods in Molecular Biology, vol. 566
DOI 10.1007/978-1-59745-562-6_2, © Humana Press, a part of Springer Science+Business Media, LLC 2009

delayed cell death cascades and a prominent neuroinflammatory response *(2–4)*. The end result is a change in baseline protein abundance levels in readily accessible body fluids that can serve as diagnostic biomarkers of injury. In fact, during the ensuing days following injury to the brain, alterations in brain function, the resulting pathology, and neurological outcome can be correlated to changes in protein abundance levels *(5–10)*. The consequence of these protein changes following injury to the brain, as related to the medical management of these patients, is an evolving field of research.

1.2. Models of Cerebral Ischemia

Several models of both global and focal cerebral ischemia are currently available for preclinical evaluation of the resultant brain injury. Models of focal brain ischemia were developed to mimic stroke in humans and generally involve the permanent or temporary occlusion of a main cerebral artery supplying the brain with blood such as the middle cerebral artery (MCA). The MCA can be directly occluded by electrocauterization or with the use of microclips following a craniectomy *(11, 12)*, although other, less-invasive techniques have been developed. Those models not involving craniectomy generally rely on an artificial embolic occlusion such as a thromboembolic implant delivered through an extracranial artery *(13–15)* or with an intraluminal filament *(16, 17)*. The most widely used model of focal brain ischemia is the intraluminal filament model of middle cerebral artery occlusion (MCAo) in rats, which will be described in detail here. Advantages of this model include its ease of use, being a less-invasive technique, and ability to produce a controlled permanent or transient ischemic injury to the brain. Our own lab has utilized the rat filament model of MCAo extensively over the past several years to study the molecular mechanisms and pathophysiological consequences associated with focal ischemic injury as well as to evaluate the preclinical efficacy of several novel therapeutics *(18–30)*.

1.3. Proteomic Analysis of Cerebral Ischemia

Protein biomarkers have important applications in the diagnosis, prognosis, and clinical research of brain injuries. Simple and rapid diagnostic tools will immensely facilitate allocation of the major medical resources required to treat brain injuries. Accurate diagnosis in acute care environments can significantly enhance decisions about patient management, including decisions whether to admit or discharge, or to administer other time consuming and expensive tests such as computed tomography (CT) and magnetic resonance imaging (MRI) scans. Development of relevant brain injury biomarkers will also improve opportunities for the conduct of clinical research including the confirmation of injury mechanism(s), diagnosis of injury type and severity level, and drug target identification. Biomarkers also possess the advantage of providing a relatively inexpensive clinical trial outcome measure that is more readily available than conventional neurological assessments or imaging modalities, thereby significantly reducing the risk and cost of human clinical trials.

Criteria for identifying useful biomarkers should include the ability to collect samples from readily accessible biological material such as CSF or blood (CSF is routinely accessible in severely brain-injured patients), to predict the magnitude of injury and resulting functional deficits, to possess a high sensitivity and specificity to the injured tissue, and to have a rapid appearance in blood with release in a time-locked sequence after injury. Ideally, biomarkers should employ biological substrates unique to the CNS and provide correlative information relevant to an injury mechanism. Utilization of protein biomarkers will also help define endpoint parameters for clinical trials that are indicative of biological activity and further to verify therapeutic efficacy. Unlike other organ-based diseases where rapid diagnosis employing biomarkers (usually involving blood tests) prove invaluable to guide treatment, there are no rapid, definitive diagnostic tests for ischemic brain injury to provide quantifiable neurochemical markers to help determine the magnitude of the injury, the anatomical and cellular pathology of the injury, or to help guide implementation of appropriate triage and medical management.

The focus of this chapter is to provide a representative overview of a preclinical brain injury protocol for the discovery and evaluation of novel biomarkers. Detailed descriptions are provided for experimental techniques involved in proteomic analysis of biological tissues obtained following a focal ischemic brain injury in rats. Additional techniques detailing various drug administration routes, surgical techniques, and placement of chronic indwelling cannulas for drug delivery through various routes of administration are also provided as relevant for the evaluation of therapeutics by neuroproteomics.

2. Materials

1. *TCA solution.* 13.3% trichloroacetic acid (TCA), 0.093% 2-mercaptoethanol, and 0.3% DDT in acetone.
2. *IPG (Immobilized pH gradient) strip rehydration buffer.* 8 M urea, 0.5% CHAPS, 0.2 mM DTT, and 0.8% IPG buffer (GE Healthcare, Piscataway, NJ). DTT and IPG buffer are added fresh prior to use (*see* **Note 1**). Small aliquots of this solution without DTT and IPG buffer can be stored at −20°C for 2–3 months.

3. *Lysis buffer A.* 7 M urea, 2 M thiourea, 4% CHAPS, 1% DTT, 0.1% SDS, 0.5% IPG buffer (pH 3–10), 1% Triton X-100, and 1/10 volume (of Lysate) of Mammalian Protease Inhibitor (Sigma, St. Louis, MO). Urea, thiourea, CHAPS, and SDS solution can be stored at −20°C for 2–3 months and rest of the chemical should be added at the time of use (*see* **Note 2**).
4. *Lysis buffer B.* 7 M urea, 2 M thiourea, 4% CHAPS, 1% DTT, 0.1% SDS, 0.5% IPG buffer (pH 3–10).
5. Nuclease Mix (GE Healthcare).
6. 2D Clean-Up Kit (GE Healthcare).
7. Mini Dialysis Kit, 1 kDa cutoff (GE Healthcare).
8. 2D Quant Kit (GE Healthcare).
9. *CyDye stock solution.* Take 2 μL of each Dye (Cy2, Cy3, and Cy5) in a vial and add 3 μL of DME (*N,N*-dimethylformamide) to each vial to make CyDye stock solutions of 400 pmol/μL. The solutions are light sensitive and should be stored in the dark at −80°C (up to 3 months).
10. *Second-dimension IPG strip equilibrium solution 1.* 50 mM Tris–HCl, pH 6.8, 6 M urea, 30% (v/v) glycerol, 2% (w/v) SDS, and 100 mM DTT. The solution without DTT can be stored at −20°C (*see* **Note 3**) for up to a year and DTT can be added at the time of use.
11. *Second-dimension IPG strip equilibrium solution 2.* 50 mM Tris–HCl, pH 6.8, 6 M urea, 30% (v/v) glycerol, 2% (w/v) SDS, and 2.5% iodoacetamide. Solution can be stored at −20°C for up to a year.
12. *Tris–glycine buffer.* 25 mM Tris–HCl, pH 8.3, 192 mM glycine, 0.1% SDS. Solution may be stored at room temperature for up to a year.
13. *Agarose sealing solution.* 100 mL Tris–glycine buffer, 0.5 g agarose, 200 μL of 0.002% (w/v) bromophenol blue. Solution may be stored at room temperature for up to a year.
14. *Coomassie blue.* BioSafe Coomassie (Bio-Rad, Hercules, CA).

3. Methods

3.1. Animal Care

All procedures described herein are currently performed at Walter Reed Army Institute of Research (WRAIR) and are approved by the WRAIR Animal Care and Use Committee. Research is conducted in compliance with the US Animal Welfare Act, *Guide for the Care and Use of Laboratory Animals* (National Research Council) and other Federal statutes and regulations relating to animals and experiments involving animals in a facility accredited by the Association for Assessment and Accreditation of Laboratory Animal Care International.

Upon arrival, all preoperative animals are housed for at least 1 week to acclimate to their new environment before use. These animals are monitored daily by our veterinary medicine staff and research personnel prior to surgery. Aseptic techniques are adhered to for all rodent recovery surgeries. Each individual surgical procedure generally requires 15–30 min to complete. Surgeons wear laboratory coats (or scrub tops as available), keeping loose sleeves away from the surgery site, as well as caps, surgical masks, sterile surgical gloves, and booties at all times during the surgery. Surgeons wash hands and arms to the elbows with antiseptic surgical preparation, and aseptically don their sterile gloves. A new pair of sterile gloves is donned before each surgical procedure if multiple surgeries are performed.

In order to assist our veterinary staff in monitoring postsurgical animals, animal cages are provided with a surgery card indicating all surgical procedures performed on an individual animal. If multiple surgeries are being performed on the same day, all instruments are wiped down and thoroughly cleaned with alcohol, sterilized for 15 s in a hot bead sterilizer and placed on a sterile pad to cool before use. All surgical sites are prepared by shaving the animal's hair from the surgical site (done away from the surgery bench site; vacuum away loose hair); this procedure takes only 30–60 s and the animal is held in the hand after appropriate anesthesia is induced. Once completed, the animal is placed on a heating pad that continuously monitors and maintains normothermic body temperature during surgery. Each surgical site is disinfected using three sequential swabs of betadine and alcohol followed by a final application of betadine solution. A sterile drape is then placed over the surgery site. When necessary, tissues are kept moist by application of sterile saline using a drop from a sterile syringe.

3.2. Anesthesia/Analgesia

All surgical procedures are performed under general anesthesia. Postoperative pain from cutaneous incisions is alleviated using subcutaneous infiltration of a local anesthetic. At the time of surgery, rats are anesthetized with an injectable (e.g., ketamine/xylazine, 70/6 mg/kg of body weight, injected intramuscular) or gas anesthesia (e.g., 2–5% isofluorane). Upon the loss of response to a pinch of the tail and hindpaws, the rats are placed in the prone or supine position on a sterile pad, depending upon the surgical procedure. Prior to all surgical cutaneous incisions, the surgical site is injected subcutaneously with 1% lidocaine with 1:100,000 epinephrine (a total of 0.25 mL delivered by a 25-G needle). Following surgery, incisions are treated with a subcutaneously injected analgesic along the margins of the incision (0.25% bupivacaine,

0.25 mL, 25-G needle). The use of additional postoperative analgesics is generally avoided to minimize conflicting interactions of analgesics with the endpoints being measured and/or the drug responses being studied.

3.3. Middle Cerebral Artery Occlusion

The MCA is the major cerebral artery that branches off the internal carotid artery (ICA) as it enters the cranial cavity in the mid-portion of the cranial floor. Therefore, it can be accessed via the ICA without craniotomy. (**Figure 1** illustrates this procedure and the type of brain injury produced by the same.) To perform MCAo, the anesthetized rat is placed on a homeothermic heating pad in the supine position.

1. Make a midline incision along the ventral cervical neck region to expose the common carotid artery and its internal and external bifurcation.
2. Place microaneurysm clips on the common carotid artery and the ICA to block the blood flow to the external carotid artery (ECA). The ECA is isolated and transected by cauterization.
3. Insert a piece of sterile 3-0 monofilament nylon suture (28 mm), with a heat-blunted tip, into the ICA through the ECA stump

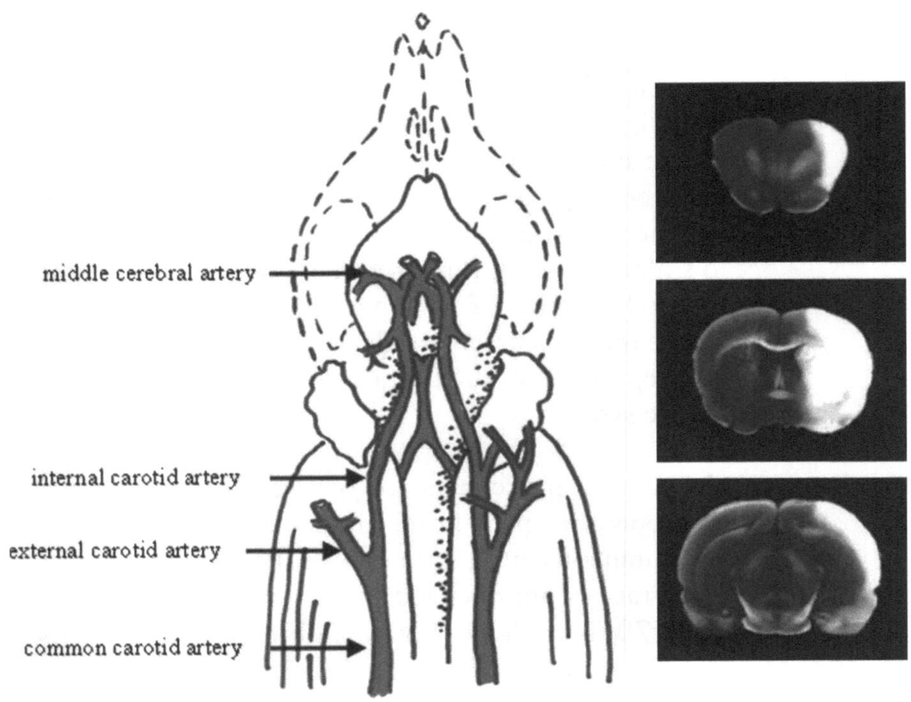

Fig. 1. The panel on the *left* represents the rat vasculature which provides blood supply to the fore brain from the common carotid artery. A nylon suture is gently inserted into the internal carotid artery via a branch of the external carotid artery and enters the brain to lodge at the origin of the middle cerebral artery. It remains in place for 2 h to produce focal ischemia. The panel on the *right* represents selected brain tissues (*Top*: rostral, *Middle*: core, and *Bottom*: caudal) with 2,3,5-triphenytetrazolium chloride (TTC) staining to demarcate the ischemic infarction (*white area*) at 24 h after transient MCAo.

(3 mm) and loosely secured with a piece of 4-0 silk suture to prevent bleeding while still allowing further insertion of the suture to the origin of the MCA.

4. Release the microaneurysm clip on the ICA to allow the advancement of the nylon suture into the ICA to a depth of 20–22 mm from the carotid bifurcation. Once it becomes lodged in the narrowing segment of the anterior cerebral artery, a slight resistance is felt indicating the blockage of the origin of the MCA.
5. Tightly secure the endovascular nylon suture by further tightening the 4-0 silk suture. The end piece of the nylon suture (3 mm) remains outside the ECA so that the suture can be pulled out at a later time point for reperfusion.
6. Close the skin incision using sterile auto clips, and allow the rat to recover from the anesthesia. The endovascular filament either remains in place permanently, or is left in place transiently (i.e., 1–2 h following occlusion) after which the rat is briefly reanesthetized (2% isoflurane), the incision is reopened and the filament is carefully retracted to allow reperfusion of blood to the brain. All surgical procedures are conducted with the aid of an operating microscope. Variations on this surgical procedure are detailed elsewhere ***(16, 17, 31–34)***.

3.4. Postsurgical Provisions

All cutaneous incisions are closed with sterile wound clips or a sterile silk suture, which will remain in place 7–10 days postsurgery or until the termination of the experiment. Immediately following surgery, the animals are placed in clean, well-ventilated, clear polycarbonate cages and observed continuously for recovery as defined by a return to the upright position and purposeful voluntary movement. Each postoperative rat is identified with a surgical card displayed on the front side of the cage with all required information detailed accordingly (i.e., anesthesia used, surgery performed, protocol numbers, point of contact, etc.). The recovery time of the animal is recorded on the card and then the animal is checked again at 30-min postrecovery and again before the end of business hours. Until recovered from anesthesia, rats are kept warm using a circulating water heating pad system. Postoperatively, food and water are available *ad libitum*. Upon recovery from anesthesia, the animals are returned to the animal holding room. In the subsequent days, the animals are checked twice daily with the observations annotated on the surgery card for each animal.

3.5. Drug Administration

All drugs administered should be pharmaceutical grade whenever possible. When pharmaceutical-grade drugs are only available in an oral tablet formulation, the research grade salt compound is used. Sterile saline or sterile water is the vehicle of choice for dissolving compounds. However, other clinically acceptable vehicles are used depending on the formulation requirement of the compound (e.g., 5% sucrose or polypropylene glycol solutions).

3.5.1. Intravenous Cannulation

An intravenous (i.v.) catheter is made from a short length of standard polyethylene (PE)-50 tubing (20 cm) with a beveled silastic tip (15 mm). The cannulas are filled with heparinized saline (20 units/mL, pharmaceutical grade) with the distal end of the cannula heat-sealed prior to implantation.

1. Under a surgical microscope, expose the right external jugular vein via a 1–2 cm cutaneous incision and isolate by careful dissection.
2. Tie off the distal end of the vein from the heart using a 4-0 silk suture, and place a microaneurysm clip on the vein proximally.
3. Make a small incision in the vessel for catheter entry between the clip and the suture.
4. Insert the silastic tip of the catheter into the vein, then remove the clip and advance the catheter for 20–30 mm approximately to the entry of the right atrium. The catheter is held in position by 4-0 silk suture.
5. After flushing with 200 μL of heparinized saline, use a steel trocar to create a subcutaneous tunnel from the incision site to a point on the dorsal surface of the neck.
6. Pass the distal end of the catheter through the tunnel, exteriorized through a stab incision on the dorsal neck.
7. Close the incision site with sterile wound clips. For i.v. infusion experiments, a plastic button is attached to the rat with a 4-0 suture at the exit of the cannula from the skin incision (the button is used to attach the animal to a metal spring/fluid swivel, allowing free movement of the animal during the experiment).

3.5.2. Intrathecal Cannulation

An intrathecal (i.t.) catheter is made of a short length of stretched PE-10 tubing (9.5 cm in length) connected to a piece of 1-cm PE-60 tubing flushed with sterile saline. During surgery, the rat is secured to a stereotaxic frame with the head flexed downward.

1. Expose the cisternal membrane and perforate with a 30-G needle.
2. Carefully insert the i.t. catheter approximately 8.5 cm into the cisterna magna to reach the L4 level of the spinal column.
3. Close the neck muscle incision along the midline using a sterile 4-0 silk suture, and close the skin incision with sterile wound clips. The distal segment of the PE-60 catheter is left to protrude from the incision for later drug injection.

3.5.3. Noncannulated, Lumbar, i.t. Bolus Injection

1. Make a dorsal midline skin incision (approximately 1 cm) immediately rostral to the pelvic girdle.
2. Dissect the muscle tissue to expose the vertebrate.
3. Using the vertebral processes as a guide, make injections using a 0.5-in. 30-G needle that is inserted between the L4–L5 vertebrae and into the subarachnoid space surrounding the cauda equina.
4. Verify the correct needle placement by monitoring the CSF flow from the needle following its insertion.
5. Close the incision using sterile wound clips.

3.5.4. Intracerebroventricular Cannulation

An intracerebroventricular (i.c.v.) catheter is made from a sterile 27-G needle tip attached to a piece of PE-20 tubing.

1. Make a 2-cm middle incision over the skull to expose the bregma.
2. Make a small burr hole stereotaxically by twisting a stainless steel needle (25 G) through the skull over the lateral cerebral ventricle(s) at 1 mm posterior and 1.5 mm lateral to the bregma.
3. Insert the catheter into the left or right lateral cerebral ventricle via the burr hole (3.5–4.0-mm deep).
4. Secure a stainless steel anchor screw to the skull approximately 2 cm posterior to the catheter.
5. Permanently fix the catheter to the skull using a small amount of dental acrylate. The animal is kept anesthetized as the dental acrylate cures and cools (about 10 min). Cold saline (4°C) is applied to facilitate cooling.

3.5.5. Direct Animal Injections

Direct injection of animals with a needle and syringe can be performed in unanesthetized, restrained rats. Intramuscular (i.m.) injections are generally delivered into the muscles of the hind limb with a 25-G needle. Subcutaneous (s.c.) injections are administered under the skin of the back. Intraperitoneal (i.p.) injections are delivered into the lower left quadrant of the abdomen. A variety of references are available describing techniques for administration of substances to rats *(35)*. Injection volumes generally range from 1 to 2 mL/kg of body weight. Needle gauge can be varied depending on the viscosity of the injected medium. If i.v. infusion is required, the injection volume can be adjusted so that it will not exceed 3 mL per day. For i.c.v. and i.t. injections, the total bolus injection volume generally does not exceed 5 μL.

3.6. Collection and Processing of Tissue

At the completion of each experiment, CSF, blood, and/or brain tissue is collected (*see* **Note 4**). At the indicated endpoint, animals are fully anesthetized and placed in a stereotaxic apparatus with the head allowed to move freely along the longitudinal axis.

1. For CSF withdrawal, flex the head so that the external occipital protuberance in the neck is prominent, and make a 1-cm dorsal midline incision over the cervical vertebrae and occiput.
2. Expose the atlanto-occipital membrane by blunt dissection and insert a sterile 25-G needle attached to sterile polyethylene tubing carefully into the cisterna magna. Approximately 100–150 μL of CSF can be collected from each rat (*see* **Note 5**).
3. Immediately following CSF collection, remove the rat from the stereotax, place on its back, and withdraw approximately 3.0 mL of blood via a direct cardiac puncture.
4. Following cardiac puncture, animals should be immediately euthanized, the skull carefully removed, and brain tissue extracted from the specific region of interest.
5. Rinse brain tissue in ice-cold PBS to remove any debris and excess blood, and snap froze in liquid nitrogen.
6. Centrifuge CSF and blood samples at 4,000 × *g* for 10 min at 4°C to clear any contaminating erythrocytes in CSF and remove blood cells.
7. Transfer cleared CSF and serum supernatants to new tubes and, along with tissue samples, store at –80°C until further processing.

3.7. Sample Purification

1. Collect the individual brain tissue in a chilled mortar (kept on dry ice) and grind slowly with a pestle into a fine powder.
2. Suspend the brain tissue powder (approximately 100 mg), CSF (600–900 μg protein), or serum (400 μL) in three volumes of freshly prepared, chilled TCA solution. Keep at 4°C overnight. Centrifuge for 15 min at 4,000 × *g* and –20°C. Wash the pellets with 2.0 mL of cold acetone (kept at –20°C). Dry and resuspend the pellet in 500 μL (approximately five-fold of dry pellet volume) of lysis buffer A.
3. Sonicate the lysate and any remaining cells in the suspension for 10 s. Add 0.5 M $MgCl_2$ to obtain final concentration of 5 mM. Then add nuclease mix (1 μL per 100 μL of lysate). Incubate on ice for 30 min. Centrifuge the suspension at 100,000 × *g* at 4°C for 1 h.
4. Clean the protein samples with the 2D Clean-up Kit.
5. Dissolve the protein pellet in 250–500 μL of lysis buffer B.
6. Dialyze the protein sample with a 1 kDa cutoff Mini Dialysis Kit against lysis buffer B at 4°C overnight (*see* **Note 6**).
7. Collect the dialyzed protein sample in a tube and measure protein concentration using the 2D Quant Kit.

3.8. Cyanine Dye Labeling for DIGE Analysis

1. Adjust the pH (with 1 M Tris buffer, pH 10) of the protein samples to 8.5. Label a sample with 1 μL of CyDye stock solution for every 100 μg of protein (*see* **Note 7**). Mix and leave on ice for 30 min in the dark. Use Cy2 to label 50 μg of protein from the experimental and control groups (used as an internal control). Use Cy3 to label 100 μg of protein from the control group and Cy5 to label 100 μg of protein from the experimental group.
2. Add 1 μL of 10 mM l-Lysine to stop the reaction. Leave on ice for 10 min in the dark (*see* **Note 8**).
3. Combine the labeled samples (300 μg of total protein) with 300 μg of unlabeled protein from each group (i.e., experimental and control samples). Mix together for a total of 900 μg of protein, and add IPG rehydration buffer to a final volume of 450 μL. Load the entire volume onto a 24-cm immobilized, nonlinear, pH 3–10 IPG strips in a strip holder (*see* **Notes 9** and **10**). IPG strips are available in different lengths, and the volume of sample to be loaded will vary depending on the length of the strip.

3.9. First-Dimensional Isoelectric Focusing

These instructions assume the use of an Amersham IPGphor isoelectric focusing (IEF) Unit. Following sample loading, IPG strips are focused with the IEF unit at 20°C under a layer of mineral oil. The voltage is initially set at 30 V for at least 12–16 h to remove salts and low-mass contaminants and to facilitate the entry of high-mass proteins into the strip. Increase the voltage slowly to 500 V over approximately 1 h, then to 1,000 V for 1 h, and finally to 8,000 V for 8 h (*see* **Note 10**). This procedure is referred to as "rehydration of strips under voltage." Alternatively, strips can be rehydrated by "passive rehydration" method in which no voltage is applied and strips are usually left overnight on bench top and next day the focusing starts with application of 500 V for 1 h, followed by 1,000 V for 1 h, and finally 8,000 V for 8 h (*see* **Note 11**).

3.10. Second-Dimensional Polyacrylamide Gel Electrophoresis

1. After the IEF procedure, IPG strips are processed immediately or can be stored at –80°C. To start second-dimensional polyacrylamide gel electrophoresis (2D-PAGE), submerge the IPG strip in 15 mL of second-dimension IPG strip equilibrium solution 1 and shake for 15 min at room temperature.
2. Next, submerge the strip in 15 mL of second-dimension IPG strip equilibrium solution 2 and shake for 15 min at room temperature.
3. Mount strips on-top of a precast slab gel. Place a 5–6 mm piece of IPG strip rehydrated with 15–20 μL of a standard molecular mass marker solution near the acidic end (marked "+") of the sample strip. Seal the strips with agarose sealing solution (*see* **Note 12**).

4. Electrophorese the gel using 1× Tris–glycine buffer in the lower (anode) chamber and 2× Tris–glycine buffer in the upper (cathode) chamber at 17 W/gel for 4–5 h at 20°C (described for use of an Ettan DALT Twelve Separation Unit from GE Healthcare).

3.11. Gel Imaging and Staining

1. Image the gel using excitation and emission wavelengths individually for each CyDye on a fluorescence imager (e.g., a Typhoon Phosphorimager from GE Healthcare).
2. Then stain the gel with Commassie blue for 2 h and then destain in water for 15 min twice.
3. Image the Commassie blue stained gel (no filters required; can be performed with any high-resolution flatbed scanner).

3.12. Spot Analysis and Selection for Mass Spectrometry

Protein spots are detected and quantitated automatically by DeCyder differential analysis software (GE Healthcare). The protein spots of interest (spots that are different between experimental and control sample) can be picked by an Ettan Robotic Spot Picker (For additional detailed information, see the user manual from GE Healthcare) and after trypsin digestion, identify the peptides by mass spectrometry as described elsewhere in this volume.

4. Notes

1. Water used to prepare all the solutions and reagents throughout the procedure should be deionized with resistivity of 18.2 MΩ cm.
2. Buffers containing Urea or Thiourea should be aliquoted in small volumes and frozen to avoid freezing and thawing.
3. Always add DTT and IPG buffer fresh on the day of use.
4. *Animal inclusion criteria.* To avoid variability in brain injury volume following MCAo, as related to incomplete or partial occlusion, several useful "inclusion criteria" can be incorporated into the protocol. (a) Cortical brain blood flow should be monitored in animals subjected to MCAo to verify a successful arterial occlusion. This can be achieved using continuous laser Doppler flow *(36)* or electroencephalography *(37)* monitoring techniques. (b) All animals subjected to MCAo should exhibit distinct acute neurological deficits contralateral to the injured brain hemisphere, which can be evaluated by a simple technique developed by Bederson et al. *(38)*. (c) The success of the MCAo injury should be verified histologically by triphenyltetrazolium chloride (TTC) staining. TTC

staining marks mitochondrial enzyme activity. Tissue with normal levels of the enzyme is stained red, whereas ischemic and infarcted tissue remained unstained due to the loss of the enzyme. This quick and easy technique has been recognized as a reliable and sensitive method for demarcation of ischemic brain infarction beyond 6-h postocclusion *(39, 40)*. Therefore, once the desired brain sections are dissected for proteomic analysis, the remaining tissue can be immediately stained with 1% TTC solution for direct visualization of presence of infarction. (d) Subarachnoid hemorrhage (SH) may occur occasionally in the intraluminal suture model of MCAo. This occurs because the nylon suture protrudes at the origin of the MCA. Animals showing SH are routinely excluded from studies.

5. The maximum amount of CSF collectable from cisterna magna is about 150 μL in normal rats. Lesser amounts may be collected following MCAo due to the development of cerebral edema/brain swelling. To compensate for this possible occurrence, the use of an *insulin syringe* is preferred, because it does not have the 50-μL "dead-volume" that regular syringes have. The cerebral edema can be severe, especially within the acute 24–48 h period post-MCAo. From our experience, when these early time points are required for CSF collection, the number of animals needed will increase by threefold.
6. Sample dialysis is recommended to remove salts and small molecular weight contaminants.
7. All CyDye working solutions (Cy2, Cy3, and Cy5) in samples should be adjusted to pH 8.5 before labeling the protein samples.
8. Samples after labeling with CyDye must be kept in the dark to prevent photobleaching of the dye.
9. For complex samples like brain tissue, serum, CSF, etc., it is recommended to use broad range nonlinear IPG strips (pH 3–10), since this will provide an even distribution of proteins over the gel.
10. For the IEF procedure, focusing longer than the suggested time will result in horizontal streaking leading to loss of resolution.
11. Avoid cleaning the strip holders with alkaline detergents and strong acids. We recommend cleaning the holders with Ettan IPG Strip Holder Cleaning solution and water, followed by autoclaving or baking.
12. Air bubbles are undesirable between the strip and the precast SDS gel when running the second dimension. Careful insertion of the top strip and slow, continuous filling of the seal gel solution can minimize bubble appearance.

Acknowledgments

Research was conducted in compliance with the Animal Welfare Act and other Federal statutes and regulations relating to animals and experiments involving animals and adheres to the principles stated in the Guide for the Care and Use of Laboratory Animals, NIH publication 85–23. Material has been reviewed by the Walter Reed Army Institute of Research. There is no objection to its presentation and/or publication. The opinions or assertions contained herein are the private views of the authors, and are not to be construed as official, or as reflecting true views of the Department of the Army or the Department of Defense.

References

1. Lipton, P. (1999) Ischemic cell death in brain neurons. *Physiol. Rev.* **79**, 1431–1568.
2. Stanimirovic, D., and Satoh, K. (2000) Inflammatory mediators of cerebral endothelium: a role in ischemic brain inflammation. *Brain. Pathol.* **10**, 113–126.
3. Won, S. J., Kim, D. Y., and Gwag, B. J. (2002) Cellular and molecular pathways of ischemic neuronal death. *J. Biochem. Mol. Biol.* **35**, 67–86.
4. Danton, G. H., and Dietrich, W. D. (2003) Inflammatory mechanisms after ischemia and stroke. *J. Neuropathol. Exp. Neurol.* **62**, 127–136.
5. Aurell, A., Rosengren, L.E., Karlsson, B., Olsson, J.E., Zbornikova, V., and Haglid, K.G. (1991) Determination of S-100 and glial fibrillary acidic protein concentrations in cerebrospinal fluid after brain infarction. *Stroke* **22**, 1254–1258.
6. Ingebrigtsen, T., and Romner, B. (2002) Biochemical serum markers of traumatic brain injury. *J. Trauma* **52**, 798–808.
7. Pineda, J.A., Wang, K.K., and Hayes, R.L. (2004) Biomarkers of proteolytic damage following traumatic brain injury. *Brain Pathol.* **14**, 202–209.
8. Siman, R., McIntosh, T.K., Soltesz, K.M., Chen, Z., Neumar, R.W., and Roberts, V.L. (2004) Proteins released from degenerating neurons are surrogate markers for acute brain damage. *Neurobiol. Dis.* **16**, 311–320.
9. Berger, R.P. (2006) The use of serum biomarkers to predict outcome after traumatic brain injury in adults and children. *J. Head Trauma Rehabil.* **21**, 315–333.
10. Sotgiu, S., Zanda, B., Marchetti, B., Fois, M.L., Arru, G., Pes, G.M., Salaris, F.S., Arru, A., Pirisi, A., and Rosati, G. (2006) Inflammatory biomarkers in blood of patients with acute brain ischemia. *Eur. J. Neurol.* **13**, 505–513.
11. Shigeno, T., Teasdale, G.M., McCulloch, J., and Graham, D.I. (1985) Recirculation model following MCA occlusion in rats. Cerebral blood flow, cerebrovascular permeability, and brain edema. *J. Neurosurg.* **63**, 272–277.
12. Buchan, A., Slivka, A., and Xue, D. (1992) The effect of the NMDA receptor antagonist MK-801 on cerebral blood flow and infarct volume in experimental focal stroke. *Brain Res.* **574**, 171–177.
13. Papadopoulos, S.M., Chandler, W.F., Salamat, M.S., Topol, E.J., and Sackellares, J.C. (1987) Recombinant human tissue-type plasminogen activator therapy in acute thromboembolic stroke. *J. Neurosurg.* **67**, 394–398.
14. Busch, E., Kruger, K., and Hossmann, K.A. (1997) Improved model of thromboembolic stroke and rt-PA induced reperfusion in the rat. *Brain Res.* **778**, 16–24.
15. Takano, K., Carano, R.A., Tatlisumak, T., Meiler, M., Sotak, C.H., Kleinert, H.D., and Fisher, M. (1998) Efficacy of intra-arterial and intravenous prourokinase in an embolic stroke model evaluated by diffusion-perfusion magnetic resonance imaging. *Neurology* **50**, 870–875.
16. Koizumi, J., Yoshida, Y., Nakazawa, T., and Ooneda, G. (1986) Experimental studies of ischemic brain edema: 1. A new experimental model of cerebral embolism in rats in which recirculation can be introduced in the ischemic area. *Japan Stroke Journal* **8**, 1–8.
17. Longa, E., Weinstein, P., Carlson, S., and Cummins, R. (1989) Reversible middle cerebral artery occlusion without craniectomy in rats. *Stroke* **20**, 84–91.

18. Phillips, J., Williams, A., Adams, J., Elliott, P., and Tortella, F. (2000) Proteasome inhibitor PS519 reduces infarction and attenuates leukocyte infiltration in a rat model of focal cerebral ischemia. *Stroke* **31**, 1686–1693.
19. Williams, A., Dave, J., Phillips, J., Lin, Y., McCabe, R., and Tortella, F. (2000) Neuroprotective efficacy and therapeutic window of the high-affinity *N*-methyl-D-aspartate antagonist conantokin-G: in vitro (primary cerebellar neurons) and in vivo (rat model of transient focal brain ischemia) studies. *J. Pharmacol. Exp. Ther.* **294**, 378–386.
20. Berti, R., Williams, A., Moffett, J., Hale, S., Velarde, L., Elliott, P., Yao, C., Dave, J., and Tortella, F. (2002) Real time PCR mRNA analysis of the inflammatory cascade associated with ischemia reperfusion brain injury. *J. Cereb. Blood Flow Metab.* **22**, 1068–1079.
21. Williams, A., Ling, G., McCabe, R., and Tortella, F. (2002) Intrathecal CGX-1007 is neuroprotective in a rat model of focal cerebral ischemia. *Neuroreport* **13**, 821–824.
22. Williams, A., and Tortella, F. (2002) Neuroprotective effects of the sodium channel blocker RS100642 and attenuation of ischemia-induced brain seizures in the rat. *Brain Res.* **932**, 45–55.
23. Yao, C., Williams, A., Cui, P., Berti, R., Hunter, J., Tortella, F., and Dave, J. (2002) Differential pattern of expression of voltage-gated sodium channel genes following ischemic brain injury in rats. *Neurotox. Res.* **4**, 67–75.
24. Williams, A.J., Ling, G., Berti, R., Moffett, J.R., Yao, C., Lu, X.M., Dave, J.R., and Tortella, F.C. (2003) Treatment with the snail peptide CGX-1007 reduces DNA damage and alters gene expression of c-fos and bcl-2 following focal ischemic brain injury in rats. *Exp. Brain Res.* **153**, 16–26.
25. Lu, X.C., Williams, A.J., Yao, C., Berti, R., Hartings, J.A., Whipple, R., Vahey, M.T., Polavarapu, R.G., Woller, K.L., Tortella, F.C., and Dave, J.R. (2004) Microarray analysis of acute and delayed gene expression profile in rats after focal ischemic brain injury and reperfusion. *J. Neurosci. Res.* 77, 843–857.
26. Williams, A.J., Berti, R., Dave, J.R., Elliot, P.J., Adams, J., and Tortella, F.C. (2004) Delayed treatment of ischemia/reperfusion brain injury: extended therapeutic window with the proteosome inhibitor MLN519. *Stroke* **35**, 1186–1191.
27. Lu, X.C., Williams, A.J., Wagstaff, J.D., Tortella, F.C., and Hartings, J.A. (2005) Effects of delayed intrathecal infusion of an NMDA receptor antagonist on ischemic injury and peri-infarct depolarizations. *Brain Res.* **1056**, 200–208.
28. Williams, A.J., Myers, T.M., Cohn, S.I., Sharrow, K.M., Lu, X.C., and Tortella, F.C. (2005) Recovery from ischemic brain injury in the rat following a 10 h delayed injection with MLN519. *Pharmacol. Biochem. Behav.* **81**, 182–189.
29. Yao, C., Williams, A.J., Lu, X.C., Hartings, J.A., Yu, Z.Y., Berti, R., Du, F., Tortella, F.C., and Dave, J.R. (2005) Down-regulation of sodium channel $Na_v1.1$ α-subunit mRNA and protein following ischemic brain injury. *Life Science* 77, 1116–1129.
30. Williams, A.J., Dave, J.R., and Tortella, F.C. (2006) Neuroprotection with the proteasome inhibitor MLN519 in focal ischemic brain injury: relation to nuclear factor kappaB (NF-kappaB), inflammatory gene expression, and leukocyte infiltration. *Neurochem. Int.* **49**, 106–112.
31. Kohno, K., Back, T., Hoehn-Berlage, M., and Hossmann, K.A. (1995) A modified rat model of middle cerebral artery thread occlusion under electrophysiological control for magnetic resonance investigations. *Magn. Reson. Imaging* **13**, 65–71.
32. Kuge, Y., Minematsu, K., Yamaguchi, T., and Miyake, Y. (1995) Nylon monofilament for intraluminal middle cerebral artery occlusion in rats. *Stroke* **26**, 1655–1657; discussion 1658.
33. Li, F., Han, S., Tatlisumak, T., Carano, R.A., Irie, K., Sotak, C.H., and Fisher, M. (1998) A new method to improve in-bore middle cerebral artery occlusion in rats: demonstration with diffusion- and perfusion-weighted imaging. *Stroke* **29**, 1715–1719; discussion 1719–1720.
34. Aspey, B.S., Taylor, F.L., Terruli, M., and Harrison, M.J. (2000) Temporary middle cerebral artery occlusion in the rat: consistent protocol for a model of stroke and reperfusion. *Neuropathol. Appl. Neurobiol.* **26**, 232–242.
35. Waynforth, H., and Flecknell, P. (1992) *Experimental and surgical technique in the rat.* Academic, London.
36. Schmid-Elsaesser, R., Zausinger, S., Hungerhuber, E., Baethmann, A., and Reulen, H.J. (1998) A critical reevaluation of the intraluminal thread model of focal cerebral ischemia: evidence of inadvertent premature reperfusion and subarachnoid hemorrhage in rats by laser-Doppler flowmetry. *Stroke* **29**, 2162–2170.
37. Lu, X., Williams, A., and Tortella, F. (2001) Quantitative electroencephalography spectral analysis and topographic mapping in a rat model of middle cerebral artery occlusion. *Neuropathol. Appl. Neurobiol.* **27**, 481–495.

38. Bederson, J., Pitts, L., Tsuji, M., Nishimura, M., Davis, R., and Bartkowski, H. (1986) Rat middle cerebral artery occlusion: evaluation of the model and development of a neurologic examination. *Stroke* 17, 472–476.

39. Bederson, J., Pitts, L., Germano, S., Nishimura, M., Davis, R., and Bartkowski, H. (1986) Evaluation of 2,3,5-triphenyltetrazolium chloride as a stain for detection and quantification of experimental cerebral infarction in rats. *Stroke* 17, 1304–1308.

40. Park, C., Mendelow, A., Graham, D., McCulloch, J., and Teasdale, G. (1988) Correlation of triphenyltetrazolium chloride perfusion staining with conventional neurohistology in the detection of early brain ischaemia. *Neuropathol. Appl. Neurobiol.* **14**, 289–298.

Chapter 3

Clinical and Model Research of Neurotrauma

András Büki, Erzsébet Kövesdi, József Pál, and Endre Czeiter

Summary

Modeling traumatic brain injury represents a major challenge for neuroscientists – to represent extremely complex pathobiological processes kept under close surveillance in the most complex organ of a laboratory animal. To ensure that such models also reflect those alterations evoked by and/or associated with traumatic brain injury (TBI) in man, well-defined, graded, simple injury paradigms should be used with clear endpoints that also enable us to assess the relevance of our findings to human observations. It is of particular importance that our endpoints should harbor clinical significance, and to this end, biological markers ultimately associated with the pathological processes operant in TBI are considered the best candidate. This chapter provides protocols for relevant experimental models of TBI and clinical materials for neuroproteomic analysis.

Key words: Fluid percussion, Impact acceleration, Traumatic brain injury, Biomarkers, Secondary injury, Diffuse axonal injury, Focal injury, Intracranial pressure

1. Introduction

The burden of traumatic brain injury (TBI) could be reduced by targeted therapy based on our understanding of the basic pathobiology caused by/operant in head injury. To provide such knowledge, experimental models should be introduced wherein various forms of injury as well as the efficacy of therapeutic interventions could be studied. The challenge associated with this very issue is based on the selection of the model itself as well as that of the endpoints used to assess injury and therapeutic efficacy.

Discouraging data indicate that despite an extreme variety of animal models introduced to study TBI, not a single experi-

Andrew K. Ottens and Kevin K.W. Wang (eds.), *Neuroproteomics*, Methods in Molecular Biology, vol. 566
DOI 10.1007/978-1-59745-562-6_3, © Humana Press, a part of Springer Science+Business Media, LLC 2009

mental therapeutic approach has been translated to better clinical care of patients *(1, 2)*. This fact was explained in detail by several authors, covering features of the most popular rodent brain (e.g., lissencephaly, size and geometry, white/gray matter ratio, etc.) *(3–5)*. Nevertheless, less emphasis was put on the usefulness of the endpoints utilized *(5)*. Theoretically, four major endpoints might be considered for TBI (1) neurological and cognitive outcome, (2) neuroimaging (3) light and electron microscopic histopathology, and (4) the analysis of cerebrospinal fluid- (CSF), brain tissue-, or serum biomarkers.

Differences between the human and experimental animal brain diminish the merit of the first two purported endpoints, the third endpoint has limited relevance for clinical conditions, leaving biomarkers as the readily assessable endpoint in both situations. Biomarkers may represent a solid basis for evaluation of therapeutic efficacy both in the experimental and the clinical setting *(6–9)*.

In concordance with the aims of this chapter, we define the models and clinical situations suitable for sample collection for neuroproteomic analysis as useful in the discovery of biomarkers. In light of the aforementioned, we provide a brief overview of the biomechanical traits of the available TBI models in **Fig. 1**, which is based on a review by Cernak et al. *(4, 10–15)*. As far as models

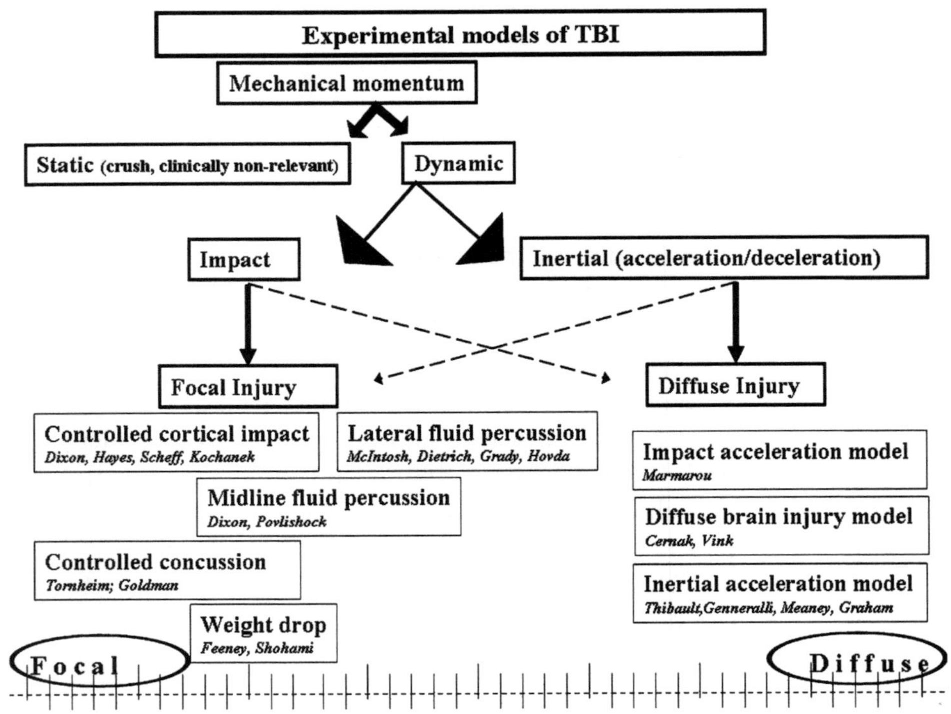

Fig. 1. Biomechanical properties of the most popular models of traumatic brain injury and their location on the scale between primarily focal and primarily diffuse injury (on the basis of the review by Cernak et al. *(4, 10–15)*.

are considered we should appreciate that the complexity of human TBI cannot be recapitulated. Still, we may mimic and identify similarity with the human condition in our models, and test efficacy of experimental therapeutic approaches. To this end, well-characterized, widely used, specific, graded, and reproducible models should be applied. To corroborate our findings, multiple models and strains should be used with rigorous comparison of findings with clinical endpoints (biomarkers provide an excellent tool for the latter premise).

Injury variety is extremely wide in humans, dependent on the type and energy of the forces evoked as well as the preinjury (premorbid) state of the patient (coagulopathy, chronic pulmonary disease, intoxication facilitating secondary injuries), and the environment (hypo/hyperthermia, compression of the body) ***(16–18)***.

General head injury classification (*see* **Table 1**) distinguishes focal injuries (due to static forces or impact-type dynamic forces) and diffuse injuries (inertial, acceleration–deceleration type dynamic forces). The complex nature of injuries sustained in real-life situations results in combined focal and diffuse injuries in most instances. This fact is reflected in the higher mortality of acute subdural hematomas (frequently associated with diffuse brain damage due to dynamic forces) than that of comparable-sized or even larger epidurals (primarily impact/focal injury) ***(16)***.

Secondary injury is another special feature of human TBI that is difficult to model. Although several classifications describe the temporal progression of sequelae associated with TBI, common thought appreciates that at the moment of the trauma, the subject sustains primary injury that is not amenable to therapeutic intervention and only the progression of this primary injury, the occurrence of secondary injuries (e.g., aspiration, blood loss, hypo- or hyperthermia concluding to impaired blood and oxygen supply of the brain), and the progression of secondary injury could and should be influenced by treatment ***(16–21)***.

In light of the aforementioned, this chapter provides a detailed description of two experimental models, which fulfill the earlier-defined criteria, and result in diffuse TBI (impact acceleration) ***(15)*** and focal/combined TBI (fluid percussion) ***(11)***. Finally, we try to summarize the most important steps and features of sample collection for neuroproteomic analysis in preclinical and clinical studies of TBI. For further clarity and simplicity our work will focus on experimental models and clinical situations of moderate/severe TBI. Although ethical issues, including waiver of consent/informed consent, will not be discussed, it is of note that an IACUC or IRB approval is a prerequisite for all experiments and trials listed in this chapter. Data collection and trials should follow national and institutional guidelines.

Table 1
Classification of closed head injury in humans[a,b]

Entity	Evoking momentum	General pathology	Cause (pathoanatomy)
Epidural, acute	Impact	Focal	Rupture of meningeal artery
Epidural, subacute/ chronic (rare)	Impact	Focal	Diploic/emissary vein rupture
Subdural, acute	**Inertial > impact**	**Focal[c]**	**Rupture of bridging vein and/or cortical (pial) artery**
Subdural, subacute/ chronic	Inertial > impact > unidentified	Focal	Rupture of a bridging vein
Traumatically evoked subarachnoid hemorrhage	**Inertial > impact**	**Focal[c]**	**Rupture of cortical (pial) artery**
Cerebral contusion coup	**Impact > Inertial**	**Focal[c]**	**Rupture of cortical (pial) artery + laceration of brain parenchyma**
Cerebral contusion-countrecoup	**Impact > Inertial**	**Focal[c]**	**Rupture of cortical (pial) artery + laceration of brain parenchyma**
Diffuse axonal injury	**Inertial**	**Diffuse**	**Axonal injury**
Diffuse neuronal somatic injury	**Inertial**	**Diffuse**	**Neuronal somatic injury**
Brain swelling	Inertial	Diffuse	Multitargetic severe primary and secondary brain injury
Hypoxic Brain Damage	Inertial	Diffuse	Multitargetic severe primary and secondary brain injury
Diffuse vascular injury	Inertial	Diffuse	Multitargetic severe primary and secondary brain injury

[a]Severe traumatic brain injury: postresuscitation Glasgow Coma Score under 9
[b]Relevance to biomarker studies is indicated with bold fonts
[c]Frequently associated with diffuse injury of brain parenchyma

2. Materials

2.1. Impact Acceleration Brain Injury in the Rat (15, 22, 23)

1. *Anesthesia.* Oxygen gas (O_2), nitrous oxide gas (N_2O), Isoflurane; anesthetizing box; laryngoscope handle – small 2AA size, Miller laryngoscope blade, size 0 and polyethylene tubing (size depends on the weight of the animal); Inspira Advanced Safety Single Animal Volume Controlled Ventilators (Harvard

Apparatus); Ohmeda Tec 4 Anesthetic Vaporizer (Ohmeda Inc., Madison, WI).

2. *Surgery*. Laboratory rat, Wistar, 365–400 g; Marmarou impact acceleration brain injury apparatus (Virginia Commonwealth University, Richmond VA); stereotaxic frame, single manipulator model; Small animal clipper; Betadine; Microsurgery instruments: straight and curved scissors, straight microtweezers, Cooper-type scissors, surgical blade Nos 15 and 11, (*see* **Note 1**); Histoacryl; Lidocain 2%.
3. *Physiological monitoring*. Rectal and temporal temperature microprobe thermometer; temperature control heating pad; pulse oximeter; disposable transducer; polyethylene tubing I.D. 1.57 mm and I.D. 0.58 mm; Heparin; Intellivue MP40 Monitor (Philips, Eindhoven, Netherlands); blood gas analyzer, disposable syringes and needles; dental acrylic.

2.2. Fluid Percussion Brain Injury in the Rat (11, 14, 24)

1. *Anesthesia*. *See* **Subheading 2.1**, **item 1**.
2. *Surgery*. *See* **Subheading 2.1**, **item 2**, and also: Fluid percussion brain injury apparatus set to produce moderate/severe TBI (*see* **Table 2**.); dental drill; a trephine with 4.8-mm diameter.
3. *Physiological monitoring*. *See* **Subheading 2.1**, **item 3**.

Table 2
Setup of the impact acceleration head injury model

mV	atm.	mV	atm.	mV	atm.
147	1	235.2	1.6	323.4	2.2
154.35	1.05	242.55	1.65	330.75	2.25
161.7	1.1	249.9	1.7	338.1	2.3
169.05	1.15	257.25	1.75	345.45	2.35
176.4	1.2	264.6	1.8	352.8	2.4
183.75	1.25	271.95	1.85	360.15	2.45
191.1	1.3	279.3	1.9	367.5	2.5
198.45	1.35	286.65	1.95	374.85	2.55
205.8	1.4	294	2	382.2	2.6
213.15	1.45	301.35	2.05	389.55	2.65
220.5	1.5	308.7	2.1	396.9	2.7
227.85	1.55	316.05	2.15	404.25	2.75

For details see text

2.3. Physiological Monitoring

1. *Arterial blood pressure.* Intra-arterial (A-) line, size: 20G/80 mm; disposable transducer; adapter cable (Philips).
2. *Cerebral perfusion pressure (CPP).* Counted by Philips IntelliVue MP 40 monitor as mean arterial blood pressure (MABP)/ICP (*see* **Note** 2).
3. *Heart rate and blood oxygen saturation (%SpO_2).* Reble SpO_2 sensor (Philips); SpO_2 extension cable (Philips).
4. *Intracranial pressure (ICP).* Intraventricular catheter *(vide supra)*; disposable transducer; adapter cable (Philips).
5. *Skin temperature (T_{skin}).* Reble skin surface temperature probe (Philips).
6. *Rectal temperature (T_{rect}).* Reble esophageal/rectal temperature probe (Philips).
7. *Brain parenchymal oxygen pressure ($P_{br}O_2$) and brain temperature (T_{br}).* Licox IP1.P Licox PMO kit (IP1 single lumen bolt and CC1.P1 PMO catheter combining both oxygen and temperature monitoring); cable connecting catheter to "Y" cable (PMO.CAB); "Y" cable to oxygen and temperature connection on CMP monitor; Licox CMP oxygen and temperature monitor (Integra NeuroSciences; Plainsboro, NJ).

2.4. Brain Tissue, CSF, and Serum-Sample Collection for Neuroproteomic Analysis in Experimental Models

1. *Blood sample collection.* Microcentrifuge tubes; disposable syringes and needles; thiopental.
2. *CSF sample collection.* Microcentrifuge tubes; disposable syringes and needles.
3. *Brain tissue collection.* Luer-type bone remover or Rongeur; microscissor with bended blades; rat-brain-blocking device; homogenizing porcelain vial; BD Vacutainer Plus Plastic Tubes (Becton Dickinson); Proteinase K (Sigma-Aldrich, St. Louis, Mo.); TRI Reagent (Sigma-Aldrich); pellet pestle; microcentrifuge tubes.
4. *Sample processing.* Centrifuge; 200–1,000 μL pipette; Cryovial, size: 2 mL 12.5 × 48 mm with color-coding caps (red = serum, yellow = CSF, green = brain tissue).

2.5. CSF and Serum-Sample Collection for Neuroproteomic Analysis in Clinical Studies

1. *Blood sample collection.* Serum separator tubes; EDTA plasma separator tubes; intra-arterial line, size: 20G/80 mm; central intravenous line; venous line, size: 18G/45 mm; disposable syringes and needles.
2. *CSF sample collection.* Serum collection tubes; CSF drainage system (Codman, Raynham, MA) connected to intraventricular catheter (*see* **Note 3**); disposable syringes and needles.
3. *Sample processing. See* **Subheading 2.4**, **item 4**.

3. Methods

3.1. Impact Acceleration Brain Injury in the Rat

1. *Anesthesia:* Put the rat in a Plexiglas anesthetic chamber for 5 min with 4% isoflurane in 70% N_2O and 30% O_2. Afterward, take the animal out of the chamber. From here you have 1–2 min for endotracheal intubation using the laryngoscope and polyethylene tubing (*see* **Note 4**). After intubation, place the rat onto the operating table and secure the tube around the nose of the animal using thread. Next, connect the respirator's hose to the tube, and maintain anesthesia. Adjust the isoflurane vaporizer to 2–3% and the flowmeter to 400–800 mL/min with 30–70% O_2–N_2O, frequency to 50–60/min, tidal volume to 1.5 mL /100 gbw (*see* **Note 5**).
2. *Surgery.* The rat is operated in the prone position; the head either rests on the operating table or fixed in the Stoelting stereotactic operating device. Clip the fur and disinfectant the area. Expose the skull between the coronal and lambdoid sutures with a midline incision. A metallic disk-shaped helmet with a diameter of 10 mm is firmly glued to this point of the skull (smooth side up). Next, place the animal in the prone position on a foam bed with the metallic helmet centered under the edge of a Plexiglas tube. The rat is prevented from falling by two belts secured to the foam bed. Brass weights weighing 450 g are allowed to fall from a height of 200 cm through the Plexiglas tube directly to the metallic disk fixed to the animal's skull *see* **Note 6**). Immediately after the injury, ventilate the animal with 100% O_2. Remove the helmet and investigate the skull for any sign of fracture, which, if found, would disqualify the animal from further evaluation. The scalp wound is sutured, with the animal remaining on artificial ventilation until spontaneous breathing recovers. Then the rat is placed in a cage until the predetermined survival point.

3.2. Fluid Percussion Brain Injury in Rat

The rat is positioned in the Stoelting apparatus, head shaved and disinfected. Following a long paramedian incision, retract the scalp and the skin laterally and expose the skull between the coronal and lambdoid sutures. Drill two burr holes (1 mm each) into the frontal and occipital bones 2 mm from bregma and lambda and 3.5 mm from the midline (*see* **Note 13**) for the insertion of fixation screws. Halfway between bregma and lambda, drill a 0.5-mm deep hole and insert the trephine; then generate a 48-mm circular craniotomy with care not to disrupt the dura and the superior sagittal sinus (*see* **Note 12**). Next the top portion of the colored hub is cut off from a 20-gauge needle, affixed over the craniotomy site using cyanoacrylate and filled with 0.9% saline.

Insert the two fixation screws into the holes in the right frontal and occipital bones (1 mm is enough). Dental acrylic is applied around the hub (injury cap) and over the screws and allowed to harden to provide stability during the injury induction.

The fluid percussion device comprised a reservoir filled with distilled water, a pendulum with variable height, and an oscilloscope. The height of the pendulum determines pressure of the fluid in the reservoir inflicting injury to the dura and via that to the brain (for details *see* **Table 2**). The right side of the reservoir ends in a piston, with a rubber ring. This end is impacted by the falling pendulum and the force of the impact travels from the rubber to the reservoir filled with water. The device is connected to an oscilloscope which shows the generated fluid pressure (and corresponding severity of injury).

When the dental cement is completely hardened, attach the triple dispenser to the injury cap, fill up with 0.9% saline, and disconnect the rat from the respirator. Hold the rat with your left hand under the forelimbs and place the rat in the vicinity of the device to connect the dispenser. Pull the pendulum to the predetermined height with your right hand and release to inflict the injury. Next the rat is disconnected from the device and reconnected to the respirator.

Following injury, the injury cap is immediately removed en bloc, the dura is investigated for any sign of disruption, which, if found, would disqualify the animal from further evaluation. Animals are monitored for spontaneous respiration and, if necessary, ventilated to ensure adequate postinjury oxygenation. Postinjury recovery times for the following reflexes are recorded: toe pinch, tail pinch, corneal blink, pinnal, and righting. Following recovery of the righting reflex (the animal returns to prone position after being placed on its back or side), animals are placed in a holding cage with a heating pad to ensure maintenance of normothermia and monitored until the appropriate time of killing.

3.3. Physiological Monitoring (See Note 8)

1. *Temperature*. Position the rat on a temperature-controlled heating pad and insert a rectal temperature probe. Clip fur and disinfect the skin. Cut with the tip of a N°11 blade a wound of 2–3 mm just above the temporalis muscle. Insert the temporal temperature probe into and underneath the muscle.
2. *Capillary blood gas saturation* ($SatO_2$). Clamp the sensor of the pulse oxymeter to the ear or to the paw of the animal. Saturation is monitored continuously and can be followed on the screen of the device (*see* **Note 8**)
3. *Arterial blood pressure* (*ABP*). (*See* **Note 9**) The rat is in the supine position with extremities attached to small pins of the operating table via rubber bands. The skin is shaved and disinfected, then incised parallel with the supposed direction

of the femoral artery, which is then identified between layers of fascia and muscles. Following dissection of the vessel a pair of 4.0 threads are positioned under its wall. The proximal one is elevated with a hemostatic clamp attached to the tip of the filament. The deep femoral artery and the distal part of the superficial artery are tied off to prevent backflow. Following such a temporary provision of hemostasis the vessel wall is cut halfway through with microscissor and the PE50 polyethylene tubing is inserted into the lumen above the proximal suture while it has been slightly released. Then the filaments are knotted and the tube is washed with heparinized saline. The distal end of the tube is connected to an external strain gauge and via that to the invasive channel of the monitor. The system is calibrated and ready to use.

4. *Blood gas analysis.* In predetermined intervals 0.1–0.2 mL of arterial blood is drained to a 2-mL sterile syringe and transferred to the lab for blood gas analysis (*see* **Note 10**).
5. *ICP monitoring.* The rat is placed and fixed in a head holder, providing maximum flexion (the incisor bar is not used in this case). Following clipping and disinfection, a midline incision of the skin is made between the top of the occiput and C-Th junction to permit an easy approach to the cisterna magna. Following a midline dissection between the muscles, the atlanto-occipital membrane is reached and identified (*see* **Note 11**). Next, puncture it using a 21-gauge needle – the appearance of clear CSF indicates a good entrance. Insert the catheter with care not to injure the medulla. To prevent the polyethylene catheter from slipping out of the cisterna magna, duracryl glue is used in conjunction with dental acrylic (PMMA). The distal end of the tube is connected to an external strain gauge and via that to the invasive channel of the monitor. The system is calibrated and ready to use.

3.4. Brain Tissue, CSF, and Serum-Sample Collection for Neuroproteomic Analysis in Experimental Models

1. *Blood sample collection.* As alluded to in **Subheading 3.3**, arterial blood samples can be drawn via the A-line from the rat, but to provide an adequate amount for analysis the animal must often be killed with an overdose of barbiturate (*see* **Note 14**). When the heart stops beating one should immediately puncture it with a 21-G needle and withdraw as much blood as is possible. To provide an optimal approach, the puncture should be performed following thoracotomy: the xyphoid process should be elevated with a forceps and the skin incised with a curved, strong scissor. Next the diaphragm is cut and the scissor is turned upward, cutting through the ribs at both sides. The anterior wall of the chest could now be elevated and the heart punctured (*see* **Note 15**). Blood samples are transferred to Cryovial with color-coding caps.

2. *CSF sample collection.* As described earlier, ICP monitoring is executed via a probe inserted into the cisterna magna. CSF samples should be withdrawn via this route with gentle suction of a small (2 mL maximum) syringe (*see* **Note 16**). Samples are stored in microcentrifuge vials with color-coded caps.
3. *Brain tissue collection.* The rat is killed with an overdose of barbiturate and decapitated with a strong Mayo Noble scissor or a Liston scissor. Next, a Luer-type bone cutter is used to remove the skull from the occipital to the frontal region with maximum care provided to preserve the dura. When all pieces of the bone are removed, cut the dura at the midline from the craniocervical- (CC-) junction to the olfactory nerves, then the CC-junction is elevated with a microscissor and the pairs of nerve roots are cut, while the brain is gently pushed up- and forward. Finally, the two huge olfactory nerves are transected, and the brain removed and put into the brain-blocking device (just taken out of the freezer). The area of interest is selected and cut, immediately transferred to a sterile microcentrifuge tube, and snap frozen in liquid nitrogen (*see* **Note 17**).
4. *Sample processing.* After the collection of the blood and CSF samples, microcentrifuge vials are centrifuged at 4, 000 × *g* for 6–8 min. The serum and the supernatant of the CSF are transferred with a pipette into Cryovials with color-coding caps (red = serum, yellow = CSF) that are stored –80°C until further analysis *(25)*.

3.5. CSF and Serum-Sample Collection for Neuroproteomic Analysis in Clinical Studies

Identification of the study population is of primary importance. Despite its complex nature, severe TBI is clinically, the best-defined condition (postresuscitation Glasgow Coma Scale under 9) with the most data on experimental and clinical therapeutic approaches. For this reason, we will focus on this population for the following study design.

1. *Selection of the patient population.* In light of the aforementioned, a relevant study population includes patients with acute subdural, epidural, traumatic subarachnoid, and/or intracerebral hematoma (cerebral contusion) (*see* **Note 18**).
2. *Selection of control group.* Age/GCS/comorbidity matched controls are optimal, where subgroups include patients with raised ICP of different origin than TBI as well as patients with extracranial injury of various severities. A third control group should include patients considered normal controls (*see* **Note 19**).
3. *Definition of sample collection schedule* (*The issue of the therapeutic window.*) The goal is to establish a reliable temporal pattern of protein accumulation in the CSF as well as serum/plasma. This temporal pattern should also be established in terms of circadian (diurnal) variations/rhythm. It is advised that the time of the first sample collection should be as close to that of the injury as possible. Frequency of sample collection should be

adjusted to the findings in the pilot phase, taking into consideration the circadian rhythm and the sensitivity of the biomarker(s) to detect adverse events; in general, 2–4 times/day is advised.

4. *Definition of the routes for sample collection.* Under clinical situations CSF and blood samples are easy to obtain. Although collection of brain tissue samples during routine surgery is feasible (lobar decompression, debridement) ethical issues (consenting), standardization of tissue removal, and thereby the value of such information do not support the application of this research modality (*see* **Note 20**).
5. *Establishment of neuromonitoring.* Basic tools of neuromonitoring include all measures that can provide data on the severity of the primary injury, and occurrence and severity of secondary brain injury. These are: A-line (MABP), central venous (CV)-line (CVP), ventriculostomy (ICP), rectal (core-) and surface temperature, Licox intracerebral (parenchymal) oxygen and temperature monitor, jugular bulb oxymetry, arterial blood gas analysis, end-tidal CO_2 monitoring (*see* **Notes 21** and **22**).
6. Blood samples are collected in serum separator tubes (for serum) and lavender top EDTA tubes (for plasma) – 5.5 mL either from an existing intra-arterial line or central intravenous line or peripheral intravenous line every 6 h (*see* **Note 23**).
7. CSF samples are collected in serum collection tubes (10 mL or as much as retrieved during a 1-h sample collection time point) from the buritrol of the CSF drainage system connected to the intraventricular catheter. It is particularly important that the buritrol of the CSF drainage system be emptied an hour before sample collection (*see* **Note 24**).
8. *Sample processing.* After the collection of the three biofluids (CSF, serum, and plasma) at a predesignated time point (according to the aforementioned time schedule), tubes should be spun in a lab centrifuge at 4,000 × *g* for 6–8 min. Next, 3 × 1 mL of plasma and serum and 5 × 1 mL of CSF are dispended with a pipette into 11 Cryovial with color-coding caps (purple = plasma, red = serum, yellow = CSF). Samples are snap frozen and stored at –80°C until further analysis.

4. Notes

1. Central sterilization facility or laboratory autoclave is suggested.
2. Philips IntelliVue MP 40 monitor uses RDE Data Viewer software (version: A.00.11). The tables in ".csv" format could be managed and edited with Microsoft Office Excel 2003 software. This combination of the monitor and software is capable

of registering biological information every minute and capable of storing it for 96 h.

3. In cases where an ICP monitor is anticipated to be left in place for longer duration (over 5 days), Bactiseal catheters are advised to reduce the risk of bacterial ventriculitis.
4. The tube should be slightly curved, and the end that enters the aditus laryngis should be cut at an angle of 45° to facilitate smooth entry. The blade of the laryngoscope should press the tongue slightly out and downward to provide adequate vision. DO NOT push the aditus with the blade and particularly do not prick it with the tube several times: this leads to severe edema precluding successful intubation. Suction is not advised for the same reason: instead use cottonoid sticks to clear the aditus. If the attempt to intubate fails, let the animal wake up and move onto another animal.
5. This setting generally works but should be continuously revised according to the physiological monitor readings and the reactions (pain) of the animal.
6. The animal should be positioned exactly at the middle of the tube. Any lateralization will interfere with the statistical analysis. To avoid a second hit due to flexibility of the foam, one should immediately grab the thread attached to the metal weights when they are moving upward following injury or, alternatively, the rebound impact could be prevented by sliding the foam bed with the animal away from the tube.
7. Although the noninvasive pulse oxymetry is usually sufficient to monitor the experiment detailed data collection is suggested for at least select animals per experiment, and in all animals in studies applying therapeutic interventions. When invasive monitoring is justified, it should be applied before infliction of trauma. Sham controls should undergo the same monitoring paradigm.
8. The reading is unreliable in hypothermia, hypoperfusion, and extreme levels of pH.
9. The comprehensive handbook of Dongen et al. provides detailed information on microsurgical – vascular interventions performed on the laboratory rat *(26)*.
10. Sodium heparin is sucked into the syringe and then immediately pushed out: the heparin film remaining on the inner surface of the syringe will prevent clot formation.
11. Injury of the occipital artery or diploic veins may lead to severe bleeding. The former may require electrocoagulation, the latter is dealt with bone wax. To achieve maximum flexion of the head, put the incisor bar of the Stoelting at the top of the nose and apply gentle pressure on it, then fix it

in place. Utilization of either a loupe or an operating microscope facilitates successful surgery.

12. The cranial bones of a rat at this weight and age are about 1.3-mm thick.
13. When craniotomy is positioned 3.5 mm over from midline, no contralateral pathology is expected *(27)*.
14. Thiopental might be used i.p. at a dose of 2.5 mg/100 gbw, alone or in conjunction with diazepam for narcosis. For overdosing the rat, a 5–7 times higher dose is suggested.
15. Intensive training of thoracotomy is important: time from cardiac arrest to cardiac puncture (as well as to craniectomy and brain tissue removal) should be minimized!
16. In the Marmarou model, acute phase withdrawal of CSF is sometimes not feasible due to intense tSAH in the cc-junction. In such cases, CSF sampling might be attempted at a later time point. Alternatively, less severe forms of injury should be tested or intraventricular drainage could be used, though it may provide an insufficient amount of CSF.
17. The tools used for homogenization depend on the purported target/subproteome; for further information the authors refer to other chapters of this book.
18. Subacute and chronic injuries harbor too many confounding variables associated with longer hospital stay/diagnostic delay primarily reflected in cumulated secondary injuries as well as extracerebral complications (pulmonary infection, etc.). Thus, the "signal-to-noise ratio" is less optimal in this group.
19. Extracranial injuries may lead to the accumulation of proteins that may interfere with characterizing the TBI neuroproteome itself. False-positive results in polytrauma cases due to such extracranial sources of proteins can diminish the diagnostic value of various biomarker candidates uncovered by neuroproteomics.
20. Serum biomarkers are superior to CSF – every effort should be made to reduce the length of the study period with CSF collection. In this early phase of the study, CSF findings may corroborate data derived from blood samples, with the latter used for subsequent studies.
21. According to the most recent issue of the Guidelines for the management of severe traumatic brain injury *(28)*, either jugular bulb oxymetry or Licox intraparenchymal oxygen pressure and temperature monitoring is advised to predict injury severity and outcome. Although jugular bulb oxymetry is cheaper, it is technically more challenging and, particularly during patient transfer, less reliable. These authors feel that Licox is more reliable, and that a "needle in the brain" is

still less invasive than a device that can interfere with venous outflow and thereby with ICP.

22. Skin temperature and rectal core temperature do not correlate properly with brain temperature; either tympanic membrane- or intraparenchymal temperature probes are advised to follow brain temperature *(29)*.
23. Description of the insertion of the A- and V-lines is beyond the scope of this chapter; any handbook of ICU specialists will provide such details. It is of note, however, that careful disinfection is mandatory just as the maintenance of the patency of these lines. To this end, heparinized normal saline is advised to rinse the taps before and after draining of the samples.
24. Prevention of ventricular infection is the most important precaution during sample collection. The length of the study with CSF sample collection should be as minimal as necessary. Disinfection (Betadine spray, sterile gloves, and syringes) is mandatory. CSF should be drained only from the tap between the buritrol and CSF collecting sack. The CSF sample discarded from the buritrol just before the 60-min sample collection period might be used for other analyses, with recognition of the 5 h storage at room temperature.

Acknowledgments

The authors thank Orsolya Farkas M.D., Ph.D. and Peter Bukovics Ph.D. for their technical help and advice.

References

1. Narayan, R. K., Michel, M. E., Ansell, B., Baethmann, A., Biegon, A., Bracken, M. B., Bullock, M. R., Choi, S. C., Clifton, G. L., Contant, C. F., Coplin, W. M., Dietrich, W. D., Ghajar, J., Grady, S. M., Grossman, R. G., Hall, E. D., Heetderks, W., Hovda, D. A., Jallo, J., Katz, R. L., Knoller, N., Kochanek, P. M., Maas, A. I., Majde, J., Marion, D. W., Marmarou, A., Marshall, L. F., McIntosh, T. K., Miller, E., Mohberg, N., Muizelaar, J. P., Pitts, L. H., Quinn, P., Riesenfeld, G., Robertson, C. S., Strauss, K. I., Teasdale, G., Temkin, N., Tuma, R., Wade, C., Walker, M. D., Weinrich, M., Whyte, J., Wilberger, J., Young, A. B., and Yurkewicz, L. (2002) Clinical trials in head injury. *J. Neurotrauma* **19**, 503–557.
2. Büki, A., and Povlishock, J. (2006) All roads lead to disconnection?–Traumatic axonal injury revisited. *Acta Neurochir. (Wien)* **148**, 181–193; discussion 193–184.
3. Biros, M. (1991) Experimental head trauma models: a clinical perspective. *Resuscitation* 22, 283–293.
4. Cernak, I. (2005) Animal models of head trauma. *NeuroRx* **2**, 410–422.
5. Kazanis, I. (2005) CNS injury research; reviewing the last decade: methodological errors and a proposal for a new strategy. *Brain Res. Brain Res. Rev.* **50**, 377–386.
6. Ottens, A. K., Kobeissy, F. H., Golden, E. C., Zhang, Z., Haskins, W. E., Chen, S. S.,

Hayes, R. L., Wang, K. K., and Denslow, N. D. (2006) Neuroproteomics in neurotrauma. *Mass Spectrom. Rev.* **25**, 380–408.

7. Pineda, J. A., Lewis, S. B., Valadka, A. B., Papa, L., Hannay, H. J., Heaton, S. C., Demery, J. A., Liu, M. C., Aikman, J. M., Akle, V., Brophy, G. M., Tepas, J. J., Wang, K. K., Robertson, C. S., and Hayes, R. L. (2007) Clinical significance of alphaII-spectrin breakdown products in cerebrospinal fluid after severe traumatic brain injury. *J. Neurotrauma* **24**, 354–366.
8. Wang, K. K., Ottens, A. K., Liu, M. C., Lewis, S. B., Meegan, C., Oli, M., Tortella, F. C., and Hayes, R. L. (2005) Proteomic identification of biomarkers of traumatic brain injury. *Expert. Rev. Proteomics* **2**, 603–614.
9. Farkas, O., Polgár, B., Szekeres-Barthó, J., Dóczi, T., Povlishock, J., and Büki, A. (2005) Spectrin breakdown products in the cerebrospinal fluid in severe head injury – preliminary observations. *Acta Neurochir. (Wien)* **147**, 855–861.
10. Cernak, I., Vink, R., Zapple, D., Cruz, M., Ahmed, F., Chang, T., Fricke, S., and Faden, A. (2004) The pathobiology of moderate diffuse traumatic brain injury as identified using a new experimental model of injury in rats. *Neurobiol. Dis.* **17**, 29–43.
11. Dixon, C., Lyeth, B., Povlishock, J., Findling, R., Hamm, R., Marmarou, A., Young, H., and Hayes, R. (1987) A fluid percussion model of experimental brain injury in the rat. *J. Neurosurg.* **67**, 110–119.
12. Lighthall, J., Dixon, C., and Anderson, T. (1989) Experimental models of brain injury. *J. Neurotrauma* **6**, 83–97.
13. Engelborghs, K., Verlooy, J., Van Reempts, J., Van Deuren, B., Van de Ven, M., and Borgers, M. (1998) Temporal changes in intracranial pressure in a modified experimental model of closed head injury. *J. Neurosurg.* **89**, 796–806.
14 McIntosh, T., Noble, L., Andrews, B., and Faden, A. (1987) Traumatic brain injury in the rat: characterization of a midline fluid-percussion model. *Cent. Nerv. Syst. Trauma* **4**, 119–134.
15. Marmarou, A., Foda, M., van den Brink, W., Campbell, J., Kita, H., and Demetriadou, K. (1994) A new model of diffuse brain injury in rats. Part I: Pathophysiology and biomechanics. *J. Neurosurg.* **80**, 291–300.
16. Mendelow, A. D., and Crawford, P. J. (2005) Primary and Secondary Brain Injury, in *Head Injury: Pathophysiology and Management* (Reily, P. L. and Bullock, R., eds.), Hodder Arnold, London, pp. 73–92.
17. Reilly, P. (2001) Brain injury: the pathophysiology of the first hours. 'Talk and Die revisited'. *J. Clin. Neurosci.* **8**, 398–403.
18. Stiefel, M., Tomita, Y., and Marmarou, A. (2005) Secondary ischemia impairing the restoration of ion homeostasis following traumatic brain injury. *J. Neurosurg.* **103**, 707–714.
19. Povlishock, J., and Pettus, E. (1996) Traumatically induced axonal damage: evidence for enduring changes in axolemmal permeability with associated cytoskeletal change. *Acta Neurochir. Suppl.* **66**, 81–86.
20. Sawauchi, S., Marmarou, A., Beaumont, A., Signoretti, S., and Fukui, S. (2004) Acute subdural hematoma associated with diffuse brain injury and hypoxemia in the rat: effect of surgical evacuation of the hematoma. *J. Neurotrauma* **21**, 563–573.
21. Koizumi, J., Yoshida, Y., Nakazawa, T., and Ooneda, G. (1986) Experimental studies of ischemic brain edema: 1. A new experimental model of cerebral embolism in rats in which recirculation can be induced in the ischemic area. *Jpn. J. Stroke* **8**, 1–8.
22. Povlishock, J., Marmarou, A., McIntosh, T., Trojanowski, J., and Moroi, J. (1997) Impact acceleration injury in the rat: evidence for focal axolemmal change and related neurofilament sidearm alteration. *J. Neuropathol. Exp. Neurol.* **56**, 347–359.
23. Rafols, J., Morgan, R., Kallakuri, S., and Kreipke, C. (2007) Extent of nerve cell injury in Marmarou's model compared to other brain trauma models. *Neurol. Res.* **29**, 348–355.
24. Thompson, H., Lifshitz, J., Marklund, N., Grady, M., Graham, D., Hovda, D., and McIntosh, T. (2005) Lateral fluid percussion brain injury: a 15-year review and evaluation. *J. Neurotrauma* **22**, 42–75.
25. Laemmli, U. (1970) Cleavage of structural proteins during the assembly of the head of bacteriophage T4. *Nature* **227**, 680–685.
26. Van Dongen, J. J., Remie, R., Rensema, J. W., and Van Wunnik, G. H. J (eds.) (1991) *Manual of Microsurgery on the Laboratory Rat.* Elsevier, Amsterdam.
27. Vink, R., Mullins, P., Temple, M., Bao, W., and Faden, A. (2001) Small shifts in craniotomy position in the lateral fluid percussion injury model are associated with differential lesion development. *J. Neurotrauma* **18**, 839–847.
28. Bratton, S. L., Chestnut, R. M., Ghajar, J., McConnell Hammond, F. F., Harris, O. A., Hartl, R., Manley, G. T., Nemecek, A., Newell, D. W., Rosenthal, G., Schouten, J., Shutter, L., Timmons, S. D., Ullman, J. S., Videtta, W., Wilberger, J. E., and Wright, D. W. (2007) X. Brain oxygen monitoring and thresholds. *J. Neurotrauma* **24**(supplement 1), S65–S70.
29. Henker, R. A., Brown, S. D., and Marion, D. W. (1998) Comparison of brain temperature with bladder and rectal temperatures in adults with severe head injury. *Neurosurgery* **42**, 1071–1075.

Chapter 4

Neuroproteomic Methods in Spinal Cord Injury

Anshu Chen and Joe E. Springer

Summary

Spinal cord injury (SCI) is a major public health problem with no known effective treatment. Traumatic injury to the spinal cord initiates a host of pathophysiological events that are secondary to the initial insult leading to neuronal dysfunction and death; yet, the molecular mechanisms underlying its dysfunction are poorly understood. Furthermore, while use of imaging methods (e.g., computed tomography scans and magnetic resonance imaging) may help define injury severity and location, they do not elucidate biological mechanisms of SCI progression. The lack of comparable biomarkers for monitoring SCI makes accurate diagnosis and evaluation of SCI progression difficult. Spinal cord contusion is an extensively used SCI model in rats that best represents the etiology of SCI in humans. In this chapter, we describe a two-dimensional (2D) gel electrophoresis-based proteomic approach to investigate the injury-related differences in the proteome and phosphoproteome of spinal cord lesion epicenter at 24 h after spinal cord contusion in rats. The purpose of this study is to elucidate the mechanisms of acute spinal cord dysfunction, as well as discover novel biomarker candidates to evaluate the biological mechanisms of SCI progression and the injury severity.

Key words: Spinal cord injury, Contusion, 2D gel electrophoresis, Rat

1. Introduction

Spinal cord injury (SCI) is a major public health problem affecting 11,000 people in the United States annually. SCI causes severe neuropathology and limited functional recovery. Following the initial insult, there is a delayed and prolonged period of secondary damage that involves a number of destructive pathophysiological and pathochemical cascades. Secondary injury may be amenable to therapeutic interventions and is characterized in part by neuronal and glial necrosis and apoptosis, increased blood–spinal barrier permeability, and neuroinflammatory responses *(1–5)*.

Andrew K. Ottens and Kevin K.W. Wang (eds.), *Neuroproteomics,* Methods in Molecular Biology, vol. 566
DOI 10.1007/978-1-59745-562-6_4, © Humana Press, a part of Springer Science+Business Media, LLC 2009

To date, the precise molecular mechanisms contributing to secondary injury remain elusive, and this incomplete understanding of disease pathogenesis has greatly impeded the development of therapeutic strategies. Furthermore, although use of imaging methods such as computed tomography scans and magnetic resonance imaging can help define the appropriate intervention for a particular patient, they do not elucidate biological mechanisms of SCI progression. In addition, the lack of comparable biomarkers for monitoring SCI makes accurate diagnosis and evaluation of SCI progression difficult.

There are several experimental SCI models available. Spinal cord contusion is the most extensively used SCI model in rats that best represents the etiology of SCI in humans. All SCI therapies tested to date in clinical trials were validated in such models *(6, 7)*. In this chapter, we describe a method and detailed procedure of spinal cord contusion using an Infinite Horizons impactor from Precision Systems and Instrumentation, which is a relatively new, computer-controlled kinetic contusion device and enables the production of consistent graded contusion-type SCI in rodents *(8–11)*.

To elucidate some of the mechanisms of spinal cord dysfunction and discover novel potential biomarker candidates to evaluate injury, it is important to study altered proteins levels and functions contributing to secondary injury *(12–14)*. Phosphorylation is one of the most significant post-translational modification of proteins, which plays an important role in various types of metabolic regulation and signal transduction *(15, 16)*. One of the gold standards for protein separation is two-dimensional (2D) gel electrophoresis *(17)*. We used a 2D gel electrophoresis-based proteomic approach to investigate injury-related differences in the proteome and phosphoproteome of the rat spinal cord lesion epicenter at 24 h after spinal cord contusion. Numerous protein spots were found to exhibit statistically significant differences in expression or relative phosphorylation levels between 24 h SCI and uninjured sham control samples.

2. Materials

2.1. SCI Modeling

1. Subjects: Long-Evans, young adult, female rats (Harlan Sprague Dawley, Indianapolis, IN) weighing approximately 200 g at the time of surgery were used in this study.
2. Surgical attire: surgical scrubs, mask, sterile surgical gloves.
3. Surgical instrument: Infinite Horizons impactor (Precision Systems and Instrumentation, Lexington, KY), glass bead

sterilizer, Plexiglas surgery board, heating pad, hair clippers, forceps, scalpels, scissors, retractors, bone rongeurs, wound closure, needle holder, 5–0 absorbable braided suture, 9-mm wound clips, 18-gauge needle, 30-mL syringe.

4. Sodium pentobarbital, betadine, 70% ethanol, buprenorphine hydrochloride.
5. Tissue isolation buffer: 215 mM mannitol, 75 mM sucrose, 0.1% (w/v) bovine serum albumin, 1 mM EGTA (ethylene glycol-bis(2-aminoethylether)-*N,N,N',N'*-tetraacetic acid), 20 mM HEPES; pH 7.2 with potassium hydroxide. Store at 4°C.
6. Acetone with 20 mM 1, 4-dithiothreitol (DTT). Prepare fresh.
7. Lysis buffer: 7 M urea, 4% (w/v) 3-[(3-cholamidopropyl) dimethylammonio]-1-propanesulfonate (CHAPS), 2 M thiourea, 40 mM Tris (base). Prepare fresh or store in 2-mL aliquots at –20°C, DTT is added just prior to use (final concentration: 40 mM).
8. 2D Quant Kit (Amersham Biosciences, Piscataway, NJ).

2.2. Two-Dimensional Gel Electrophoresis

1. PeppermintStick phosphoprotein molecular weight standards (Molecular Probes, Eugene, OR).
2. Immobiline DryStrip (24 cm, pH 3–10, nonlinear, Amersham Biosciences).
3. Rehydration buffer: 8 M urea, 2% (w/v) CHAPS, 0.5% (v/v) immobilized pH gradient (IPG) buffer (Amersham Biosciences), 0.002% (w/v) bromophenol blue. Store in 1.5-mL aliquots at –20°C, DTT is added just prior to use (final concentration: 20 mM).
4. 1.5 M Tris–HCl, pH 8.8: prepare stock solution in water, and filter through a 0.45-μm filter. Store at 4°C.
5. Sodium dodecyl sulfate (SDS): prepare 20% (w/v) SDS stock solution in water and filter through a 0.45–μm filter. Store at room temperature.
6. SDS equilibration buffer: 50 mM Tris–HCl of pH 8.8, 6 M urea, 30% (v/v) glycerol, 2% (w/v) SDS, trace bromophenol blue, store in 40-mL aliquots at –20°C. Add DTT (final concentration 1% (w/v)) or iodoacetamide (final concentration 2.5% (w/v)) just prior to use.
7. Ammonium persulfate: prepare 10% (w/v) solution in water and immediately store in 0.2-mL aliquots at –20°C.
8. Gel solution recipe: 12.5% (w/v) acrylamide using 30% acrylamide/bis solution (37.5:1 with 2.6% C), 0.375 M Tris–HCl of pH 8.8, 0.1% (w/v) SDS, 0.05% (w/v) ammonium persulfate, 0.033% (v/v) TEMED (*N,N,N',N'*-tetramethylethylenediamine).

9. Water-saturated 1-butanol: mix 1-butanol with water at a ratio of 70:30 in a glass bottle, shake vigorously for several minutes, and wait until separated completely into two layers. Use the top layer. Store at room temperature.
10. Gel storage solution: 0.375 M Tris–HCl of pH 8.8, 0.1% (w/v) SDS. Store at 4°C.
11. SDS electrophoresis buffer: prepare 4X stock with 100 mM Tris base (do not adjust pH), 768 mM glycine, 0.4% (w/v) SDS. Store at room temperature.
12. Agarose sealing solution: add 0.25 g agarose, trace bromophenol blue, and 100 mL 1X SDS electrophoresis buffer into a glass beaker. Microwave until agarose is dissolved. Prepare just prior to use.
13. Pro-Q Diamond phosphoprotein gel stain (Molecular Probes, Eugene, OR).
14. Fixing solution: 50% (v/v) methanol, 7% (v/v) acetic acid. Store at room temperature.
15. Washing solution: 10% (v/v) methanol, 7% (v/v) acetic acid. Store at room temperature.
16. SYPRO Ruby protein gel stain (Molecular Probes, Eugene, OR).
17. Amersham Biosciences Ettan IPGphor II isoelectric focusing system, Ettan DALT*six* gel caster, Ettan DALT*six* large vertical system, thermostatic circulator, Typhoon 9,400 laser scanner.

2.3. Data Analysis

ImageMaster 2D platinum software (Amersham Biosciences).

3. Methods

3.1. SCI Modeling

The surgical procedures assume the use of Infinite Horizons impactor. The procedures are easily adaptable to other contusion devices.

3.1.1. Preoperative Procedures

1. Turn on the heating pad and set to medium head under the Plexiglas surgery board (*see* **Note 1**).
2. Turn on the sterilizer and allow it to warm up. Sterilize surgical instruments by dipping them into the sterilizer for a few seconds and then place them in the sterile surgical area to cool (*see* **Note 2**).
3. Anesthetize rats with 0.15–0.20 mL sodium pentobarbital (40 mg/kg, ip) (*see* **Note 3**).

4. Shave the back hair using hair clippers, removing as much hair as possible from the area of the back where the surgery will be performed (*see* **Note 4**).
5. Prepare the skin for surgery with betadine and then with 70% ethanol over the surgical area.

3.1.2. Operative Procedures

1. Make a dorsal midline incision with a scalpel through the skin but not into the muscle, above the vertebral column and at the level of thoracic segment 6 (T_6)–T_7 to T_{12}–T_{13}. This will need to be about an inch long and open wide enough to perform SCI.
2. Make incisions with a scalpel along each side of the vertebral column to separate the muscle and to form a "trench" for the clamps to hold the rat in the impactor (*see* **Note 5**).
3. Palpate the ribs to determine where the level of T_{13} is (floating and last rib), and then count the intervertebral spaces rostral to T_{10}.
4. Make small incisions with a scalpel to separate the intervertebral spaces at T_{11}–T_{10}, and T_{10}–T_9, and excise the muscle and connective tissue in this area with scissors.
5. Using rongeurs, clip off the spinous processes of T_9, T_{10}, T_{11}, and remove the deep muscle to prepare to remove the body of the T_{10} vertebrae (a dorsal laminectomy).
6. Carefully remove the T_{10} vertebral body using rongeurs until the spinal cord is exposed enough for the impactor rod (*see* **Note 6**).
7. Stabilize the vertebral column of the rat by clamping the vertebrae at T_9 and T_{11} into the Infinite Horizons impactor (keeping the spinal cord in a horizontal position) (*see* **Note 7**).
8. Perform the injury with the Infinite Horizons impactor. A stainless steel-tipped probe (2.5 mm diameter) is rapidly lowered onto the dorsal surface of the spinal cord until an impact force of 150 kdynes is achieved. A data acquisition program subsequently displays the actual force applied to the spinal cord, the maximum spinal cord displacement, and the velocity of the probe at the time of peak force and peak displacement.
9. Remove the rat from the clamps.
10. Clean the injury/surgery site and close the muscles of the incision with 5–0 absorbable braided suture, and then the skin using 9-mm wound clips.

3.1.3. Postoperative Procedures

1. Place rats into an incubator or on a heating pad (33°C) for 2–3 h to recover from anesthesia, and then return to their home cages in the colony room after the rats regain consciousness.

2. Manually express bladders of the spinally injured rats twice daily until euthanized (*see* **Note 8**).

3.1.4. Spinal Cord Tissue Dissection

1. At 24 h after surgery, rats are deeply anesthetized with an overdose of sodium pentobarbital (100 mg/kg, ip) and then decapitated (*see* **Note 9**).
2. Attach an 18-gauge needle to a 30-mL syringe filled with ice-cold tissue isolation buffer. Insert the needle tip into the most caudal region of the vertebral column, and then inject buffer to force the spinal cord out of the vertebral column through the rostral foramen by the pressure of the injection (*see* **Note 9**).
3. Rapidly excise a 5-mm segment of spinal cord that contains the injury epicenter, or the identical region from the spinal cord of the sham control rat.
4. Freeze the dissected spinal cord tissue in liquid nitrogen, and store in –70°C immediately.

3.1.5. Protein Extraction and Quantification

1. Homogenize the lesion epicenter tissue (~3 mg) in 1 mL ice-cold acetone with 20 mM DTT and precipitate proteins at –20°C for 2 h.
2. Remove acetone supernatant by centrifugation, and then further remove residual acetone in protein pellets by lyophilization.
3. Homogenize the protein pellets in ice-cold lysis buffer, sonicate at 100 W for 0.5 min, and centrifuge at 12,000 *g* for 1 h.
4. Measure protein supernatant concentrations in triplicate by 2D Quant Kit.

3.2. Two-Dimensional Gel Electrophoresis

A 2D gel electrophoresis-based proteomic approach is used to investigate injury-related differences in the proteome and phosphoproteome of the rat spinal cord lesion epicenter at 24 h after contusion. Briefly, spinal cord tissue protein samples (sham control or 24 h SCI) are separated by 2D gel electrophoresis. Gels are stained with Pro-Q Diamond phosphoprotein gel stain for phosphorylation level and then stained for total protein expression using SYPRO Ruby protein gel stain. ImageMaster 2D platinum software is used to quantify protein abundances. The ratio of Pro-Q Diamond and SYPRO Ruby signals is used as a quantitative method for determining the relative phosphorylation level of each protein spot (**Fig. 1**).

3.2.1. First Dimension: Isoelectric Focusing

These instructions assume the use of an Ettan IPGphore II System. They are easily adaptable to other formats.

1. Add 1 μg PeppermintStick phosphoprotein molecular weight standards and rehydration buffer to each protein sample (500 μg, 450 μL final volumes), and then rehydrate into IPG strips (Immobiline DryStrip) for 20 h.

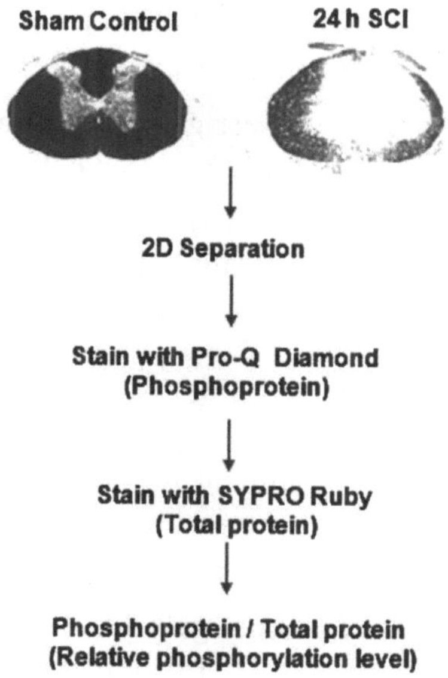

Fig. 1. Differential proteomic platform. Spinal cord tissue protein samples (sham control or 24 h SCI) are separated by 2D gel electrophoresis. Gels are stained with Pro-Q Diamond phosphoprotein gel stain for phosphoprotein expression and then imaged with a Typhoon 9,400 laser scanner. Gels are then stained for total protein expression using SYPRO Ruby protein gel stain and imaged again. ImageMaster 2D platinum software is used to quantify protein abundances. The ratio of Pro-Q Diamond and SYPRO ruby signals is used as a quantitative method for determining the relative phosphorylation level of each protein spot.

2. Isoelectric focus proteins using an Ettan IPGphore II System at 20°C for a total of 70 kVh (step and hold to 500 V for 0.5 kVh, step and hold to 1,000 V for 1 kVh, and step and hold at 8,000 V for 68.5 kVh).

3.2.2. Preparing SDS-PAGE Gels and Electrophoresis Unit

1. Cast 12.5% homogenous SDS-PAGE gels (255 × 205 × 1.0 mm) using an Ettan DALT*six* gel caster. Assemble the gel cassette, and then make up 500 mL gel solution without TEMED and ammonium persulfate; degas for 30 min. Add TEMED and ammonium persulfate to the gel solution, mix briefly, and then immediately pour the gel. Overlay each gel with water-saturated 1-butanol (2 mL), and allow a minimum of 2 h for polymerization (*see* **Note 13**).
2. Pour 4.5 L 1X SDS electrophoresis buffer (lower chamber buffer) into the Ettan DALT*six* Electrophoresis unit, switch on the thermostatic circulator, and adjust the temperature to 25°C (*see* **Note 14**).

3.2.3. Equilibrating IPG Strips

1. Equilibrate the focused IPG strips in equilibration buffer with 1% DTT for 15 min at room temperature.
2. Equilibrate the focused IPG strips again in equilibration buffer with 2.5% iodoacetamide for 15 min at room temperature.

3.2.4. Second Dimension: SDS-PAGE

1. Place the IPG strips on top of 12.5% SDS-PAGE gels, and then seal the IPG strips in place with 0.25% agarose solution.
2. Insert gels into Ettan DALT*six* Electrophoresis unit, and use 1 L 2X SDS electrophoresis buffer as upper chamber buffer.
3. Carry out the second-dimensional SDS-PAGE at 5 W/gel for 30 min followed by 17 W/gel until the bromophenol blue dye front had run off the base of the gel (for about 5 h).

3.2.5. Protein and Phosphoprotein Gel Stain

1. Stain gels with Pro-Q Diamond phosphoprotein gel stain for protein phosphorylation levels, and then image at excitation/emission wavelengths 532/560 nm with a Typhoon 9,400 laser scanner. Representative Pro-Q stained gel images are shown in **Fig. 2**.
2. Stain gels for total protein expression levels using SYPRO Ruby protein gel stain and image again at excitation/emission wavelengths 457/610 nm. Representative SYPRO Ruby stained gel images are shown in **Fig. 2**.

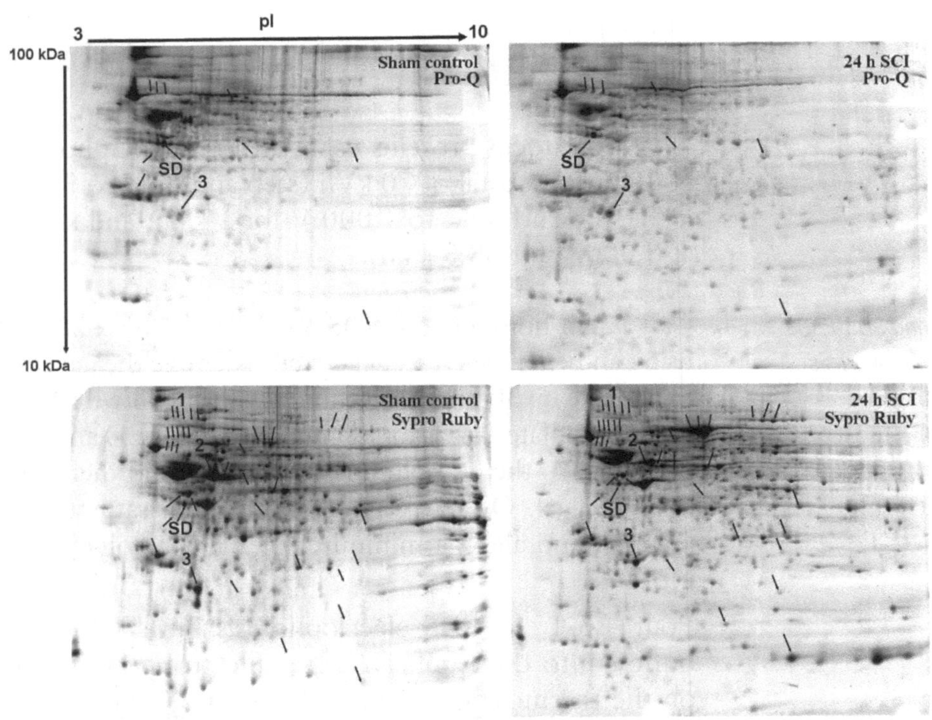

Fig. 2. 2D gel analysis of injury-related differences in the proteome and phosphoproteome of rat spinal cord tissue. Representative Pro-Q and SYPRO Ruby (sham control and 24 h SCI) stained gel images are shown. The spot images were mapped by isoelectric point (*pI*), ranging from 3 to 10, and molecular weight (MW), ranging from 100 to 10 kDa. Ovalbumin was loaded onto each gel as an internal standard (SD). Numerous protein spots (indicated by arrows) were found to exhibit statistically significant differences in expression levels or relative phosphorylation level between the 24 h SCI and uninjured sham control samples.

3.3. Data Analysis

1. Quantify protein abundances by ImageMaster 2D platinum software.
2. Use the ratio of Pro-Q Diamond and SYPRO ruby signals as a quantitative method for determining the relative phosphorylation level of each protein spot (**Fig. 1**).
3. Calculate *p*-values with an unpaired Student's *t* test for each comparison.
4. Calculate changes in expression and relative phosphorylation levels, whereby a fold increase is calculated from the direct (SCI/sham control) abundance ratio, and a fold decrease equals 1/ratio (**Table 1**).

4. Notes

1. Rats should be kept warm during surgery. Rats lose heat rapidly when under general anesthesia, and if heat is not supplied, they can easily die from hypothermia. It is important to maintain body temperature during anesthesia by providing a heat source.

Table 1
Examples of identified differentially regulated proteins

Protein name[a]	Average fold change
Spinal cord injury-specific proteins	
1. Heat-shock protein, HSP 90-alpha (HSP 86)	N/A
Expression level	
2. Glial fibrillary acidic protein	−2.00*
Relative phosphorylation level	
3. Ubiquitin carboxyl-terminal hydrolase isozyme L1	6.26**

ImageMaster 2D platinum software was used to quantify protein abundances. The ratio of Pro-Q Diamond and SYPRO ruby signals was used as a quantitative method for determining the relative phosphorylation level of each protein spot. Average fold abundance increase was calculated from the direct (SCI/sham Control) abundance ratio, and a fold decrease was equal to 1/abundance ratio.

[a] Numbers refer to spot numbers in **Fig. 2**.

*$p < 0.05$; **$p < 0.01$ unpaired Student's *t* test ($n = 6$).

2. All surgical procedures should be performed under aseptic conditions. Begin surgery with sterile instruments and handle instruments aseptically. Surgical instruments should be cleaned prior to submersion into the glass beads of a sterilizer. Surgeons should wash and dry their hands before aseptically donning sterile surgical gloves.
3. Rats are allowed access to food and water *ad libitum*. Fasting can cause hypoglycemia and dehydration, which can cause anesthetic complications and death. Animal's vital signs should be closely monitored throughout the surgery.
4. Hair removal should be performed in an area separate from where the surgery is to be performed.
5. Avoid making a deep incision, as this could puncture the pleural cavity (collapsing a lung and killing the rat).
6. Do not touch the spinal cord while exposing it during the laminectomy procedure.
7. This injury parameter (150 kdynes) produces a "mild/moderate" SCI. Rigid stabilization of the spinal column is very important for the production of a consistent graded contusion-type SCI.
8. If the rats need to survive for more than 24 h after SCI, the general condition of the animal should be closely monitored postoperatively. Rats should have their bladders manually expressed twice daily until micturition returns. Prophylactic antibiotic treatment should be provided. The skin and fur should be washed and dried as necessary. Analgesics (buprenorphine hydrochloride: 0.01–0.05 mg/kg, sc) should be administered every 6–12 h for 3–5 days. Rats with survival periods of at least 1 week should be examined for early signs of autophagia, and preventative strategies should be implemented.
9. Decapitate rats and remove spinal cord as soon as possible (in about 1 min).
10. Unless stated otherwise, all solutions should be prepared in double distilled water with a resistivity of 18.2 MΩ.cm at 25°C, and a total organic content of less than five parts per billion. This standard is referred to as "water" in this text.
11. Acrylamide is a neurotoxin when unpolymerized, so use of gloves and a mask is required at all times. TEMED is corrosive and highly flammable and should be handled in a fume hood.
12. 1-Butanol has significant, unpleasant smell and should be handled in a fume hood.
13. TEMED is used in conjunction with ammonium persulfate to accelerate acrylamide polymerization. 1-Butanol generates a significant, unpleasant smell; cast gels in a fume hood.

14. Adequate cooling to keep the buffer precisely at 25°C is essential to prevent heat-induced damage to the gels.

Acknowledgments

The authors would like to thank Dr. Melanie L. McEwen and Dr. Rangaswamy Rao Ravikumar for technical assistance. This work was supported by PHS grant NS46380 and an endowment from Cardinal Hill Rehabilitation Hospital.

References

1. Grossman, S. D., Rosenberg, L. J., and Wrathall, J. R. (2001) Temporal-spatial pattern of acute neuronal and glial loss after spinal cord contusion. *Exp. Neurol.* **168**, 273–282.
2. Springer, J. E. (2002) Apoptotic cell death following traumatic injury to the central nervous system. *J. Biochem. Mol. Biol.* **35**, 94–105.
3. Bareyre, F. M., and Schwab, M. E. (2003) Inflammation, degeneration and regeneration in the injured spinal cord: insights from DNA microarrays. *Trends Neurosci.* **26**, 555–563.
4. Ahn, Y. H., Lee, G., and Kang, S. K. (2006) Molecular insights of the injured lesions of rat spinal cords: Inflammation, apoptosis, and cell survival. *Biochem. Biophys. Res. Commun.* **348**, 560–570.
5. Ling, X., and Liu, D. (2007) Temporal and spatial profiles of cell loss after spinal cord injury: Reduction by a metalloporphyrin. *J. Neurosci. Res.* **85**, 2175–2185.
6. Pearse, D. D., and Bunge, M. B. (2006) Designing cell- and gene-based regeneration strategies to repair the injured spinal cord. *J. Neurotrauma* **23**, 438–452.
7. Onifer, S. M., Rabchevsky, A. G., and Scheff, S. W. (2007) Rat models of traumatic spinal cord injury to assess motor recovery. *ILAR J.* **48**, 385–395.
8. Scheff, S. W., Rabchevsky, A. G., Fugaccia, I., Main, J. A., and Lumpp, J. E., Jr. (2003) Experimental modeling of spinal cord injury: Characterization of a force-defined injury device. *J. Neurotrauma* **20**, 179–193.
9. Cao, Q., Zhang, Y. P., Iannotti, C., DeVries, W. H., Xu, X. M., Shields, C. B., and Whittemore, S. R. (2005) Functional and electrophysiological changes after graded traumatic spinal cord injury in adult rat. *Exp. Neurol.* **191**, S3–S16.
10. Ravikumar, R., McEwen, M. L., and Springer, J. E. (2007) Post-treatment with the cyclosporin derivative, NIM811, reduced indices of cell death and increased the volume of spared tissue in the acute period following spinal cord contusion. *J. Neurotrauma* **24**, 1618–1630.
11. McEwen, M. L., Sullivan, P. G., and Springer, J. E. (2007) Pretreatment with the cyclosporin derivative, NIM811, improves the function of synaptic mitochondria following spinal cord contusion in rats. *J. Neurotrauma* **24**, 613–624.
12. Denslow, N., Miche, M. E., Temple, M. D., Hsu, C. Y., Saatman, K., and Hayes, R. L. (2003) Application of proteomics technology to the field of neurotrauma. *J. Neurotrauma* **20**, 401–407.
13. Wang, K. K., Ottens, A., Haskins, W., Liu, M. C., Kobeissy, F., Denslow, N., Chen, S., and Hayes, R. L. (2004) Proteomics studies of traumatic brain injury. *Int. Rev. Neurobiol.* **61**, 215–240.
14. Ottens, A. K., Kobeissy, F. H., Fuller, B. F., Liu, M. C., Oli, M. W., Hayes, R. L., and Wang, K. K. (2007) Novel neuroproteomic approaches to studying traumatic brain injury. *Prog. Brain Res.* **161**, 401–418.
15. Cohen, P. (1982) The role of protein phosphorylation in neural and hormonal control of cellular activity. *Nature* **296**, 613–620.
16. Cohen, P. (1992) Signal integration at the level of protein kinases, protein phosphatases and their substrates. *Trends Biochem. Sci.* **17**, 408–413.
17. Gygi, S. P., Corthals, G. L., Zhang, Y., Rochon, Y., and Aebersold, R. (2000) Evaluation of two-dimensional gel electrophoresis-based proteome analysis technology. *Proc. Natl. Acad. Sci. USA* **97**, 9390–9395.

Chapter 5

Modeling Substance Abuse for Applications in Proteomics

Scott E. Hemby and Nilesh Tannu

Summary

The ability to model aspects of human addictive behaviors in laboratory animals provides an important avenue for gaining insight into the biochemical alterations associated with drug intake and the identification of targets for medication development to treat addictive disorders. The intravenous self-administration procedure provides the means to model the reinforcing effects of abused drugs and to correlate biochemical alterations with drug reinforcement. In this chapter, we provide a detailed methodology for rodent intravenous self-administration and the isolation and preparation of proteins from dissected brain regions for Western blot analysis and high-throughput proteomic analysis. Examples of cocaine-induced alterations in the abundances of ionotropic glutamate receptor subunits in reinforcement-related brain regions are provided.

Key words: Cocaine, Self-administration, Glutamate, Reinforcement, Infrared immunoblotting

1. Introduction

A generally accepted tenet in drug abuse research is that certain drugs can function as reinforcing stimuli that contribute to their abuse liability in humans, and can be modeled in animal. The ingestion of drugs, such as cocaine, exerts maladaptive effects on various biochemical substrates in several brain regions, which in turn lead to further intake, drug craving, binge administration, and relapse. A significant amount of research investigating the neurobiology of drug abuse is conducted in animal models (rodent and nonhuman primate) that closely resemble characteristics of human drug intake. Various behavioral models have been developed, and are commonly used for studying the effects of abused drugs on brain

Andrew K. Ottens and Kevin K.W. Wang (eds.), *Neuroproteomics,* Methods in Molecular Biology, vol. 566
DOI 10.1007/978-1-59745-562-6_5, © Humana Press, a part of Springer Science+Business Media, LLC 2009

biochemistry including, self-administration, place conditioning, locomotor activity and sensitization, and intracranial self-stimulation, to name a few. Although each model plays an important role in helping us understand the effects of drugs on behavior and biochemistry, the self-administration paradigm provides the ability to directly link such changes to the *reinforcing* effects of the drug. Generally, the self-administration paradigm involves the emission of specific behavior(s) by the animal or human (e.g., lever-press; nose-poke) that is maintained by drug administration (e.g., intravenous, oral, or intracranial). There are several advantages of the self-administration paradigm, including that substances abused by humans can function as positive reinforcing stimuli under laboratory conditions, the ability to generate clear dose-effect curves *(1, 2)* and the homology of brain regions between species (rodent, nonhuman primate and human) that contribute to the reinforcing effects of the drugs.

The advent of high-throughput screening technologies has produced a paradigm shift in the manner in which scientists are able to detect and identify molecular mechanisms related to disease. Proteomic analysis strategies allow the simultaneous assessment of thousands of genes and proteins of known and unknown function, thereby enabling a global biological view of addictive disorders. A major challenge in proteomic biology is to understand the function of proteins in the context of human disease, such as in addictive disorders. Broad scale evaluations of protein expression are well suited to the study of drug abuse given the multigenic nature of drug addiction, the anatomical and cellular complexity of the brain compared with other tissues (including the vast representation of expressed proteins in the brain), and the relatively limited knowledge of the molecular pathology of these disorders *(3–5)*.

2. Materials

2.1. Intravenous Catheter for Rat Self-Administration

1. The majority of equipment for catheterization can be produced in the lab, or is commercially available through Med Associates Inc., Coulbourne Instruments and other manufacturers.
2. Backplate and polyvinyl chloride tubing catheter *(6, 7)*. Catheters are composed of 25-ga polyethylene tubing with two-2 mm sections of 20-ga polyethylene tubing placed over the smaller tubing at 2.5 cm and 12 cm from the proximal end. Between the points, the smaller tubing is looped to prevent crimping. Two 10-cm strands of suture are tied at each of these points for anchoring the catheter to the muscle.

3. Spring leash to protect the catheter and provide points of connection with backplate and swivel.
4. Single channel swivel to enable side-to-side motion of the animal while connecting the syringe.
5. Syringe infusion pump. Investigators should take into account the motor speed of the syringe pump and the diameter of syringe in determining the volume of drug that will be delivered through the catheter.
6. Operant chamber with response lever(s), stimulus lights, and tone generator. Commercially available from Med Associates, Inc. and Coulbourn Instruments.
7. Bacteriostatic heparinized saline (1.7 units/mL).
8. Drugs for anesthesia (pentobarbital and thiopental) as well as for self-administration, which require state and federal government licensure for the investigator.

2.2. Protein Isolation, Fractionation, and Quantification (see Note 1)

1. Phosphate buffered saline (10×).
2. TSE buffer: 10 mM Tris, pH 7.5, 300 mM sucrose, 1 mM EDTA.
3. CLB buffer: 100 mM HEPES, 10 mM NaCl, 1 mM KH_2PO4, 5 mM $NaHCO_3$, 1 mM $CaCl_2$, 0.5 mM $MgCl_2$.
4. Commercial protease inhibitor cocktail (e.g., Halt Protease Inhibitor Cocktail, Pierce Biotechnology) or 10 mM HEPES, 10 mM NaCl, 1 mM KH_2PO_4, 5 mM $NaHCO_3$, 1 mM $CaCl_2$, 0.5 mM $MgCl_2$, 5 mM EDTA, and the following protease inhibitors: 1 mM phenylmethylsulfonylfluoride, 10 mM benzamidine, 10 μg/mL aprotinin, 10 μg/mL leupeptin, and 1 μg/mL pepstatin. Commercial phosphatase inhibitor (e.g., Halt Phosphatase Inhibitor Cocktail, Pierce Biotechnology/Thermo Scientific) (*see* **Note 2**).
5. Suspension buffer: 20 mM Tris-HCl, pH 8.0, 1 mM ETDA, with protease inhibitor cocktail.
6. Sucrose buffer: 10 mM Tris-HCl, pH 7.5, 300 mM sucrose, 1 mM EDTA, 0.1% NP40 with protease inhibitor cocktail.

2.3. Infrared Western Blotting

1. Precast polyacrylamide gels. Store at 4°C.
2. Running buffer (1× TGS buffer): 25 mM Tris-HCl, pH 8.3, 192 mM glycine, 0.1% SDS. Store at room temperature.
3. Transfer buffer (1× TG buffer): 25 mM Tris-HCl, 192 mM glycine, 20% methanol. Store at room temperature.
4. Wash buffer: 1× PBS, 0.1% Tween-20. Store at room temperature.
5. Transfer membrane: Pure nitrocellulose membrane (0.2 μm).
6. 3MM chromatography paper.

7. Primary antibodies: NR1, 1:300 mouse monoclonal (MAB363, Chemicon: recognizes NR1–1a, NR1–1b, NR1–2a, NR1–2b splice variants); NR2A, 1:300 mouse monoclonal (MAB1526, Chemicon); NR2B, 1:1,000 rabbit polyclonal (07–351; Millipore); neuronal tubulin, 1:1,000 mouse monoclonal (05–559, Millipore). Polyclonal antibodies for AMPA and kainate receptor subunits are purchased from Millipore and used at a dilution of 1:1,000.
8. Secondary antibodies: IRdye 800 conjugated affinity purified (Rockland Immunochemicals, Gilbertsville, PA); Alexa Fluor 680 (Invitrogen, Carlsbad, CA). Store at 4°C.
9. LiCor Odyssey Infrared Imaging System (Li-Cor Biosciences, Lincoln, NE).
10. Blocking buffer: Odyssey blocking buffer (LiCor). Store at 4°C.
11. Pre-stained molecular markers: Precision Plus Protein Western C standards (Bio-Rad, Hercules, CA). Store at 4°C.

2.4. Protein Processing for Proteomics

1. Sample clean-up kits such as 2D clean-up kit (GE HealthCare); ReadyPrep 2-D Cleanup Kit (Bio-Rad) and 2-D Sample Prep Kit (Pierce Biotechnology).
2. Sample buffer: 30 mM Tris-HCl, 2 M thiourea, 7 M urea, and 4% CHAPS, pH 8.5. (*see* **Note 3**).
3. 2D-Quant kit (GE HealthCare).
4. CyDye DIGE Fluor minimal dyes 2, 3, and 5 (GE HealthCare).
5. >99.5% pure dimethylformamide (DMF, USB Corporation) and 10 mM lysine.
6. Rehydration buffer: 2 M thiourea, 7 M urea, 2% dithiothreitol (DTT), 4% CHAPS and 2% Pharmalyte (*see* **Note 3**).
7. Destreak rehydration buffer (GE HealthCare).

2.5. Two-Dimensional Polyacrylamide Gel Electrophoresis (2D-PAGE)

1. EttanTMIPGphorTMapparatus(GEHealthCare);alternative equipment is PROTEAN IEF System (Bio-Rad), Multiphor II Horizontal Electrophoresis Unit (PerkinElmer) or UniPhor Horizontal Electrophoresis Unit(Sigma-Aldrich).
2. Immobiline DryStrips: 240 ×3 × 0.5 mm^3, linear pH ranges 4–7/6–9/3–10 (GE Healthcare); alternative sources are Bio-Rad, Sigma-Aldrich, and Isogen Lifesciences.
3. Ettan Dalt II System (GE HealthCare); alternative equipment is PROTEAN Plus Dodeca Cell (Bio-Rad).
4. Precast 8–15% gradient SDS-PAGE (2,400 × 2,000 × 1 mm; Jule Inc.).

5. Mass spectrometer: Applied Biosystems 4700/4800 Proteomics Analyzer (MALDI-TOF/TOF); most other mass spectrometers are capable of analysis; however, MALDI-TOF-TOF has an advantage of higher throughput and provides detailed peptide sequence analysis.

3. Methods

Historically, the abuse liability of cocaine is attributed to the direct effects of the drug on dopamine uptake blockade, yielding elevated extracellular dopamine concentrations that occur in discrete areas of the brain, specifically the nucleus accumbens (NAc), ventral tegmental area (VTA), prefrontal cortex – data from human subjects *(8–10)* as well as animal models *(2, 11–15)*. More recently, attention has focused on significant alterations in glutamatergic transmission in the VTA and NAc following cocaine administration in rodents and humans, which has been associated with the neuroplasticity of cocaine addiction *(16–21)*. A further delineation of the neural contributions and alterations of addictive behaviors has been postulated recently, in which the dysregulation of prefrontal glutamatergic projections to NAc is identified as an essential component. Briefly stated, prefrontal cortical dopamine alterations lead to preferential responding for drug-related stimuli, whereas accumbal-glutamatergic alterations underlie the unmanageable aspects of drug-seeking behaviors *(22, 23)*.

In many respects, substance abuse can be seen as a disease of synaptic dysregulation and pathology. Several studies have demonstrated significant morphological and electrophysiological disturbances in mesolimbic brain structures reflecting significant synaptic modification following cocaine administration – effects that are mediated predominantly by ionotropic glutamate receptors. Since subunit composition determines the functional properties of ionotropic glutamate receptors *(24)*, alterations in the abundance of ionotropic glutamate receptor subunits in specific brain regions are correlated with changes in neuronal excitability and synaptic strength, principles underlying long-term biochemical and behavioral effects of cocaine that, in turn, may affect subsequent drug intake.

Our group and others have examined the effects of self-administered cocaine on glutamate dysregulation in rats, nonhuman primates, and in postmortem tissue from cocaine overdose victims.

In this chapter, we provide an overview of methods used for rodent self-administration as well as biochemical procedures used to explore changes in glutamate receptor abundances in brain regions following self-administration.

3.1. Intravenous Catheter Implantation for Rat

1. Rats are anesthetized by administration, and surgery is undertaken using aseptic surgical procedures.
2. A 3-cm region is shaved on the right region of the underside of the neck along with a 5 × 5-cm region on the back, 6 cm from the base of the neck and proceeding caudally. Both areas are hand shaved to remove remaining fur, and swabbed with a tincture of iodine.
3. A 3-cm incision is made in the shaved area of the back, 1.5 cm lateral from the midline, and covered with sterile gauze. Following, a 2.5-cm incision is made in the shaved area of the neck above the jugular vein. The jugular vein is exposed and cleared of fascia.
4. A trocar (3-mm diameter) is inserted at the base of the neck incision, and is guided subcutaneously around the caudal region of the right forelimb to exit at the center of the back incision. The distal end of the polyethylene catheter is inserted through the trocar to exit at the back. Holding the proximal end of the catheter, the trocar is removed through the back incision.
5. The jugular vein is reexposed, and a 23-ga needle is used to puncture the vein to provide a point of entry for the proximal end of the catheter. The catheter is inserted as the needle is withdrawn, and extends to just outside the right atrium. The catheter is anchored to muscle near the point of entry into the vein.
6. The incision is sutured, and treated topically with Neosporin antibiotic powder. The distal end of the catheter is guided through a Teflon shoulder harness. The harness provides a point of attachment for a spring leash connected to a single channel swivel at the opposing end. The catheter is threaded through the leash for attachment to the swivel.
7. The fixed end of the swivel is connected to a syringe by polyethylene tubing. The syringe is placed in a computer controlled motor driven syringe pump. An infusion of thiopental is administered as needed to assess catheter patency.
8. Immediately following surgery, an analgesic should be administered. Rats should be monitored every 30 min after surgery, until conscious, and a minimum of three times per day for 2 days following surgery. After this time, rats are observed a minimum of two times per day for the remainder of the experiment. The health of the rats should be monitored according to the guidelines issued by the Institutional Animal Care and Use Committee and the National Institute of Health.

3.2. Rat Self-Administration Procedures

1. These instructions utilize the Med Associates Drug Self-Administration Chamber for rats, but the general instructions can be adapted to other operant apparatus.
2. Rats are transferred from their home cage to an operant conditioning chamber enclosed in a sound attenuated box. Each chamber contains a response lever, a house light, a tone generator, a ventilation fan, and a syringe pump located outside the sound attenuated box.
3. Computer software is used for programming of the self-administration session and for data collection (Med Associates). Extraneous noise is masked by the ventilation fan and white noise generator in the room. Prior to each session, the swivel and catheter should be flushed with heparinized saline before connecting the catheter to the infusion pump via a 20-ga Luer hub and 28-ga male connector.
4. Responding is normally engendered under a fixed ratio one (FR1) schedule of reinforcement, whereby one response on the lever results in the infusion of the drug. Upon completion of the response requirement, a drug infusion is delivered and a time-out is in effect. During the time-out, the lever light is extinguished, the house light illuminated, and a tone generated. The end of the time-out is signaled by illumination of the lever light and extinguishing of the house light and tone. During the time-out period, responses on levers are recorded, but have no scheduled consequence. Following stable responding, the schedule of reinforcement can be adjusted according to the study design.
5. It is noted that numerous studies have described self-administration in nonhuman primates, but the behavioral and procedural complexity requires significant training, expertise, and resources that are beyond the scope of this chapter.

3.3. Necropsy and Dissection

1. Following the completion of experimental studies, subjects should be euthanized according to the guidelines set forth by the Institutional Animal Care and Use Committee. We recommend the most expedient and humane method of sedation. Please keep in mind that sedatives may affect certain proteins of interest (e.g., GABA receptors). Ensure complete sedation (unresponsive to tactile and painful stimuli). The following steps pertain to necropsy and dissection for both rat and monkey.
2. Following intracardial perfusion with ice cold PBS, craniotomy is performed exposing the brain. Following removal of bone and dura, the brain is removed, rinsed with 1× PBS (4°C) and placed immediately in a brain matrix at 4°C. (*see* **Note 4**)
3. The brain is blocked in the plane of interest (e.g., coronally). From the rostral to caudal aspects, the brain blocks are

removed and placed caudal side down on an aluminum plate. At this point, the regions of interest can be dissected immediately or the blocks can be stored at –80°C for later use.

4. Upon dissecting the regions of interest from rodent, monkey, or human samples, tissue is pulverized using a metal mortar and pestal (kept on dry ice) in the presence of liquid nitrogen. Following evaporation of the liquid nitrogen, the pulverized tissue is stored in Eppendorf tubes and kept at –80°C.

3.4. Protein Isolation

1. Pulverized tissue samples are Dounce homogenized in the presence of RIPA lysis buffer and protease inhibitors. Crude protein homogenates can be used for analysis or processed further using a variety of fractionation protocols. Commonly, we use a fractionation protocol, which yields membrane, cytosolic and nuclear protein fractions.
2. Homogenates are centrifuged at 7,500 × *g* for 5 min. The supernatant is removed and the pellet (nuclei and debris) are resuspended in Suspension Buffer and centrifuged at 7,500 × *g* for 5 min.
3. Repeat twice and resuspend pellet in the solution and store at –20°C (nuclear fraction).
4. Centrifuge supernatant at 25,000 × *g* for 30 min at 4°C. Following, the supernatant containing the cytosolic fragment is removed and stored at –20°C (cytosolic fraction).
5. Resuspend pellet in Sucrose Buffer and centrifuged at 5,000 × *g* for 5 min at 4°C. The supernatant is discarded, and the pellet is resuspended in the buffer and washed three times before resuspension in the buffer and protease inhibitors and storing the samples at –20°C (membrane fraction).
6. Protein concentrations of samples are calculated using the Bicinochoninic Acid (BCA) Protein Assay Kit (Pierce, Rockford, IL).
7. Mix reagents and prepare serial dilutions of standards 0–2 μg/μL as described in the BCA Assay protocol and incubate for 30 min at 37°C.
8. Samples are quantified using spectrophotometer, and concentrations are determined according to the standard curve.

3.5. SDS-Polyacrylamide Gel Electrophoresis

1. These instructions use the BioRad Ready Gel System, but are easily adaptable to other formats. Similarly, these instructions are specific for infrared immunoblotting using the LiCor Odessey imaging system. However, visualization of proteins using chemiluminescence can also be performed.
2. Dilute samples in 1.5 mL centrifuge tubes with Laemmeli sample buffer to achieve the final protein concentrations. Place

tubes in a heat block (95°C) for 5 min and then immediately on ice. After cooling, centrifuge samples briefly to collect contents in bottom of tube.

3. Prepare a Tris-HCl gel of the percent polyacrylamide according to the protein(s) of interest. Carefully remove and discard the adhesive film on the bottom of the gel and the inserted comb.
4. Add 1× Running Buffer to the chamber, covering the wells. Load identical concentrations of proteins along with prestained molecular weight marker.
5. Connect the assembly to a power supply. The gel can be run at 25–30 V or at 90–100 V if a cooling unit is available. Bromophenol blue will be the leading dye front, and gel electrophoresis should be stopped when it reaches the anodic end of the gel.

3.6. Immunoblot Analysis

1. A tray containing transfer buffer needs to be of sufficient size to lay out the transfer cassette with the requisite foam and two pieces of 3MM Whatman paper. One fiber pad is placed on each side of the open cassette followed by a piece of 3MM paper. A section of nitrocellulose membrane cut slightly larger than the size of the separating gel is saturated with transfer buffer. The gel is removed from the gel unit and placed on top of one sheet of the saturated 3MM paper. The saturated nitrocellulose membrane is placed on top of the gel followed by the remaining piece of 3MM paper. A long thin cylinder (such as a pipette) is rolled over the 3MM paper to remove any bubbles. Afterwards, the remaining fiber pad is placed in top of the 3MM paper and the cassette is closed.
2. The transfer cassette is placed into the transfer tank with the nitrocellulose membrane between the gel and the anode **(Fig. 1)**. This is extremely important, as an incorrect orientation will result in the proteins being electrophoresed into the buffer instead of being transferred onto the nitrocellulose.
3. A refrigerated circulating water bath is used to maintain temperature between 5 and 10°C.
4. The lid to the transfer tank is secured and the power supply is activated at 25 V for 12 h or 65 V for 3 h.
5. Following completion of the transfer, the cassette is removed and disassembled by removing the top fiber pad and 3MM paper. The nitrocellulose membrane is removed, placed in a small tray, and rinsed with transfer buffer for 5 min. The colored molecular markers should be clearly visible on the membrane.
6. The nitrocellulose membrane is incubated with 50% LiCor Buffer/50% PBS for 1 h at room temperature on a rocking platform.

Fig. 1. Immunoblotting preparation showing location of fiber pad, Whatmann paper, nitrocellulose, and gel.

7. The blocking buffer is discarded, and the membrane is incubated with the desired primary antibody/antibodies, diluted in 50% LiCor Buffer/50% Wash Buffer for 12–15 h at 4°C on a rocking platform. In addition to the antibody of interest, we also include a neuronal tubulin antibody to ensure equal protein loading and efficiency of transfer (*see* **Note 5**).
8. The primary antibody is removed, and the membrane is rinsed with Wash Buffer four times for 5 min each to remove excess primary antibody. Add secondary antibody in 50% Odyssey buffer/50% Wash buffer. Monoclonal antibodies are visualized with goat anti-mouse AlexFluor680 conjugated secondary antibody (A21076, Molecular Probes), and polyclonal antibodies were visualized with goat anti-rabbit IRDye800 conjugated secondary antibody (610–132–121, Rockland Immunochemicals) diluted 1:15,000. Incubate for 2 h at room temperature on a rocking platform. Be sure to cover the incubating tray with aluminum foil to prevent bleaching of the conjugated fluorophore on the secondary antibody.
9. The incubation buffer is discarded and the membrane is rinsed three times for 15 min each with Wash Buffer, followed by two rinses with 1× PBS two times for 15 min each. The membrane is scanned on the LiCor Odyssey infrared scanner or stored at 4°C for later analysis. Membrane should be protected from light as noted earlier.
10. Membranes are scanned with the Licor Odyssey infrared scanner, and signals are quantified with Odyssey version 1.2 software. Signal intensities for proteins of interest were reported as percent control relative to tubulin **(Fig. 2)**.

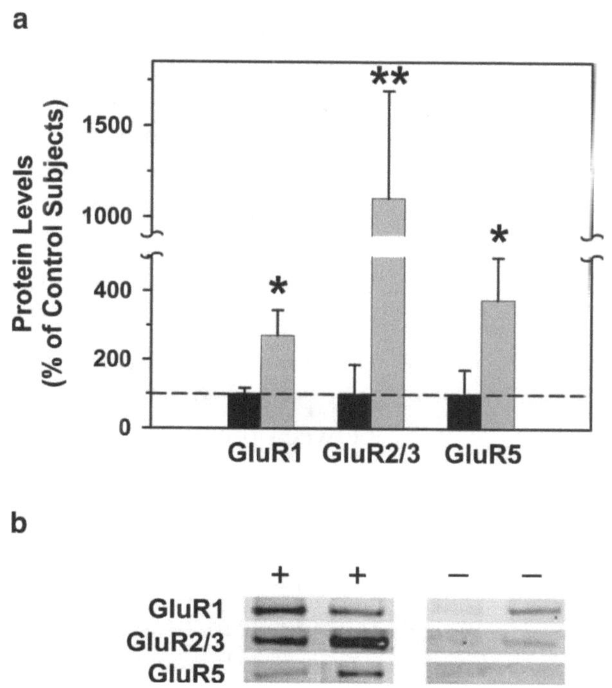

Fig. 2. (**a**) Representative Western blot analysis of proteins isolated from the nucleus accumbens of monkeys following 18 months of cocaine self-administration compared with controls. Protein was isolated from membrane fractions, and levels of ionotropic glutamate receptor subunit proteins GluR1, GluR2/3, and GluR5 were evaluated (5 μg) following separation on 10% SDS-PAGE. Data are expressed as mean (± S.E.M.) of the percent of control values per amount of protein loaded. *Asterisks* indicate a significant difference ($P < 0.05$). (**b**) Representative bands from two cocaine self-administration monkeys (+) and two control subjects (–) for each subunit are shown. Reprinted with permission from the Journal of Neurochemistry.

3.7. Protein Processing for Proteomics

1. Precipitate the protein sample using the 2D clean-up kit according to the manufacturer's recommendations at 20°C overnight.
2. The next day, pellet the sample by centrifugation at 13,400 × g for 5 min at 4°C and air-dry the pellet for 2 min.
3. Determine the protein concentration using the 2D-Quant kit, which is compatible with the reagent concentration in the sample buffer. The protein concentration should be in the range of 5–10 mg/ml (concentrate or dilute the samples as necessary).
4. Check the pH of all the samples and make sure that it is between 8 and 9 during cyanine dye labeling. Bring each sample up to 450 mL with Rehydration Buffer and add 50 mL of Destreak rehydration buffer (*see* **Notes 6** and **7**).

3.8. Two-Dimensional Polyacrylamide Gel Electrophoresis

1. Perform the IEF according to GE Healthcare setup using the 24-cm, pH 4–7 NL Immobiline DryStrips on an Ettan IPGphor apparatus. The protocol can be adapted on most isoelectric focusing equipment (*see* **Note 8**).

2. Equilibrate the IEF DryStrips to reduce the disulfide bonds by gently rocking them in 10 mL of reducing buffer/strip for 10 min. Immediately after this, alkylate the –SH groups of proteins by gently rocking the proteins in 10 mL of alkylating buffer/strip for 10 min. The SDS in the buffers also helps the proteins to acquire a negative charge, which drives their migration under the electrical current. Before proceeding to the next step, rinse the IEF strip in the SDS electrophoresis running buffer (*see* **Note 9**).
3. Overlay 0.6% agarose solution on the top of the glass plate of a precast 8–15% gradient SDS-PAGE. Place the IEF strip between the glass plates, and push it with a thin plastic spacer ensuring that the IEF strip rests on the SDS-PAGE. The proteins are then separated on the basis of their molecular weight at 4 W overnight until the bromophenol blue dye front reaches the bottom of the gel (*see* **Note 10**).
4. The gels can be scanned by varied scanners; Typhoon 9400 scanner (GE Healthcare); FLA 5100 Imaging System (FUJIFILM) or Ettan DIGE imager (GE HealthCare) (*see* **Note 11**).
5. The gel images can be analyzed by DeCyder (GE HealthCare), Progenesis SameSpots (Nonlinear Dynamics) – the most automated of the softwares available for DIGE analysis – or Delta2D (DECODON).
6. The differentially regulated protein spots are analyzed by mass spectrometry for protein identification **(Fig. 3)**.

4. Notes

1. All solutions should be prepared in double distilled water with resistivity up to 18.2 MΩ cm and total organic content less than 1 part per billion or HPLC grade water.
2. Different combinations of protease inhibitors can be used depending on the proteins and moieties of interest.
3. Confirm that all the solutions containing urea are prepared fresh and have not been heated above 37°C to prevent protein carbamylation and subsequent formation of charge trains on the 2D gel.
4. The determination of the appropriate plane of dissection and the width of the sections are based on the region(s) of interest. Commercially available brain matrices for rodents and monkeys provide consistency of sectioning.
5. Antibodies should be tested across a range of protein concentrations to determine linearity of antigen to signal for each species as well as each brain region of interest.

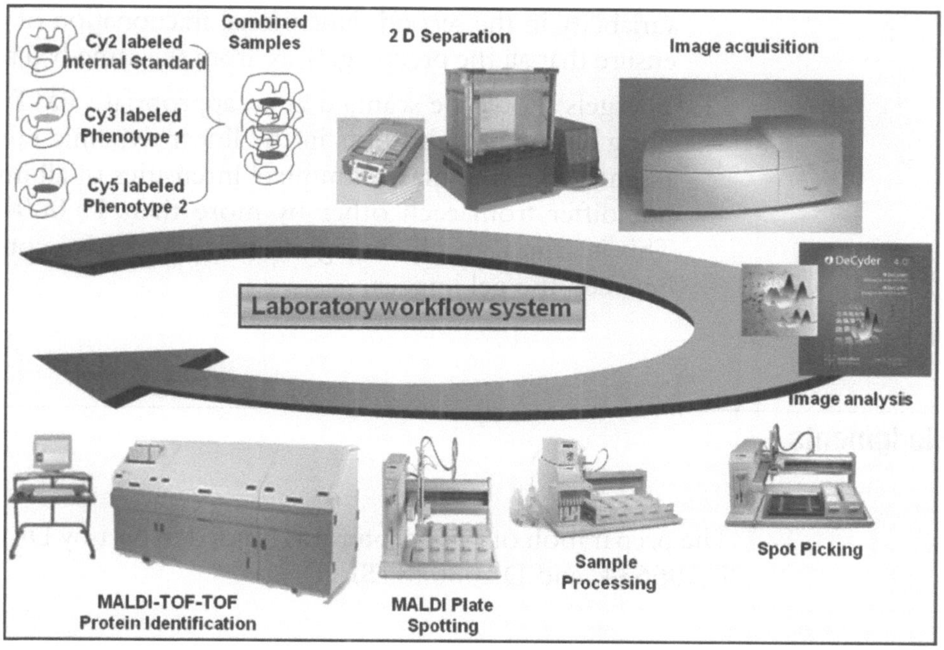

Fig. 3. 2D-DIGE work-flow for comparative expression-proteomics. Samples are labeled with fluorescent dyes (Cy 2, 3, and 5) and combined together prior to the IEF. An optimized protocol based on the tissue of interest is used for the IEF and second dimension separation of proteins. The three gel images are scanned by Typhoon™ scanner (GE Healthcare), so that the maximum pixel intensities are within the linear dynamic range and at the same time are consistent across the entire set of gels in the experiment. The differentially regulated protein spots, after the image analyses by DeCyder™ analysis software, are picked by robotic picker. These spots are robotically processed, including in-gel digestion, and prepared for acquiring mass spectra (MS and MS/MS) by MALDI-TOF-TOF. The acquired mass spectra are searched against the protein database, using the MASCOT search engine, for the species of interest to obtain protein identification.

6. The optimal labeling of the sample of interest should be achieved by a preliminary study. The goal is to achieve labeling for less abundant proteins at the same time maintaining the most abundant proteins in the linear dynamic range for quantitative analyses. A range of ratios for protein concentration: CyDye amount (50 mg:100 pmol to 50 mg:400 pmol) should be tested.
7. Prepare the rehydration buffer by freshly adding DTT and IPG buffer.
8. The length of the pH strip, its pH range, as well as the IEF setup should be empirically determined to provide the best possible resolution for your sample of interest.
9. Ensure fresh DDT and iodoacetamide in the reducing and alkylating buffers, respectively. To minimize protein loss, do not exceed the stipulated alkylation and reduction times.
10. Make certain that low-fluorescence glass plates with a reference marker are used. This is critical for the background pixel values of the scanned images to be as low as possible. To avoid

variability in the second dimensional fractionation of proteins, ensure that all the precast gels are from the same batch.

11. The gels should be scanned at an appropriate PMT value to bring all the protein spots in the linear dynamic range. It is essential that the maximum pixel intensities of all images do not differ from each other by more than 10,000–15,000. This is crucial to obtain significant quantitative comparison between the gel images.

Acknowledgments

The preparation of this chapter was funded in part by DA012498, DA06634, and DA03628 (SEH).

References

1. Hemby, S. E. (1999). Recent advances in the biology of addiction. *Curr. Psychiatry Rep.* **1**, 159–165.
2. Hemby, S. E., Johnson, B. A., and Dworkin, S. I. (1997). Neurobiological basis of drug reinforcement, in *Drug Addiction and its Treatment: Nexus of Neuroscience and Behavior* (Johnson, B. A., and Roache, J. D. eds., Lippincott-Raven Publishers, Philadelphia, pp. 137–169.
3. Tannu, N., Mash, D. C., and Hemby, S. E. (2007). Cytosolic proteomic alterations in the nucleus accumbens of cocaine overdose victims. *Mol. Psychiatry* **12**, 55–73.
4. Tannu, N. S., and Hemby, S. E. (2006). Methods for proteomics in neuroscience. *Prog. Brain Res.* **158**, 41–82.
5. Tannu, N. S., and Hemby, S. E. (2006) Two-dimensional fluorescence difference gel electrophoresis for comparative proteomics profiling. *Nat. Protoc.* **1**, 1732–1742.
6. Weeks, J. R. (1962). Experimental morphine addiction: method for automatic intravenous injections in unrestrained rats. *Sci.* **138**, 143–144.
7. Weeks, J. R. (1972). Long-term intravenous infusions, in *Methods in Psychobiology, Vol.* 2 (Myers, R. D. ed.), Academic Press, New York, pp. 155–168.
8. Breiter, H. C., Gollub, R. L., Weisskoff, R. M., Kennedy, D. N., Makris, N., Berke, J.D., Goodman, J.M., Kantor, H.L., Gastfriend, D.R., Riorden, J.P., Mathew, R.T., Rosen, B.R., and Hyman, S.E. (1997). Acute effects of cocaine on human brain activity and emotion. *Neuron.*, **19**, 591–611.
9. Kilts, C. D., Gross, R. E., Ely, T. D., and Drexler, K. P. (2004). The neural correlates of cue-induced craving in cocaine-dependent women. *Am. J. Psychiatry* **161**, 233–241.
10. Kilts, C. D., Schweitzer, J. B., Quinn, C. K., Gross, R. E., Faber, T. L., Muhammad, F., Ely, T. D., Hoffman, J. M., and Drexler, K. P. (2001) Neural activity related to drug craving in cocaine addiction. *Arch. Gen. Psychiatry* **58**, 334–341.
11. Hemby, S. E., Co, C., Dworkin, S. I., and Smith, J. E. (1999). Synergistic elevations in nucleus accumbens extracellular dopamine concentrations during self-administration of cocaine/heroin combinations (Speedball) in rats. *J. Pharmacol. Exp. Ther.* **288**, 274–280.
12. Hemby, S. E., Co, C., Koves, T. R., Smith, J. E. and Dworkin, S. I. (1997). Differences in extracellular dopamine concentrations in the nucleus accumbens during response-dependent and response-independent cocaine administration in the rat. *Psychopharmacology (Berl).* **133**, 7–16.
13. Pettit, H. O., Ettenberg, A., Bloom, F. E., and Koob, G. F. (1984). Destruction of dopamine in the nucleus accumbens selectively attenuates cocaine but not heroin self-administration in rats. *Psychopharmacology* **84**, 167–173.

14. Pettit, H. O., Pan, H. T., Parsons, L. H., and Justice, J. B., Jr. (1990). Extracellular concentrations of cocaine and dopamine are enhanced during chronic cocaine administration. *J. Neurochem.* **55**, 798–804.
15. Zito, K. A., Vickers, G., and Roberts, D. C. (1985). Disruption of cocaine and heroin self-administration following kainic acid lesions of the nucleus accumbens. *Pharmacol. Biochem. Behavior* **23**, 1029–1036.
16. Churchill, L., Swanson, C. J., Urbina, M., and Kalivas, P. W. (1999). Repeated cocaine alters glutamate receptor subunit levels in the nucleus accumbens and ventral tegmental area of rats that develop behavioral sensitization. *J. Neurochem.* **72**, 2397–2403.
17. Fitzgerald, L. W., Ortiz, J., Hamedani, A. G., and Nestler, E. J. (1996). Drugs of abuse and stress increase the expression of GluR1 and NMDAR1 glutamate receptor subunits in the rat ventral tegmental area: Common adaptations among cross-sensitizing agents. *J. Neurosci.* **16**, 274–282.
18. Tang, W. X., Fasulo, W. H., Mash, D. C., and Hemby, S. E. (2003). Molecular profiling of midbrain dopamine regions in cocaine overdose victims. *J. Neurochem.* **85**, 911–924.
19. Ungless, M. A., Whistler, J. L., Malenka, R. C., and Bonci, A. (2001). Single cocaine exposure in vivo induces long-term potentiation in dopamine neurons. *Nat.* **411**, 583–587.
20. White, F. J., Hu, X. T., Zhang, X. F., and Wolf, M. E. (1995). Repeated administration of cocaine or amphetamine alters neuronal responses to glutamate in the mesoaccumbens dopamine system. *J. Pharmacol. Exper. Therapeutics* **273**, 445–454.
21. Zhang, X. F., Hu, X. T., White, F. J., and Wolf, M. E. (1997). Increased responsiveness of ventral tegmental area dopamine neurons to glutamate after repeated administration of cocaine or amphetamine is transient and selectively involves AMPA receptors. *J. Pharmacol. Exper. Therapeutics* **281**, 699–706.
22. Kalivas, P. W., McFarland, K., Bowers, S., Szumlinski, K., Xi, Z. X., and Baker, D. (2003). Glutamate transmission and addiction to cocaine. *Ann. N. Y. Acad. Sci.* **1003**, 169–175.
23. Kalivas, P. W., Volkow, N., and Seamans, J. (2005) Unmanageable motivation in addiction: A pathology in prefrontal-accumbens glutamate transmission. *Neuron.* **45**, 647–650.
24. Borges, K., and Dingledine, R. (200). Molecular pharmacology and physiology of glutamate receptors, in *Glutamate and addiction* (Herman, B. H., Frankenheim, J., Litten, J. Z., Sheridan, P. H., Weight, F. F., and Zukin, S. R., eds.), Humana, Totawa, NJ, pp. 3–22.

Chapter 6

Protein Aggregate Characterization in Models of Neurodegenerative Disease

Andrew T.N. Tebbenkamp and David R. Borchelt

Summary

A pathological hallmark of many neurodegenerative diseases is the presence of protein aggregates. Transgenic mice that recapitulate this pathology are a valuable resource to isolate these proteins for detailed study. One aspect of our research program is to characterize and quantify aggregates β-amyloid (Aβ) peptides, superoxide dismutase 1 (SOD1), and huntingtin (htt) that comprise pathologic lesions found in Alzheimer's disease, familial amyotrophic lateral sclerosis, and Huntington's disease, respectively. In this chapter, we describe methods, based on sequential detergent extraction and ultracentrifugation, to isolate and analyze these protein aggregates. These methods have been applied to human tissues to some extent, but have been highly useful in studies involving transgenic mouse models of these diseases.

Key words: Protein aggregation, Neurodegenerative disease, Detergent solubility, Protein misfolding

1. Introduction

Postmortem analyses of Alzheimer's disease (AD) brains show neurodegeneration of the basal forebrain and hippocampus. Neurons in these regions accumulate intracellular aggregates composed of hyperphosphorylated tau (neurofibrillary tangles), and are intermixed with extracellular aggregates (plaques), composed mainly of β-amyloid (Aβ). Transgenic mice that overexpress mutant forms of amyloid precursor protein (APP), the precursor protein for Aβ peptides *(1)*, recapitulate human amyloid pathology and provide a resource for study of amyloid formation and clearance *(2)*.

Andrew K. Ottens and Kevin K.W. Wang (eds.), *Neuroproteomics,* Methods in Molecular Biology, vol. 566
DOI 10.1007/978-1-59745-562-6_6, © Humana Press, a part of Springer Science+Business Media, LLC 2009

Of the total number of cases of amyotrophic lateral sclerosis (ALS), only 1–2% of the cases are due to mutations in the enzyme superoxide dismutase 1 (SOD1). Disease-causing mutations exist in catalytic and noncatalytic regions of the enzyme suggesting alterations in activity do not contribute to pathogenesis *(3)*. In SOD1-linked cases of human ALS, and in transgenic mice that express mutant forms of SOD1, aggregated forms of mutant SOD1 accumulate as disease symptoms worsen *(4–8)*. Biochemical detection of aggregates has been critical in assessing the overall contribution of mutant SOD1 aggregation in disease; although tissues from both humans and mice can contain histologically visible inclusions that are immunoreactive for mutant SOD1, quantification of these aggregates is labor intensive and sometimes not representative of overall protein aggregate levels. Moreover, biochemical isolation of such aggregates is critical in determining the role of protein modification in the misfolding of the mutant protein.

Huntington's disease (HD) results from a CAG repeat expansion (>36), encoding a polyglutamine tract, in exon 1 of the huntingtin (*htt*) gene. In brain tissues of affected humans and transgenic mice that express all or part of mutant htt, protein aggregates accumulate in the cytoplasm and nucleus of neurons. The principal component of these aggregates is an N-terminal fragment of htt that contains the polyglutamine tract *(9)*. In addition to htt, these aggregates are immunoreactive for a number of other proteins and the entrapment of proteins in htt aggregates has been suggested as a potential mechanism of toxicity *(10)*. Biochemical isolation of these aggregates from transgenic mouse models is essential in advancing understanding of pathogenic mechanisms.

The methods described later have been utilized to varying degrees in each of the disease settings. The filter-trap assay, described below, is an adaptation of a procedure developed in the laboratory of Dr. Erich Wanker to detect and quantify aggregates of mutant huntingtin protein *(11, 12)*. The filter trap assay has proven to be very useful in quantifying the levels of aggregates in tissues, particularly the levels of Aβ amyloid and tau *(13)*. However, this method is not conducive for detailed biochemical characterization; instead methods involving detergent extraction, centrifugation/sedimentation, and immunoblotting provide much more information. Moreover, such methods are more suitable for subsequent analyses such as mass spectroscopy. The methods described below are adaptations of methodology developed to study prion proteins *(14–16)*.

2. Materials

2.1. Filter Trap Assay

1. Phosphate-buffered saline (PBS) with 5 mM EDTA; with freshly added cocktails of protease inhibitors. (Mammalian cell cocktail, Sigma, St. Louis, MO).
2. Stock solution of 10% sodium-dodecyl sulfate (SDS) in dH_2O to be diluted to 1% working solution. Can be stored at room temperature (RT).
3. Cellulose acetate membrane, pore size 0.22 μm, pre-wet with PBS/1% SDS (Schleicher and Schuell, Keene, NH).
4. Blocking solution: 5% nonfat dry milk powder (w/v) in 10 mM Tris-buffered saline (TBS). Can be stored at 4°C for approximately 1 week.
5. Appropriate primary antibody [e.g., polyclonal anti-Aβ antibody (Zymed Laboratories) diluted 1:600 in blocking solution].
6. Washing solution (TBST): 0.1% Tween-20 (v/v) in 10 mM TBS, can be stored at RT.
7. HRP-conjugated protein A (Sigma) diluted 1:5,000 in blocking solution.
8. Chemiluminescence substrates (PerkinElmer, Boston, MA).

2.2. Differential Detergent Extraction and Centrifugation

1. Lysis buffer (1× TEN): 10 mM Tris-HCl pH 8.0, 1 mM EDTA pH 8.0, and 100 mM NaCl.
2. Buffer A: 1× TEN, 1% Nonidet P40, proteinase inhibitor cocktail 1:100 dilution (P8340, Sigma).
3. Buffer B: 1× TEN, 0.5% Nonidet P40.
4. Buffer C: 1× TEN, 0.5% Nonidet P40, 0.5% deoxycholic acid, 0.25% SDS.
5. Buffer D: 1× TEN, 0.5% Nonidet P40, 0.5% deoxycholic acid, 2% SDS.

2.3. SDS-PAGE and Western Blot

1. Precast 18% Tris-glycine (TG) polyacrylamide gel (Invitrogen, Carlsbad, CA).
2. 4× Laemmli sample buffer: 0.25 M Tris-HCl, pH 6.8, 8% SDS, 40% glycerol, 0.05% Bromophenol Blue, 20% 2-mercaptoethanol, fill to desired volume with dH_2O.
3. Bicinchoninic acid (BCA) protein concentration assay (Pierce, Rockford, IL).
4. Running buffer: TG-SDS buffer, powder (Amresco, Solon, OH).
5. Transfer buffer: 20% methanol in running buffer, store at 4°C.

6. Optitran nitrocellulose membrane (Schleicher and Schuell, Keene, NH).
7. Blocking solution (*see* **Subheading 2.1, item 4**).
8. Appropriate primary antibodies (e.g., anti-Aβ, anti-SOD1, anti-htt).
9. Secondary antibodies (Kirkegaard & Perry Laboratories, Gaithersburg, MD).
10. Washing solution (*see* **Subheading 2.1, item 6**).

3. Methods

Strategies to separate soluble proteins from insoluble proteins are useful in detecting and quantifying protein aggregates. These strategies also aid in determining whether these proteins modify the solubility of any interacting partners.

3.1. Filter Trap Assay (*see* Note 1)

1. Homogenize brain tissue in 5 volumes of PBS/EDTA/protease inhibitor cocktail using sonication for 60 s.
2. Dilute with 10 volumes of previous solution, centrifuge briefly (3,000 × *g*) for upto 5 minutes, and discard the pellet.
3. Dilute supernatant with stock 10% SDS to a final concentration of 1%.
4. Make serial dilutions (1:2) with 1× PBS/1% SDS and spot 100 μL of each aliquot onto a 0.22 μm cellulose acetate filter (pre-wet with PBS/1%SDS) sealed within a dot blot apparatus, followed by vacuum filtration.
5. Wash wells five times for 10 min each with PBS, and then place the membrane in blocking solution with anti-Aβ antibody diluted 1:600.
6. Wash membrane three times for 10 min each in TBST followed by incubation in HRP-conjugated protein A (Sigma) diluted 1:5,000 in blocking solution for 1 h at RT.
7. Wash membrane three times for 10 min each in TBST before visualizing with chemiluminescence substrates. Accurate quantification is accomplished by imaging luminescence in a gel documentation system, such as manufactured by Fuji, Kodak, or BioRad.

3.2. Differential Detergent Extraction and Centrifugation (*see* Note 2)

1. Homogenize tissue at a 10:1 volume to weight ratio in 1× TEN.
2. Mix homogenate 1:1 with Buffer A, sonicate (Microson XL2000; Misonix, Farmingdale, NY – 2W at 22.5 kHz) for

30 s, and centrifuge at >100,000 × *g* for 5 min in an Airfuge® (Beckman Coulter, Inc, Fullerton, CA). The supernatant (S1) is saved and used for analysis as "soluble fraction."

3. Resuspend pellet (P1) in Buffer B (volume equal to original supernatant), sonicate as described earlier, and centrifuge >100,000 × *g* for 5 min in an Airfuge® to obtain pellet P2.
4. Resuspend pellet P2 in Buffer C, sonicate, and centrifuge >100,000 × *g* for 5 min to obtain pellet P3. Pellet P3 can be resuspended in Buffer D for storage.

3.3. SDS-PAGE and Western Blot

1. Determine protein concentration of sample to be loaded by using BCA assay.
2. Standard procedures for SDS-PAGE, electrotransfer, and immunoblotting are followed. Samples can be prepared for SDS-PAGE by mixing with 4× Laemmli buffer to a final concentration of 1×; with or without reducing agent, allowing for analysis of disulfide cross linking; with or without boiling, allowing for analysis of SDS-resistant oligomeric structures.
3. Immunoblots are analyzed by incubation in the appropriate antiserum and dilutions determined empirically. Primary antibody incubations less than 3 h can be performed at room temperature. Times longer than 3 h should be incubated at 4°C.
4. Incubate the membrane with secondary antibody diluted in blocking solution, rinse in TBST, and visualize using chemiluminescence.

4. Notes

1. The filter trap assay method described here is adapted from Scherzinger et al., 1997 *(11)* and has been specifically optimized for analysis of amyloid peptide levels in the brains of transgenic mice that model AD. Other variations on this theme have been utilized for detection of aggregates of mutant htt *(see* **ref.** *12)*, tau protein in Alzheimer's *(13)*, and mutant SOD1 *(5)*.
2. The differential detergent extraction and centrifugation procedure described earlier has been optimized for analysis of aggregates formed by mutant SOD1 in spinal cords of transgenic mice *(6, 7)*. Decreasing or increasing the SDS concentration in buffers in different steps can alter the stringency of the assay; enhancing or reducing the sedimentation of material less tightly bound aggregates. The method has also

been utilized to detect mutant SOD1 aggregates formed in HEK 293 cells transiently transfected with pEF.Bos expression vectors *(6)*. With minimal adaptations the procedure can be scaled up to generate sufficient material for further characterization – such as mass spectroscopy (B. F. Shaw, A. Durazo, H.L. Lelie, G. Xu, E.B. Gralla, A.M. Nerissian, A. Tiwari, L.J. Hayward, D.R. Borchelt, J.S. Valentine, J.P. Whitelegge, manuscript in preparation).

References

1. Weidemann, A., König, G., Bunke, D., Fischer, P., Salbaum, J. M., Masters, C. L., and Beyreuther, K. (1989). Identification, biogenesis, and localization of precursors of Alzheimer's disease A4 amyloid protein. *Cell* **57**,115–126.
2. Jankowsky, J. L., Savonenko, A., Schilling, G., Wang, J., Xu, G., and Borchelt, D. R. (2002). Transgenic mouse models of neurodegenerative disease: opportunities for therapeutic development. *Curr Neurol Neurosci Rep* **2**,457–464.
3. Valentine, J. S., Hart, P. J. (2003). Misfolded CuZnSOD and amyotrophic lateral sclerosis. *Proc Natl Acad Sci USA* **100**,3617–3622.
4. Wang, J., Xu, G., Gonzales, V., Coonfield, M., Fromholt, D., Copeland, N. G., Jenkins, N. A., and Borchelt, D. R. (2002). Fibrillar inclusions and motor neuron degeneration in transgenic mice expressing superoxide dismutase 1 with a disrupted copper-binding site. *Neurobiol Dis* **10**,128–138.
5. Wang, J., Xu, G., and Borchelt, D. R. (2002). High molecular weight complexes of mutant superoxide dismutase 1: Age- dependent and tissue-specific accumulation. *Neurobiol Dis* **9**,139–148.
6. Wang, J., Slunt, H., Gonzales, V., Fromholt, D., Coonfield, M., Copeland, N. G., Jenkins, N. A., and Borchelt, D. R. (2003). Copper-binding-site-null SOD1 causes ALS in transgenic mice: aggregates of non-native SOD1 delineate a common feature. *Hum Mol Genet* **12**,2753–2764.
7. Wang, J., Xu, G., Li, H., Gonzales, V., Fromholt, D., Karch, C., Copeland, N. G., Jenkins, N. A., and Borchelt, D. R. (2005). Somatodendritic accumulation of misfolded SOD1-L126Z in motor neurons mediates degeneration: {alpha}B-crystallin modulates aggregation. *Hum Mol Genet* **14**,2335–2347.
8. Wang, J., Xu, G., Slunt, H. H., Gonzales, V., Coonfield, M., Fromholt, D., Copeland, N. G., Jenkins, N. A., and Borchelt, D. R. (2005). Coincident thresholds of mutant protein for paralytic disease and protein aggregation caused by restrictively expressed superoxide dismutase cDNA. *Neurobiol Dis* **20**, 943–952.
9. Schilling, G., Klevytska, A., Tebbenkamp, A. T., Juenemann, K., Cooper, J., Gonzales, V., Slunt, H., Poirer, M., Ross, C. A., and Borchelt, D. R. (2007). Characterization of huntingtin pathologic fragments in human Huntington disease, transgenic mice, and cell models. *J Neuropathol Exp Neurol* **66**, 313–320.
10. Nucifora, F. C., Jr., Sasaki, M., Peters, M. F., Huang, H., Cooper, J. K., Yamada, M., Takahashi, H., Tsuji, S., Troncoso, J., Dawson, V. L., Dawson, T. M., and Ross, C. A. (2001). Interference by huntingtin and atrophin-1 with CBP-mediated transcription leading to cellular toxicity. *Science* **291**,2423–2428.
11. Scherzinger, E., Lurz, R., Turmaine, M., Mangiarini, L., Hollenbach, B., Hasenbank, R., Bates, G. P., Davies, S. W., Lehrach, H., and Wanker, E. E. (1997). Huntingtin-encoded polyglutamine expansions form amyloid-like protein aggregates in vitro and in vivo. *Cell* **90**,549–558.
12. Scherzinger, E., Sittler, A., Schweiger, K., Heiser, V., Lurz, R., Hasenbank, R., Lehrach, H., and Wanker, E. E. (1999). Self-assembly of polyglutamine-containing huntingtin fragments into amyloid-like fibrils: Implications for Huntington's disease pathology. *Proc Natl Acad Sci USA* **96**,4604–4609.
13. Xu, G., Gonzales, V., and Borchelt, D. R. (2002). Rapid detection of protein aggregates in the brains of Alzheimer patients and transgenic mouse models of amyloidosis. *Alzheimer Dis Assoc Disord* **16**,191–195.
14. Bolton, D. C., McKinley, M. P., and Prusiner, S. B. (1982) Identification of a protein that purifies with the scrapie prion. *Science* **218**,1309–1311.

15. Meyer, R. K., McKinley, M. P., Bowman, K. A., Braunfeld, M. B., Barry, R. A., and Prusiner, S. B. (1986). Separation and properties of cellular and scrapie prion proteins. *Proc Natl Acad Sci USA* **83**,2310–2314.
16. McKinley, M. P., Meyer, R. K., Kenaga, L., Rahbar, F., Cotter, R., Serban, A., and Prusiner, S. B. (1991). Scrapie prion rod formation in vitro requires both detergent extraction and limited proteolysis. *J Virol* **65**,1340–1351.
17. Shaw BF, Lelie HL, Durazo A, Nersissian AM, Xu G, Chan PK, Gralla EB, Tiwari A, Hayward LJ, Borchelt DR, Valentine JS, Whitelegge JP. Detergent-insoluble aggregates associated with amyotrophic lateral sclerosis in transgenic mice contain primarily full-length, unmodified superoxide dismutase-1. *J Biol Chem.* 2008 Mar 28;283(13):8340–50. Epub 2008 Jan 11. PubMed PMID: 18192269; PubMed Central PMCID: PMC2276386.

Part II

Sub-Proteome Separations and Neuroproteomic Analysis

Chapter 7

Sub-Proteome Processing: Isolation of Neuromelanin Granules from the Human Brain

Florian Tribl

Summary

The sub-proteome analysis of organelles is a field of high relevance for molecular biology, because it provides detailed insights into the protein composition of cellular compartments. This approach not only results in a catalogue of organellar proteins, but in fact holds the potential to uncover the enzymatic armament engaged in biochemical reactions and to identify novel mechanisms of organelle biogenic pathways. Knowledge about protein localization may be a first step towards extensive functional analyses of specific target proteins engaged in development, aging, or disease. Moreover, several disorders of the human brain include aberrant protein function in specific compartments. Thus, a closer look at cellular organelles will allow for advancing our current perceptions of pathogenic processes. This chapter aims to provide a methodological workflow given by the isolation of neuromelanin granules from the human midbrain. This approach encompasses several modular steps that can easily be adjusted to any other organelle of interest and follows the sequence of (1) organelle isolation, (2) isolation quality controls by transmission electron microscopy and Western immuno blotting, and (3) gel-based protein separation towards protein identification by mass spectrometry.

Key words: Human brain, Neuromelanin, Lysosome-related organelle, Organelle isolation, Density gradient, Transmission electron microscopy, Subcellular proteomics

1. Introduction

Neuromelanin (NM) granules are pigmented organelles in the human brain that give name to a brain area termed *substantia nigra pars compacta* (SN; latin, *black substance*). Macroscopically, the granules appear as a brown to black area in the brain stem because of the insoluble NM pigment *(1)*. The SN massively degenerates in Parkinson's disease (PD), which gives rise to

Andrew K. Ottens and Kevin K.W. Wang (eds.), *Neuroproteomics*, Methods in Molecular Biology, vol. 566
DOI 10.1007/978-1-59745-562-6_7, © Humana Press, a part of Springer Science+Business Media, LLC 2009

severely disabling movement symptoms *(2)*. NM granules are suggested to play an important role in the neurodegenerative events of PD: redox-active iron is bound to NM and thereby retained within this compartment, but in PD it is thought to be increasingly released into the cytosol *(3)*. Additionally, the protein α-synuclein, which has a tendency to aggregate into insoluble Lewy bodies, was recently demonstrated to be attached to NM granules in PD *(4, 5)*.

Because the SN is pigmented in primates, but not in rodents, NM granules thus have escaped from a detailed investigation to clarify their origin. Recently, however, subcellular proteomics enabled the successful isolation of NM granules from the human brain and resulted in their identification as lysosome-related organelles *(6)*. These findings underscore the potential of subcellular proteomics and encourage implementing this approach to tackle open questions in brain research (*see* **Note 1**) *(7)*.

Processing of sub-proteomes starts with the dissection of the target organelle from tissue or cells. After release of the cellular organelles, the organelle of interest needs to be separated from the residual compartments and contaminating cellular matter, e.g. by density gradient centrifugation *(8, 9)*. Next, the outcome of an organelle preparation needs to be thoroughly validated, e.g. by inspection on the morphological level by transmission electron microscopy to estimate the amount of contaminating organelles present *(9–11)*. Alternatively, an inspection on the molecular level using Western immunoblotting allows for estimating the degree of enrichment of the organelle of interest and may uncover specific sources of contamination on the basis of organelle marker proteins. Finally, the subcellular proteome is separated, e.g. by one-dimensional SDS-polyacrylamide gel electrophoresis (1D SDS-PAGE), to further reduce the complexity of the sample. Additionally, prefractionation by 1D SDS-PAGE offers a convenient and well-established platform to start the sample preparation for downstream protein identification by peptide-based mass spectrometry.

2. Materials

2.1. Isolation of Organelles

1. Separation buffer: 10 mM HEPES, 10% glucose, pH 6; store at 4°C.
2. Isolation buffer: 10 mM HEPES, 1 mM EDTA, 100 mM KCl, 10% (m/v) sucrose, pH 7.5; store at 4°C.
3. Washing buffer: 10 mM HEPES, 250 mM NaCl, 0.01% (v/v) Triton X-100, pH 7.5; store at 4°C (*see* **Note 2**).

4. Protease inhibitor cocktail (PIC) supplied in dimethyl sulfoxide (Sigma-Aldrich, Munich, Germany). Store in aliquots of 100 μL at –20°C.
5. Sucrose gradient solutions: Prepare dilutions of 1, 1.2, 1.4, and 1.6 M sucrose in 10 mM HEPES, pH 7.5. Store at 4°C (*see* **Note 3**).
6. Percoll (Fluka, Buchs, Switzerland): Prepare an 80% (v/v) solution and store at 4°C.
7. A syringe equipped with a 26-gauge needle.
8. Sieving mesh and a Petri dish.
9. A plate cooler set at 4°C.

2.2. Transmission Electron Microsopy

1. Prepare 2% (v/v) glutardialdehyde in 0.1 M phosphate buffered saline (PBS), pH 7.4 at 4°C (*see* **Note 4**).
2. Prepare 2% (w/v) OsO_4 in 1.5% (v/v) glutardialdehyde, in 0.1 M phosphate buffered saline (PBS), pH 7.4 at 4°C.
3. Dry 100% ethanol (EtOH) over $CuSO_4$.
4. Prepare the following aqueous EtOH solutions: 30% (v/v), 50% (v/v), 70% (v/v), 80% (v/v), 90% (v/v), 96% (v/v).
5. Have on hand 1,2-epoxypropane, EPON™ epoxy resin (Sigma-Aldrich, Munich, Germany), lead citrate, and uranyl acetate.

2.3. One-Dimensional SDS-Polyacrylamide Gel Electrophoresis

1. Novex 10–20% Tricine gels (Invitrogen, Carlsbad, CA) (*see* **Note 5**).
2. Invitrogen XCell II™ Mini Gel Electrophoresis System.
3. Tricine running buffer (1×): Prepare 700 mL with 10× Tricine SDS Running Buffer stock (Invitrogen).
4. Invitrogen Tricine SDS Sample Buffer (2×).
5. Invitrogen NUPAGE® Sample Reducing Agent (10×).

2.4. Western Immunoblot

1. Nitrocellulose membranes.
2. XCell II™ blot module (Invitrogen) and sponges.
3. Invitrogen Tris-glycine transfer buffer (25×).
4. TBS-T 10× stock: Prepare stock with 1.37 M NaCl, 27 mM KCl, 250 mM Tris-HCl, pH 7.3, and 1% (v/v) Tween-20.
5. TBS-T: Dilute 100 mL of 10× stock with 900 mL water for use.
6. Blocking solution: Prepare 5% (w/v) nonfat dry milk in TBS-T (*see* **Note 6**).
7. Primary antibody: Dilute selected antibody in blocking solution (*see* **Note 7**).
8. Blot washing buffer: TBS-T.

9. Secondary antibody: goat anti-mouse IgG (Cell Signaling, Frankfurt/Main, Germany) diluted 1:1,000 (v/v) in 10 mL of TBS-T.
10. Enhanced chemiluminescence (ECL-system™, Boehringer Ingelheim, Ingelheim, Germany) (*see* **Note 8**).

2.5. Stripping and Reprobing Immunoblots

1. Stripping solution: 62.5 mM Tris-HCl, pH 6.7, 100 mM β-mercaptoethanol and 2% SDS (w/v). Warm to a working temperature of 50°C and add β-mercaptoethanol to a final concentration of 100 mM (*see* **Note 9**). Store at room temperature.
2. Wash buffer: 0.1% (w/v) bovine serum albumin in TBS-T.

2.6. Gel Staining with Colloidal Coomassie Brilliant Blue G-250

1. Fixing solution: 50% (v/v) methanol and 2% (v/v) phosphoric acid.
2. Staining solution: 34% (v/v) methanol, 2% (v/v) phosphoric acid, 17% (m/v) ammonium sulphate.
3. Coomassie Brilliant Blue G-250 powder.

2.7. In-Gel Digestion with Trypsin

1. Muffled quartz reaction tubes (Sigma, St. Louis, MO).
2. Destain solution A: 10 mM NH_4HCO_3, pH 7.8 (*see* **Note 10**).
3. Destain solution B: Prepare 1:1 (v/v) mixture of 10 mM NH_4HCO_3, pH 7.8, and acetonitrile (HPLC grade).
4. Trypsin solution: Reconstitute lyophilized trypsin (Promega, Madison, WI) in 10 mM NH_4HCO_3, pH 7.8, to a final concentration of 0.05 µg/µL. Store reconstituted trypsin at 4°C.

2.8. Sample Preparation for Mass Spectrometry Analysis

1. Extraction solution: Prepare a 0.1% 1:1 (v/v) mixture of trifluoroacetic acid and acetonitrile mixed.
2. Separate quartz tubes.

3. Methods

3.1. Isolation of Organelles

The protocol described here is optimized for the isolation of NM granules from human brain being not affected by neurodegenerative or psychiatric disorders. If it is the aim to isolate a different target compartment, this protocol should be regarded as a concept rather than a protocol **(Fig. 1)**. A comprehensive collection of protocols for most organelles is available from “A Practical Approach Series” *(12)*. All steps can easily be adapted and optimized for the given structure, e.g. by variation of the density media. If organelles are to be isolated from cultured cells, one should pelletize the cells appropriately and directly start at **step 11**.

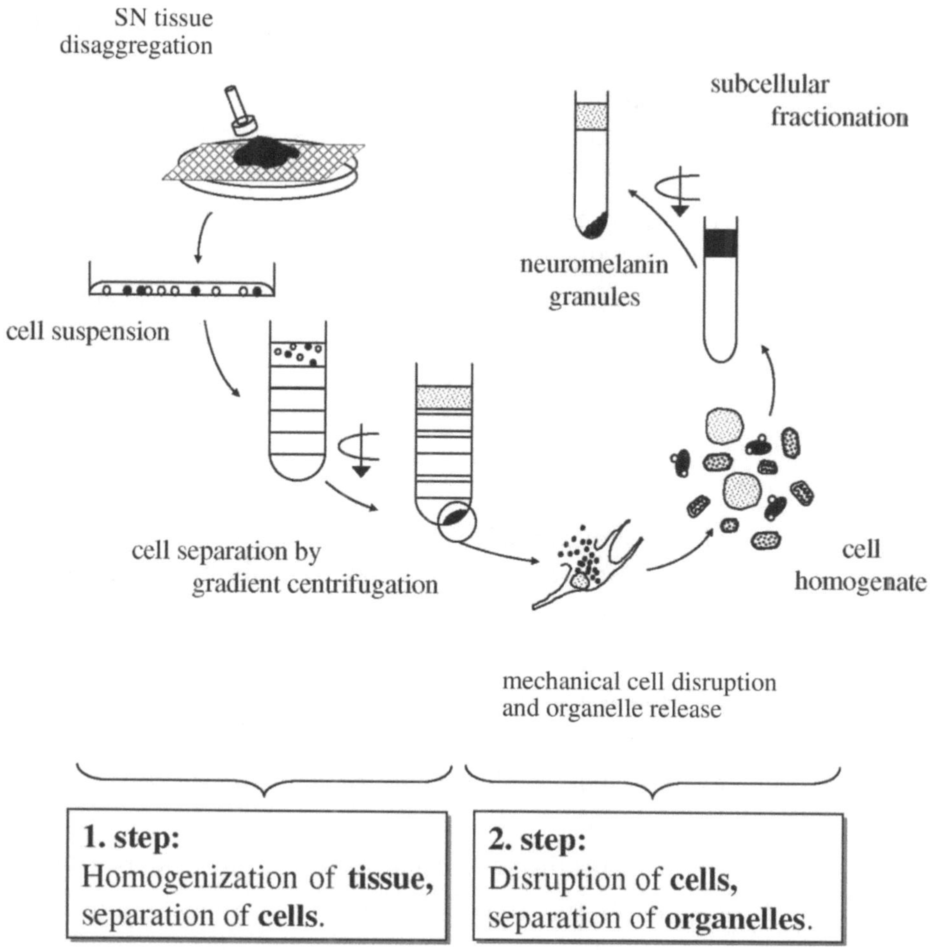

Fig. 1. Neuromelanin granule isolation strategy. Tissue is disaggregated by sieving through a mesh, and the resulting suspension is fractionated in a discontinuous sucrose gradient (1. step). The target cells are collected and homogenized by mechanical disruption. The cell homogenate containing several types of cellular organelles is layered on top of an 80% Percoll cushion. Finally, pigmented neuromelanin granules are isolated according to their density following centrifugation through Percoll (2. step; reproduced form **ref. *6*** with the permission of the American Society for Biochemistry and Molecular Biology).

1. Assemble all the materials required for the treatment and termination protocol: a styropor box with ice to maintain the buffer solutions at 4°C; thaw the PIC at room temperature; set a cooling plate to 4°C; cool a centrifuge to 4°C; prepare a syringe equipped with a 26-gauge needle.
2. Place a Petri dish on the cooler plate and cover it with a sieving mesh. Alternatively, standard sieving grids could be used.
3. Supply the separation buffer with thawed PIC 1:100 (v/v) (*see* **Note 11**).
4. Thaw the tissue in the PIC-containing separation buffer.
5. Place the tissue on the covered Petri dish and pass it through the sieving device, e.g., with a plugger of a syringe.
6. Combine the disconnected tissue suspensions and place on ice.

7. Prepare a discontinuous sucrose gradient in a 50 mL centrifuge tube: apply step-wise 5 mL layers of the 1–1.6 M sucrose solutions, starting with the solution of highest density, and carefully overlaying with solutions of decreasing density (*see* **Note 12**).
8. Apply the sample on top of the density gradient by gently allowing it to flow over the top layer.
9. Centrifuge the sample through the sucrose gradient at 4,000 × *g* for 45 min at 4°C.
10. Cellular matter of different density collects as bands at the density interfaces. Melanized cell bodies are collected as a dark brown pellet at the bottom of the 50 mL centrifuge tube.
11. Isolate the pellet by carefully removing the supernatant and resuspend in 1 mL of PIC-containing "isolation buffer."
12. Carefully homogenize the pellet by aspirating 10 times through a 26-gauge needle while avoiding foam formation.
13. Layer the cell homogenate on top of a 5 mL 80% Percoll cushion in a 15 mL centrifuge tube.
14. Centrifuge the sample through the Percoll cushion at 4,000 × *g* for 15 min at 4°C.
15. Discard the supernatant and carefully save the organelle pellet.
16. Wash the organelle pellet by resuspending it in washing buffer and centrifuge again at 4,000 × *g* for 15 min at 4°C (*see* **Note 13**).

3.2. Transmission Electron Microscopy

1. Fix the organelle pellet overnight in 2% glutardialdehyde at 4°C *(11)*.
2. Incubate for 30 min in 2% OsO_4 solution in a dark place.
3. Remove the fixation solutions by a brief rinse with water.
4. Incubate the sample in water three times for 15 min each.
5. Dehydrate the specimen with increasing concentrations of EtOH: start with 30% (v/v) EtOH for 10 min, then apply the 50, 70, 80, 90, 96% solutions for 10 min, respectively.
6. Working in a fume hood, incubate the dried dample in 1,2-epoxypropane (2 × 15 min), and remove the supernatant.
7. Prepare fresh a 1:1 (v/v) mixture of 1,2-epoxypropane and EPON™ epoxy resin.
8. For embedding, incubate the sample in the mixture of 1,2-epoxypropane and EPON™ epoxy resin overnight.
9. Remove the supernatant and incubate the sample in EPON™ epoxy resin 2 h, then repeat once, remove the supernatant and again add fresh EPON™ epoxy resin.

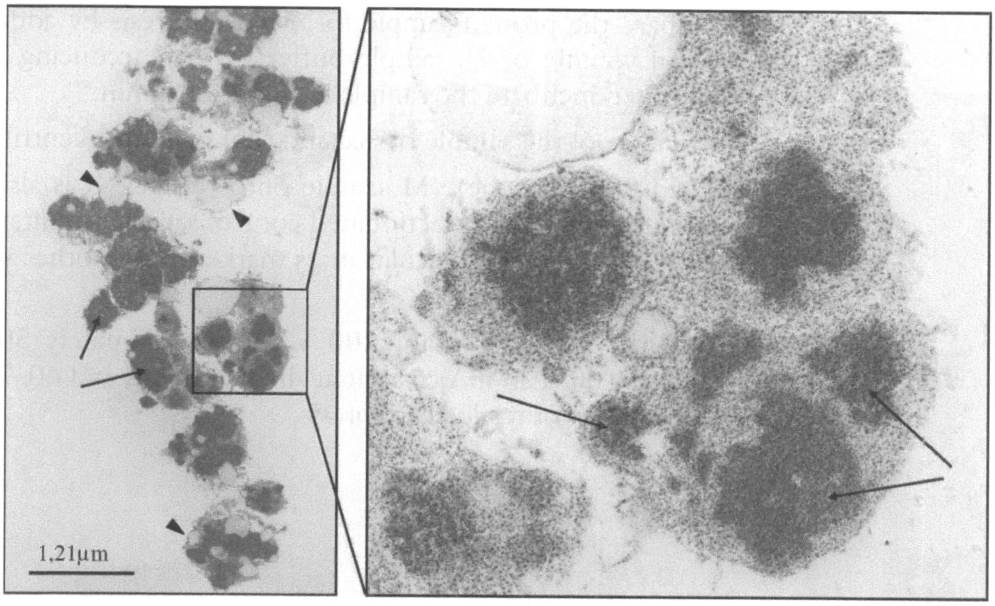

Fig. 2. Quality assessment of the organelle preparation on a morphological level by transmission electron microscopy. Isolated neuromelanin granules are virtually free of contaminating organelles and retain their morphological appearance after isolation. Dark areas represent the electron-dense neuromelanin, which is embedded in a protein matrix. Lipidic bulbs and are characteristic for neuromelanin granules still attached to the organelles (Reproduced from **ref. *6*** with the permission of the American Society for Biochemistry and Molecular Biology).

10. Polymerize the EPON™ epoxy resin in an incubator at 55–65°C for 48 h.
11. Prepare thin sections with an ultramicrotome and mount them on copper or nickel grids.
12. Contrast the specimen with lead citrate and uranyl acetate according to Reynolds *(13)*. An example of a transmission electron microscopic visualization of isolated NM granules is given in **Fig.** 2.

3.3. SDS-Polyacrylamide Gel Electrophoresis (SDS-PAGE)

1. These instructions assume the use of precast gels, e.g, 10–20% gradient Tricine gels, and the XCell II™ Mini Gel Electrophoresis System from Invitrogen. The procedure is adaptable to other formats, including self-cast minigels.
2. Bring the sample buffer to room temperature to completely dissolve the SDS. Set a heater to 90°C.
3. Remove the gel cassette from its package, briefly rinse with water, peel off the tape from the bottom and remove the comb.
4. Pour approximately 100 mL of the 1× Tricine SDS running buffer into the electrophoresis cell. Then insert the gel into the cell. Fill the upper chamber with the 1× Tricine SDS running buffer. Check that the sample wells are covered with buffer. Remove air bubbles with a pipette, if necessary.

5. Prepare the protein sample for electrophoresis by adding an equal volume of 2× sample buffer with the reducing agent, and then incubate the sample at 90°C for 5 min.
6. Briefly cool the sample on ice and spin in a table centrifuge.
7. Load 20 μL of the NM sample onto the gel. It is also suggested to run an unfractionated control sample simultanously (**Fig. 3**). Load a molecular mass marker into another well to discern protein masses.
8. Run the gel at constant 200 V for approximately 50 min. Expect 100–120 mA current at the beginning and 60–70 mA at the end of the electrophoresis.

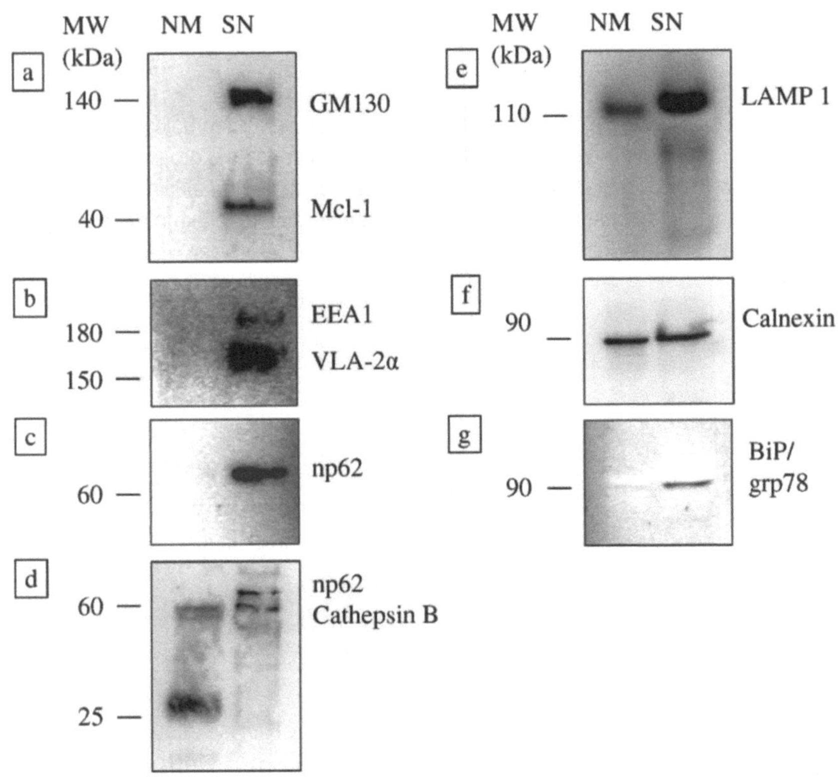

Fig. 3. Quality assessment of the organelle preparation on a molecular level by Western immuno blotting. Probing for organellar marker proteins enables the visual assessment of isolated neuromelanin granules (NM) purity when compared with total *substantia nigra pars compacta* tissue (SN). Marker proteins for (**a**) the Golgi network (GM130) and mitochondria (Mcl-1), (**b**) early endosomes (EEA-1) and the plasma membrane (VLA-2a), and (**c**) the nucleus (np62) are all not detected in the NM preparation. Markers for (**d, e**) late endosomes and lysosomal marker proteins (cathepsin B, LAMP1) are, however, present at a lesser amount in NM. (**f**) Calnexin, a protein predominantly found in the endoplasmic reticulum, is found in pigmented organelles as a melanogenic chaperone and is abundant in the NM preparation. (**g**) The endoplasmic reticulum marker protein BiP/grp78 is absent, however, confirming that the calnexin is from the association with NM (Reproduced from **ref. *6*** with the permission of the American Society for Biochemistry and Molecular Biology).

3.4. Western Immunoblot

Quality control for organelle preparations on a molecular level is crucial to evaluate the degree of enrichment. Several organelle marker proteins, which are specific for a given cellular compartment, are commercially available and allow for visualizing the purity of the sample (**Fig. 3**).

1. Remove the gel from the plastic cassette and cut off the wells with a knife. Then cover the gel with a few mL of 1× transfer buffer.
2. Wet the precut nitrocellulose membranes in 1× transfer buffer. The membranes should be completely covered with buffer.
3. Presoak 6–7 sponges in 1× transfer buffer: apply slight pressure to remove air and facilitate complete uptake of buffer.
4. Prepare the XCell II™ blot module and place three sponges in the blotting cassette.
5. Prepare the "gel sandwich" as follows: wet a sheet of Whatman paper by briefly dipping into 1× transfer buffer and place the gel onto it. Cover the gel with the nitrocellulose membrane and put a second wet Whatman paper onto the nitrocellulose membrane. Press out residual air bubbles to allow uniform protein transfer onto the membrane. Then lay the "gel sandwich" onto the sponges, cover it with the remaining sponges and tightly close the cassette.
6. The cassette is placed into the XCell II™ blot module. Perform the protein transfer onto the membrane at 70 V for 90 min. Alternatively, the transfer can be performed overnight at 4°C and 35 V.
7. The transfer efficacy can be evaluated by dipping the membrane into a Ponceau Red solution that reversibly visualizes protein as reddish bands. Stained membranes can either be briefly unstained in TBS or directly placed into 30 mL of blocking solution.
8. Blocking of unspecific antibody binding sites is accomplished for 1 h at room temperature on a rocking platform.
9. Discard the blocking solution. Dilute the primary antibody for probing. Gently pour the antibody solution onto the membrane and incubate for 1 h at room temperature on a rocking platform. Alternatively, incubate with the primary antibody overnight at 4°C without a rocking platform.
10. Then remove the primary antibody and wash the membrane 3 × 15 min with 30 mL of TBS-T (*see* **Note 13**).
11. Dilute the secondary antibody in TBS-T. Discard the "blot washing buffer" and incubate the membrane with the secondary antibody for 1 h at room temperature on a rocking platform.

12. Finally, remove the secondary antibody and wash the membrane 3 × 15 min with 30 mL TBS-T.
13. While waiting for the last washing step, warm 1–2 mL of each ECL solution to room temperature and prepared a 1:1 mixture.
14. Briefly rinse the membrane in TBS, and then pour the mixed ECL solution onto the membrane and disperse evenly.
15. Finally, detect the chemoluminescence in a gel imaging system.

3.5. Stripping and Reprobing Immunoblots

In most cases, organelle preparations are available in limited amounts; thus, stripping of the blotted membranes and reprobing with a different antibody is advisable. Sequentially, the membranes can be probed, stripped, and probed again.

1. Warm the stripping solution to 50°C (*see* **Note 9**).
2. Incubate the membrane for 50 min with gentle agitation.
3. Discard the stripping solution and wash the membranes in 3 × 100 mL of washing buffer for 10 min. The membrane is now ready to be blocked and reprobed with another primary antibody, beginning at **step 8** of **Subheading 3.4**.

3.6. Gel Staining with Colloidal Coomassie Brilliant Blue G-250

1. Colloidal Coomssie Brilliant Blue (CBB) G-250 is a sensitive stain that is compatible with downstream mass spectrometric analyses *(14)*. Fix the gel in 100 mL of fixing solution for 1 h on a rocking platform at room temperature.
2. Discard the fixing solution and wash 3 × 5 min with water. Then add 100 mL of staining solution and allow the gel to equilibrate for 30 min on the rocking platform.
3. Sprinkle a small amount (covering the tip of a small spatula) of CBB G-250 directly on top of the staining solution, spreading over the gel, and leave overnight. Do not dissolve CBB G-250 (*see* **Note 14**).
4. Discard the staining solution and briefly rinse with water. Wash the gel for 3 h, occasionally exchanging the water to reduce the background staining.

3.7. In-Gel Digestion with Trypsin

1. For in-gel digestion, dissect the whole lane of separated proteins into 3–4 mm high gel slices, or dissect a protein band of interest. Then dissect the gel slice into approximately 1 mm gel cubes to enlarge the surface-to-volume ratio for effective protease uptake. Use a scalpel cleaned with EtOH for dissection. Transfer each gel cube into a quartz reaction tube (*see* **Notes 15 & 16**).
2. For washing a gel cube, add 15–20 µL of destain solution A using a 50 µL HAMILTON™ syringe with a blunt platinum needle (*see* **Note 17**). The gel should be completely covered by the washing solution. Incubate for 10 min at room temperature.

3. Remove destain solution A and add 15–20 µL of destain solution B. Incubate for 10 min, then remove and discard destain solution B.
4. Repeat **steps 2** and **3** twice. This procedure is sufficient to destain most gel cubes effectively.
5. Place the quartz tubes with the washed gel cubes in a vacuum concentrator and dry the gel cubes for 10 min. The gel cubes should then have a white appearance.
6. Add 2 µL of trypsin solution onto a dried gel cube and incubate at 37°C overnight.

3.8. Sample Preparation for Mass Spectrometry Analysis

1. To extract the peptides generated by in-gel digestions, add 15–20 µL of the extraction solution and place the samples into an ultrasound bath. Add ice to the water to cool the samples during 15 min of sonication.
2. Remove the extraction solution and transfer it to a fresh quartz tube.
3. Repeat **step 1** and combine the extraction solutions in the same quartz tube.
4. Place the quartz tubes into a vacuum concentrator and evaporate for 5 min.

The peptide samples are then compatible for mass spectrometric analysis *(15)*. Details of the mass spectrometric methods for identification of peptides and proteins are beyond the scope of this chapter. The reader is referred to descriptions in other chapters of this volume.

4. Notes

1. Research on human tissue requires the approval of an ethics committee.
2. Note that the washing buffer in the last step should not contain potassium if a subsequent step includes the application of SDS, e.g. in 1D-SDS-PAGE. In the presence of potassium, SDS turns into potassium dodecyl sulfate, which has reduced solubility.
3. Sucrose gradient solutions may be stored at 4°C for at least 4 weeks.
4. A comprehensive protocol collection for electron microscopy of organelles is provided by Nigel *(11)*.
5. Keep in mind that gels should be stored at 4°C. Expired precast gels should not be used to avoid abnormal protein migration.

6. Nonfat dry milk is recommended to be dissolved properly in TBS-T by stirring thoroughly at room temperature for at least 30 min. The "blocking solution" can be stored at 4°C for at least 2 days.
7. Primary antibodies to probe for organelle marker proteins were taken from the "Organelle Sampler Kit" purchased from BD Biosciences and diluted to the recommended ratio.
8. The mixed ECL solution can be stored at 4°C and reused at least once on the following day.
9. β-mercaptoethanol is toxic! Work under a fume hood and use caution when handling, e.g., while preparing the stripping solution.
10. Destain solutions A and B should be prepared freshly before each use.
11. It is not recommended to store buffers containing the protease inhibitor cocktail for reuse on the following day.
12. Sucrose gradients may be generated by using a peristaltic pump to form sharper density interfaces.
13. Primary antibody solutions can be stored at 4°C for a few days and reused for subsequent experiments. The signal intensity, however, may decline and thus requires a prolonged exposure in a gel imaging system or on film.
14. CBB G-250 should not be dissolved in the staining solution but rather be dispersed in form of clumps on the surface of the liquid. The rational is that colloidal particles, and not single dye molecules, should stain the proteins on the surface of the gel to yield higher sensitivity and to increase the signal-to-background ratio ***(14)***.
15. Labeling on the quartz reaction tubes should be additionally protected with an adhesive tape to prevent the loss of the label in the waterbath (*see* **Subheading 3.8**, **step 1**).
16. The adsorption of peptides to the surface of a quartz reaction tube is significantly less than to the surface of a microcentrifuge tube.
17. Platinum needles are recommended since less peptide is adsorbed to the surface than with plastic pipette tips.

Acknowledgments

The author would like to thank Prof. Peter Riederer, Prof. Manfred Gerlach, and Prof. Gerhard Bringmann for constant encouragement, and Prof. Katrin Marcus and Prof. Helmut E Meyer for training in proteomics. The author is grateful to Dr. Esther Asan for assistance in transmission electron microscopy, and to Dr. Thomas Arzberger and Dr. Thomas Tatschner for their expertise in human brain tissue asservation. This work was supported by the Austrian Academy of Sciences, by the *BrainNet Europe II*, by BMBF Grant 031U102F, the Deutsche Parkinson Vereinigung, and the Fond der Chemischen Industrie.

References

1. Fedorow, H., Tribl, F., Halliday, G., Gerlach, M., Riederer, P., and Double, K. L. (2005) Neuromelanin in human dopamine neurons: comparison with peripheral melanins and relevance to Parkinson's disease. *Prog. Neurobiol.* **75**, 109–124.
2. Hirsch, E., Graybiel, A. M., and Agid, Y. A. (1988) Melanized dopaminergic neurons are differentially susceptible to degeneration in Parkinson's disease. *Nature* **334**, 345–348.
3. Fasano, M., Bergamasco, B., and Lopiano, L. (2006) Is neuromelanin changed in Parkinson's disease? Investigations by magnetic spectroscopies. *J. Neural. Transm.* **113**, 769–774.
4. Fasano, M., Giraudo, S., Coha, S., Bergamesco, B., and Lopiano, L. (2003) Residual substantia nigra neuromelanin in Parkinson's disease is crosslinked to alpha-synuclein. *Neurochem. Int.* **42**, 603–606.
5. Halliday, G. M., Ophof, A., Broe, M., Jensen, P. H., Kettle, E., Fedorow, H., Cartwright, M. I., Griffiths, M. I., Sheperd, C. E., and Double, K. L. (2005) alpha-Synuclein redistributes to neuromelanin lipid in the substantia nigra early in Parkinson's disease. *Brain* **128**, 2645–2664.
6. Tribl, F., Gerlach, M., Marcus, K., Asan, E., Tatschner, T., Arzberger, T., Meyer, H. E., Bringmann, G., and Riederer, P. (2005) "Subcellular Proteomics" of neuromelanin granules isolated from the human brain. *Mol. Cell Proteomics* **4**, 945–957.
7. Tribl, F., Marcus, K., Bringmann, G., Meyer, H. E., Gerlach, M., and Riederer, P. (2006) Proteomics of the human brain: Sub-proteomes might hold the key to handle brain complexity. *J. Neural. Transm.* **113**, 1041–54.
8. Hinton, R. H., and Mullock, B. M. (1997) Isolation of subcellular fractions, in *Subcellular Fractionation: A Practical Approach* (Graham, J. M., and Rickwood, D., eds.), Oxford University Press, Oxford, England, pp. 31–70.
9. Huber, L. A., Pfaller, K., and Vietor, I. (2003) Organelle proteomics: implications for subcellular fractionation in proteomics. *Circ. Res.* **92**, 962–968.
10. Dreger, M. (2003) Subcellular proteomics. *Mass Spectrom. Rev.* **22**, 27–56.
11. Nigel, J. (1997) Electron microscopy of organelles, in *Subcellular Fractionation: A Practical Approach* (Graham, J. M., and Rickwood, D., eds.), Oxford University Press, Oxford, England, pp. 303–328.
12. Graham, J. M., and Rickwood, D. (eds.) (1997) *Subcellular Fractionation. A Practical Approach*, Oxford University Press, Oxford, England.
13. Reynolds, E. S. (1963) The use of lead citrate at high pH as an electron-opaque stain in electron microscopy. *J. Cell Biol.* **17**, 208–212.
14. Neuhoff, V., Arold, N., Taube, D., and Erhardt, W. (1988) Improved staining of proteins in polyacrylamide gels including isoelectric focusing gels with clear background at nanogram sensitivity using Coomassie Brilliant Blue G-250 and R-250. *Electrophoresis* **9**, 255–262.
15. Schäfer, H., Nau, K., Sickmann, A., Erdmann, R., and Meyer, H. E. (2001) Identification of peroxisomal membrane proteins of Saccharomyces cerevisiae by mass spectrometry. *Electrophoresis* **22**, 2955–2968.

Chapter 8

Proteomic Analysis of Protein Phosphorylation and Ubiquitination in Alzheimer's Disease

Stefani N. Thomas, Diane Cripps, and Austin J. Yang

Summary

Posttranslational modifications such as phosphorylation and ubiquitination serve, independently or together, as gatekeepers of protein transport and turnover in normal and disease physiologies. Aberrant protein phosphorylation is one of the defining pathological hallmarks of more than 20 different neurodegenerative disorders, including Alzheimer's disease (AD). The disruption of the phosphorylation of neurotransmitter receptors has been implicated as one of the causal factors of impaired memory function in AD *(1–3)*. Another feature of AD is the aberrant accumulation of proteins that are normally degraded by the ubiquitin proteasome system upon being conjugated to ubiquitin. Thus, elucidating the protein targets of phosphorylation and ubiquitination that can serve as AD biomarkers will aid in the development of effective therapeutic approaches to the treatment of AD. This chapter provides details pertaining to the qualitative and quantitative liquid chromatography tandem mass spectrometry-based analysis of an affinity purified, phosphorylated, and ubiquitinated protein, paired-helical filament tau.

Key words: Alzheimer's disease (AD), Phosphorylation, Ubiquitin, Tau, Liquid chromatography tandem mass spectrometry (LC-MS/MS), Selected reaction monitoring (SRM)

1. Introduction

Phosphorylation and ubiquitination are two of the many posttranslational modifications of the microtubule-associated protein tau that are critical to the molecular pathogenesis of neurodegeneration in Alzheimer's disease (AD) *(4–13)*. A distinct hierarchical pattern of tau phosphorylation has been shown to correlate with the progression of AD neuropathology *(14, 15)*. Paired helical filaments (PHFs) of hyperphosphorylated tau aggregates comprise the degradation-resistant core of intraneuronal neurofibrillary tangles in AD and other tauopathies. The accumulation

Andrew K. Ottens and Kevin K.W. Wang (eds.), *Neuroproteomics,* Methods in Molecular Biology, vol. 566
DOI 10.1007/978-1-59745-562-6_8,

of degradation-resistant PHF tau is linked to the impaired function of the ubiquitin proteasome system *(16)*. Although PHF tau is ubiquitinated, it is not subsequently degraded by the proteasome system *(6, 17–19)*. Hence, the determination of the precise phosphorylation and ubiquitination sites of tau that specifically correlate with the progression of AD could serve as effective biomarkers.

The identification of phosphorylation sites typically relies upon the use of antibodies, thereby limiting the investigator's ability to comprehensively evaluate all possible phosphorylation sites of a protein of interest. Similarly, the detection of ubiquitinated proteins has conventionally been achieved with antibody-based detection methods, the use of which provides information about the general ubiquitination status of a given protein, but not the specific site(s) of ubiquitin conjugation. Here we describe a sensitive and quantitative method employing liquid chromatography tandem mass spectrometry (LC-MS/MS) to evaluate tau phosphorylation and ubiquitination. Although the details of the methods described herein pertain to the analysis of tau, these methods can be applied to the analysis of any phosphorylated and/or ubiquitinated protein.

2. Materials

2.1. SDS-PAGE and in-Gel Trypsin Digestion

1. Precast Tris-glycine 4–20% polyacrylamide gel (Bio-Rad).
2. Running buffer (10×): 250 mM Tris-HCl, 1.92 M glycine, 1% (w/v) SDS. Store at room temperature.
3. Sample buffer (4×): 250 mM Tris-HCl pH 6.8, 40% glycerol, 8% SDS (w/v), 200 mM DTT, 0.008% bromophenol blue. Store at –20°C (*see* **Note 1**).
4. Precision plus dual color standard (Bio-Rad).
5. Scalpels (*see* **Note 2**).
6. Gel-loading pipette tips.
7. Microcentrifuge tubes rinsed with 50% methanol.
8. Coomassie blue stain or Silver Stain (Silver Quest Silver Staining Kit, Invitrogen) (*see* **Note 3**).
9. Water (Burdick and Jackson, HPLC grade).
10. Digestion buffer: 50 mM Ammonium bicarbonate (Sigma) made from 100 mM Ammonium bicarbonate stock solution. Check to ensure pH is between 7.5 and 9.0.
11. Acetonitrile (Burdick and Jackson, HPLC grade).
12. Methanol (Burdick and Jackson, HPLC grade).

13. TCEP ("Bond-breaker neutral pH TCEP solution", Pierce/ Thermo Fisher Scientific): Prepare 10 mM solution in 100 mM ammonium bicarbonate. Prepare 10 mM solution fresh.
14. Iodoacetamide (Sigma): 55 mM prepared in 100 mM ammonium bicarbonate. Prepare fresh.
15. TPCK-treated trypsin (Promega) (*see* **Note 4**).
16. Formic acid (Sigma).

2.2. In-Solution Trypsin Digestion

1. Digestion buffer: 100 mM ammonium bicarbonate (Sigma) in water; pH between 7.5 and 9.0 is acceptable.
2. Centrifugal filter devices: (Microcon) for buffer exchange and sample concentration.
3. Desalting spin columns: (Pierce/Thermo Fisher Scientific).
4. Methanol: HPLC grade.
5. TCEP: ("Bond-breaker neutral pH TCEP solution", Pierce/ Thermo Fisher Scientific).
6. Iodoacetamide (Sigma): 1 M stock prepared in 100 mM ammonium bicarbonate; freshly prepared.
7. DTT (Sigma): 1 M stock prepared in 100 mM ammonium bicarbonate. Stock solution can be stored in aliquots at -20°C.
8. TPCK-treated trypsin: (Promega).
9. Glacial acetic acid: HPLC grade.

2.3. Immobilized Metal Affinity Chromatography for Phosphopeptide Isolation

1. Phosphopeptide isolation kit (Pierce/Thermo Fisher Scientific).
2. Water (Burdick and Jackson, HPLC grade).
3. Ammonium bicarbonate (Sigma). Prepare 0.1 M stock in water and ensure pH is in the range of 8.7–9.1.
4. Acetonitrile (Burdick and Jackson, HPLC grade).
5. Glacial acetic acid, HPLC grade.

2.4. Liquid Chromatography

1. Solvent "A" (aqueous buffer): 2% acetonitrile (Burdick and Jackson – HPLC grade), 0.1% formic acid (Fluka) in water (Burdick and Jackson – HPLC grade).
2. Solvent "B" (organic buffer): 95% acetonitrile (Burdick and Jackson – HPLC grade), 0.1% formic acid (Fluka) in water (Burdick and Jackson – HPLC grade).
3. C18 reversed phase microcapillary column: 15 cm length, 75 µm i.d., 5 µm particles with 300 Å pores (Micro-Tech Scientific).
4. XTreme Simple LC system (Micro-Tech Scientific).

2.5. Mass Spectrometry

1. LTQ linear ion trap mass spectrometer equipped with a nano-spray ion source (Thermo Electron).
2. PicoTip™ spray tips: uncoated, 360 µm fused silica tubing o.d., 20 µm tip o.d., 10 µm tip i.d. (New Objective).
3. Ultra high-grade purity helium.

2.6. Mass Spectrometry Data Analysis

1. Xcalibur™ for mass spectrometer instrument control and data analysis (Thermo Electron) (*see* **Note 5**).
2. Bioworks software using the SEQUEST algorithm (Thermo Electron).

3. Methods

Davies et al. were among the first groups to publish data suggesting that various forms of phosphorylated tau could be utilized as functional AD biomarkers *(14, 20–23)*. Hyperphosphorylated PHF-tau is largely degradation resistant, although it is known to be ubiquitinated. However, the exact temporal correlation between tau phosphorylation, ubiquitination, and neurodegeneration in AD is not known. We have recently applied a series of functional proteomic and LC-MS/MS analyses to determine the extent of phosphorylation and ubiquitination in PHF-tau isolated from human AD brain *(24)* using an antibody that recognizes a conformational variant of tau that represents an early stage of PHF generation in AD *(25, 26)*. Our data indicate that this conformation of PHF-tau is phosphorylated on at least 30 amino acid residues and is ubiquitinated at Lys-254, Lys-311, and Lys-353. These ubiquitination events occur as a combination of both mono-ubiquitination and poly-ubiquitination.

The methods detailed in this chapter are written based on the analysis of PHF-tau; however, these methods can be extrapolated and utilized for the study of any purified protein that is phosphorylated and/or ubiquitinated. The starting amount of purified PHF-tau used for our studies was 100 µg.

3.1. In-Gel Trypsin Digestion

1. If proteins other than the protein of interest are present in the purified protein sample, it might be desirable to separate the purified protein via SDS-PAGE. Use 50 µg protein for SDS-PAGE analysis. Following electrophoresis, wash the gel three times for 10 min per wash with deionized water to remove any residual SDS.
2. Proceed to Coomassie blue staining or Silver staining. For Silver staining methods, overnight fixation helps to decrease background staining. After staining is complete, wash the gel for 10 min with deionized water.

3. Capture an image of the gel. Excise protein band using a clean scalpel and cut into 1 × 1 mm^2 cubes. Transfer gel pieces to microcentrifuge tube washed with 50% methanol (*see* **Note 6**). If using Silver stain, destain gel pieces before proceeding to next step.
4. Add 100% methanol for 5 min to dehydrate gel pieces. Add sufficient volume of methanol to cover gel pieces.
5. Remove 100% methanol and add 30% methanol for 5 min to rehydrate gel pieces (*see* **Note 7**).
6. Remove 30% methanol and wash gel pieces twice for 10 min per wash with water (HPLC-grade).
7. Wash gel pieces three times for 10 min per wash with 30% acetonitrile in 100 mM ammonium bicarbonate.
8. Dry gel pieces in vacuum centrifuge (speed vac) for 15 min. Gel pieces will become opaque when dry.
9. Add 10 mM TCEP (enough volume to cover gel pieces) and let incubate for 1 h at 60°C to reduce protein disulfide bonds.
10. Briefly centrifuge gel pieces and remove liquid.
11. Add 55 mM iodoacetamide (enough volume to cover gel pieces) and let incubate for 45 min at room temperature in the dark to alkylate cysteine residues.
12. Briefly centrifuge gel pieces and remove liquid.
13. Wash gel pieces for 15 min with 100 mM ammonium bicarbonate.
14. Remove liquid. Shrink gel pieces with 100% acetonitrile.
15. Remove liquid. Dry gel pieces in vacuum centrifuge (speed vac) for 15 min. It is critical to ensure that gel pieces are completely dry to facilitate the absorption of trypsin into the gel pieces in the next step.
16. Rehydrate the gel pieces in trypsin solution at 4°C (on ice) for 45 min. Trypsin solution: 1.5 ng/μL trypsin in 50 mM ammonium bicarbonate. Add adequate volume of trypsin solution to completely cover gel pieces.
17. Remove any remaining trypsin solution and add sufficient volume of digestion buffer to completely cover gel pieces.
18. Let incubate at 37°C overnight.
19. Briefly centrifuge gel pieces and transfer supernatant to another microcentrifuge tube washed with 50% methanol.
20. Add 50 mM ammonium bicarbonate to cover gel pieces and incubate at 37°C with shaking for 15 min.
21. Add volume of acetonitrile equal to volume of 50 mM ammonium bicarbonate utilized in **step 20** and incubate at 37°C with shaking for 15 min.

22. Centrifuge gel pieces and add supernatant to the supernatant collected in **step 19**.
23. Extract peptides: Add acetonitrile/5% formic acid (50:50) to gel pieces and incubate at 37°C with shaking for 15 min.
24. Remove supernatant and combine with supernatant collected in **steps 19** and **22**.
25. Repeat **step 23**.
26. Remove supernatant and combine with supernatant collected in **steps 19**, **22**, and **24**.
27. Dry down gel extract and reconstitute in volume of liquid chromatography Solvent "A" that is compatible with the volume of the sample loop in the liquid chromatography system that will be utilized to analyze the sample (*see* **Note 8**).

3.2. In-Solution Trypsin Digestion

1. Use at least 50 μg protein for in-solution digestion. Protein sample should ideally be free of any detergents prior to in-solution digestion and LC-MS/MS (*see* **Note 9**). Perform buffer exchange with 100 mM ammonium bicarbonate using Microcon centrifugal filter devices. Be aware of the molecular weight cut-off (MWCO) for the filter devices. Optimum volume of the sample following buffer exchange is the volume that yields a protein concentration of ~0.1–1.0 μg/μL.
2. Add methanol to 40% to denature protein (*see* **Note 10**).
3. If protein does not contain any cysteine residues, skip this step and proceed to **step 4**.
   (a) Add TCEP to 5 mM final concentration to reduce protein disulfide bonds. Vortex and incubate for 30 min at 37°C.
   (b) Alkylate cysteine residues by addition of iodoacetamide to 10 mM final concentration. Vortex and incubate for 1 h at room temperature in the dark.
   (c) Stop alkylation reaction by adding DTT to 20 mM final concentration. Vortex and incubate for 1 h at room temperature.
   (d) Rid samples of any excess iodoacetamide using a protein desalting spin column (Pierce/Thermo Fisher Scientific).
4. Add trypsin to sample at a ratio of 1 trypsin:50 protein. Let incubate for 2 h at 37°C.
5. Repeat **step 4**.
6. Add trypsin to sample at a ratio of 1 trypsin:50 protein and let incubate overnight at 37°C.
7. Stop enzymatic digestion by acidifying with addition of acetic acid to 5% final volume. Dry sample in speed vac.

3.3. Immobilized Metal Affinity Chromatography for Phosphopeptide Isolation

1. Resuspend dried, digested peptide sample in 50 μL 5% acetic acid in water. Ensure that sample pH is <3.5. If the sample pH is >3.5, a significant amount of nonspecific binding may occur, consequently decreasing the selectivity of the gallium-chelated resin for binding to phosphopeptides.
2. Follow the procedure detailed in the Pierce phosphopeptide isolation kit (product 89853) product information sheet.
3. Do not combine final elution fractions. Dry down elution fractions in speed vac and reconstitute peptides in liquid chromatography Solvent "A".

3.4. Liquid Chromatography

Utilize an LC method with the following parameters:

1. Flow rate of 0.4 μL/min.
2. 30 min equilibration with Solvent "A".
3. Sample loading time (load at 95% Solvent "A") equivalent to the time it would take to load a sample with a volume that is 2.5× the sample loop volume.
4. For the efficient separation of hydrophilic phosphopeptides, utilize the following gradient: 5–25% Solvent "B" in 60 min followed by 25–90% Solvent "B" in 10 min. Otherwise, utilize a gradient of 5–40% Solvent "B" in 60 min followed by 40–90% Solvent "B" in 10 min.
5. Condition column with 90% solvent "B" for 15 min.

3.5. Mass Spectrometry

1. Set spray voltage to 2.0 kV and heated capillary temperature to 200°C.
2. Create a data dependent MS/MS instrument method with the following parameters:
   (a) Full scan range: 400–1,800 m/z.
   (b) Five most abundant ions selected for MS/MS fragmentation.
   (c) Dynamic exclusion enabled – exclude ions for 30 s after being detected twice within 30 s (*see* **Note 11**).
   (d) Minimum MS signal: 500 counts, miimum MS/MS signal: 100 counts.
   (e) Activation time: 30 ms (120 ms for MS/MS/MS methods; *see* **step 3**.).
   (f) Collision energy: 35% (24% for MS/MS/MS methods; *see* **step 3**.).
3. For the analysis of phosphopeptides, use a data-dependent MS/MS/MS (MS3) method whereby an MS3 scan will be triggered if, among the three most abundant ions in the MS/MS scan, a neutral loss of 98, 49, or 32.7 Da is detected. These neutral loss masses correspond to the loss of phosphoric acid

(H_3PO_4) on singly, doubly, and triply charged precursor ions, respectively (*see* **Note 12**). **Figure 1** is an illustration of the neutral loss data-dependent algorithm that is utilized for the identification of phosphopeptides.

3.6. Mass Spectrometry Data Analysis of Protein Phosphorylation and Ubiquitination

1. Conduct a SEQUEST search against the appropriate organism database using a threshold of 100, monoisotopic mass type, and automatic charge state determination. Include the following differential modifications: +57.02 on Cys (if protein was reduced and alkylated using iodoacetamide); +16.0 on Met (oxidation); +80.0 on Ser, Thr, and Tyr (phosphorylation); −18.0 on Ser, Thr (neutral loss of phosphoric acid) (*see* **Note 12**); +114 on Lys (–GG from the c-terminus of ubiquitin that remains conjugated to a Lys residue of an ubiquitinated protein following trypsin digestion).
2. Manually inspect spectra passing a threshold of cross-correlation vs. charge state values of 1.5 for +1 ions, 2.0 for +2 ions, and 2.5 for +3 ions to verify that all major fragment ions are identified and, in the case of serine and threonine phosphopeptides, that phosphorylated residues that are identified with a differential modification of −18 are from MS3 scans.

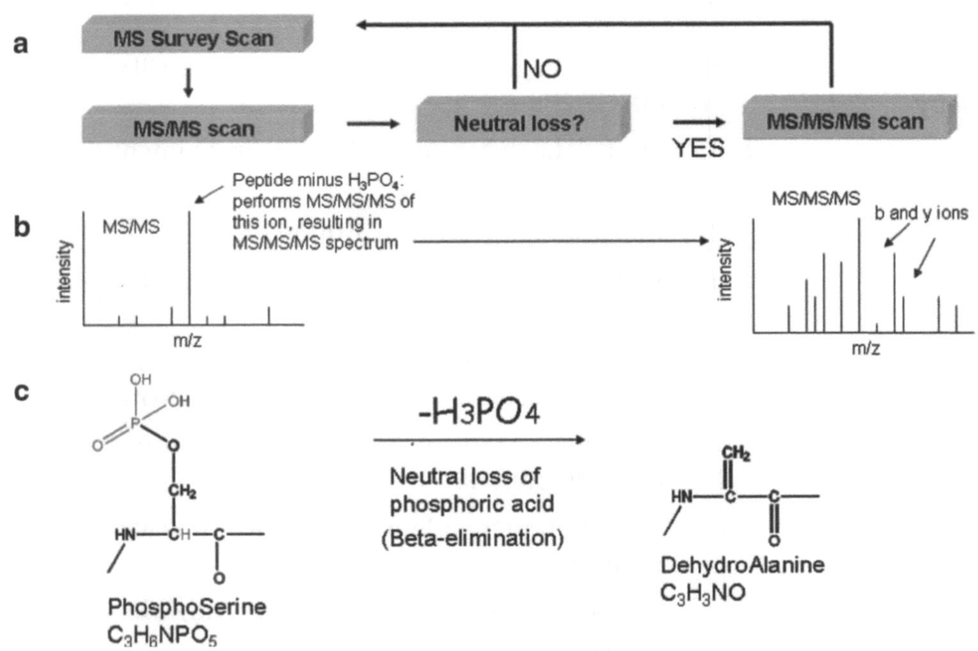

Fig. 1. Mass spectrometry data-dependent algorithm for the identification of phosphopeptides. (**a**) Mass spectrometer instrument scan cycle. (**b**) Schematic representation of an MS/MS spectrum in which a peptide has undergone a neutral loss of phosphoric acid and the resultant MS/MS/MS spectrum. (**c**) Structure of phosphoserine amino acid residue which, upon a neutral loss of phosphoric acid by a beta-elimination reaction, is converted to dehydroalanine.

3.7. Quantification of Posttranslational Modifications by Selected Reaction Monitoring

3.7.1. Evaluation of External Standard Peptides for Use in SRM Method

1. After performing qualitative LC-MS/MS analysis (*see* **Subheadings 3.4–3.6**) to determine possible phosphorylation and/or ubiquitination sites on the protein of interest, synthesize peptides to be utilized as external standards for performing quantification by selected reaction monitoring (SRM). Adhere to the following considerations and guidelines when selecting suitable peptides for quantitative analysis:
   (a) Perform a BLAST search to ensure that the peptide selected for quantification is unique to the protein of interest.
   (b) Select peptides with a length of ~6–13 amino acid residues. Peptides of this length are efficiently retained on reversed phase microcapillary columns and elute with a peak width that is amenable for quantification.
   (c) If possible, avoid peptides containing reactive and labile residues such as cysteine and methionine that readily undergo oxidation during synthesis and while in typical solvents. Tryptophan is prone to alkylation during synthesis. Such modifications will reduce the actual amount of external standard peptide and will result in erroneous quantification, namely over-estimation.
   (d) Synthesize both the modified (phosphorylated or ubiquitinated) and unmodified versions of the peptide standards to deduce the stoichiometry or extent of posttranslational modification at specific amino acid sites. However, be mindful that sites of posttranslational modification that occur in proximity to the proteolytic site could impede proteolysis.
2. Utilize standard Fmoc chemistry to synthesize the external peptide standards (*see* **Note 13**).
3. Verify the stock amount of the synthesized peptide by quantitative amino acid analysis (*see* **Note 13**).
4. Assess the purity of the synthesized peptide by reversed phase chromatography (*see* **Note 13**).
5. Confirm the identity of the synthesized peptide by infusing the peptide into the mass spectrometer and conducting MS/MS fragmentation.

3.7.2. SRM Method for Quantification of Post-translational Modifications

Although triple quadrupole mass spectrometers are ideal for SRM-based quantification, linear ion trap mass spectrometers are practical alternatives *(27)*. A schematic of the SRM method used for quantitative analysis is presented in **Fig. 2**.

1. Select an external synthetic peptide MS/MS product ion with an m/z greater than that of the precursor ion.

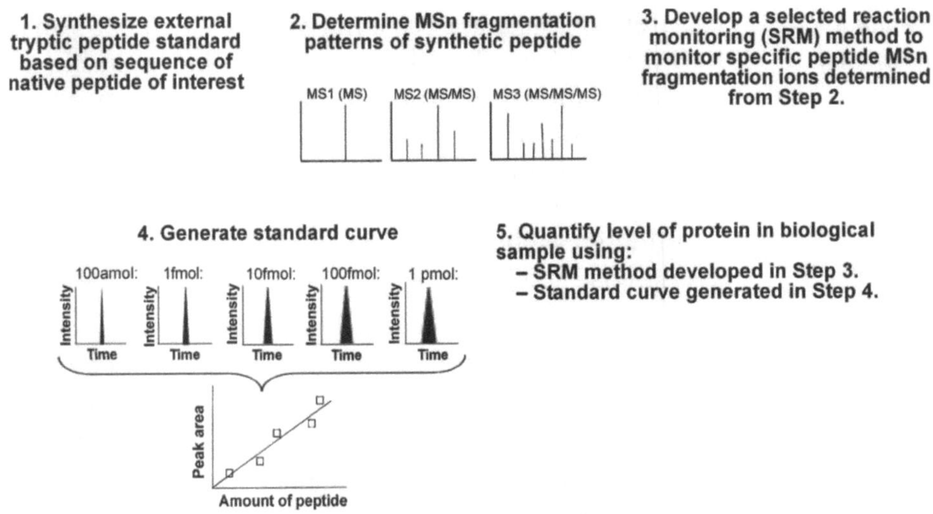

Fig. 2. Selected reaction monitoring (SRM) method used for quantitative mass spectrometry-based protein quantification.

2. Create an SRM instrument method whereby the precursor-to-product ion transition selected in **step 1** is monitored using a scan range that minimizes the duty cycle of the mass spectrometer.
3. Analyze the external peptide standard using the LC method detailed in **Subheading 3.4** (the method can be significantly shortened, so long as the peptide elutes in a distinct peak) and the SRM method created as outlined above.
4. Utilize Qual Browser (Xcalibur™ software) to display only the scans that contain distinctive product ions for each spectrum and integrate the resultant peaks using the ICIS peak detection algorithm with 15-point Boxcar smoothing.
5. Repeat **steps 3** and **4** using various amounts of external standard peptide spanning at least 1 order of magnitude (e.g., 100 amol, 1 fmol, 10 fmol, 100 fmol, 1 pmol) to calculate an equation for the line of best-fit based on the standard curve of the peak area vs. amount of peptide.
6. Analyze the sample of interest using the SRM method developed in **step 2** above.
7. Repeat **step 4** and determine the amount of the protein of interest in the sample by using the integrated peak area of the monitored peptide and the line of best-fit calculated from the SRM analysis of the external peptide standards. For further information about the effective development and optimization of SRM methods, please refer to *(28)*.

4. Notes

1. Any variation of the standard Laemmli buffer will work. β-mercaptoethanol, a typical component of Laemmli buffers that is included to reduce protein disulfide bonds, is a common source of keratin contamination. We, therefore, prepare our sample buffer with the reducing agent DTT instead.
2. The use of scalpels as opposed to razor blades is preferred so as to minimize the chances of gloves making contact with the gel when excising gel bands and cutting the bands into smaller pieces. Avoid using latex gloves as these are a common source of keratin contamination.
3. There are alternative commercially available mass spectrometry compatible silver staining kits. Mass spectrometry compatible silver staining methods are those that do not involve the use of glutaraldehyde to fix the proteins in the gel.
4. TPCK-trypsin from any other commercial source may also be used. TPCK treatment of trypsin inactivates chymotryptic activity and modification by reductive methylation or acetylation prevents autolytic activity.
5. Xcalibur version 1.4 was used for the studies from which these methods were derived; however, any later version of the software is adequate.
6. All steps for in-gel digestion should be performed in a clean laminar flow hood to minimize keratin contamination. Wipe down all surfaces in the hood with 50% methanol prior to beginning any work.
7. For all steps of the in-gel digestion protocol that entail removing solutions from gel pieces, use a gel-loading pipette tip.
8. Digested samples can be desalted using a C18 Zip Tip (Millipore) or any other C18-based spin column, cartridge or tip. However, when analyzing phosphopeptides, it should be noted that phosphate groups generally render peptides more hydrophilic and there is a risk of the phosphopeptides not binding to the C18 resin used for desalting, consequently resulting in sample loss.
9. Perform acetone precipitation if detergents are present in the protein sample at or above the following concentrations: SDS – 0.05%; Triton X-100, CHAPS, NP-40, Tween 20 and Octyl glucopyranoside – 1%. The procedure for acetone precipitation is as follows: add 4× sample volume of cold acetone (−20°C); vortex and incubate for 1 h at −20°C; centrifuge for 10 min at 12,000 × *g*; decant supernatant while being careful to not dislodge the protein pellet; resuspend protein pellet in 100 mM ammonium bicarbonate.

10. This method is intended for use with purified protein samples. Thus, the use of stronger denaturants, or acid-labile surfactants should not be required for efficient enzymatic digestion. However, if desired, the enzymatic digest can be performed in the presence of RapiGest™ (Waters), which is a mass spectrometry-compatible acid-labile surfactant that has been demonstrated to generate quantitative and qualitative improvements in peptide and protein identifications. A detailed protocol for in-solution enzymatic digestion using RapiGest™ is available in *(29)*.
11. The optimum duration for dynamic exclusion should be empirically determined based on the average chromatographic peak width. Using a dynamic exclusion duration of 30 s with an LC system that generates peak widths of ~1 min permits the acquisition of MS/MS data for a peptide as it begins to elute as a peak, at the apex of the peak and at the peak tail.
12. Phosphoserine and phosphothreonine residues are converted to dehydroalanine and 2-aminodehydrobutyric acid, respectively, following a neutral loss of phosphoric acid (H_3PO_4) upon collision induced dissociation (CID). Phosphotyrosine residues rarely undergo this neutral loss upon CID.
13. Most universities have a Molecular Biology Resource Core Facility or a Biopolymer Core Facility that offers these services to investigators for a fee.

Acknowledgments

The authors thank Peter Davies for providing the affinity-purified PHF-tau samples that we analyzed utilizing the methods detailed in this chapter. This work was supported by National Institutes of Health grants MH59786 and AG25323 (A.Y.).

References

1. Sze C, Bi H, Kleinschmidt-DeMasters BK, Filley CM, Martin LJ. (2001) *N*-Methyl-D-aspartate receptor subunit proteins and their phosphorylation status are altered selectively in Alzheimer's disease. *J Neurol Sci* **182**, 151–159.
2. Zhao D, Watson JB, Xie CW. (2004) Amyloid beta prevents activation of calcium/calmodulin-dependent protein kinase II and AMPA receptor phosphorylation during hippocampal long-term potentiation. *J Neurophysiol* **92**, 2853–2858.
3. Palop JJ, Chin J, Bien-Ly N, et al. (2005) Vulnerability of dentate granule cells to disruption of arc expression in human amyloid precursor protein transgenic mice. *J Neurosci* **25**, 9686–9693.
4. Braak H, Braak E, Strothjohann M. (1994) Abnormally phosphorylated tau protein related to the formation of neurofibrillary

tangles and neuropil threads in the cerebral cortex of sheep and goat. *Neurosci Lett* **171**, 1–4.

5. Lim J, Lu KP. (2005) Pinning down phosphorylated tau and tauopathies. *Biochim Biophys Acta* **1739**, 311–322.
6. Gong CX, Liu F, Grundke-Iqbal I, Iqbal K. (2005) Post-translational modifications of tau protein in Alzheimer's disease. *J Neural Transm* **112**, 813–838.
7. Lee G, Thangavel R, Sharma VM, et al. (2004) Phosphorylation of tau by fyn: Implications for Alzheimer's disease. *J Neurosci* **24**, 2304–2312.
8. Yoshimura Y, Ichinose T, Yamauchi T. (2003) Phosphorylation of tau protein to sites found in Alzheimer's disease brain is catalyzed by Ca2+/calmodulin-dependent protein kinase II as demonstrated tandem mass spectrometry. *Neurosci Lett* **353**, 185–188.
9. Liu F, Zaidi T, Iqbal K, et al. (2002) Role of glycosylation in hyperphosphorylation of tau in Alzheimer's disease. *FEBS Lett* **512**, 101–106.
10. Buee L, Bussiere T, Buee-Scherrer V, Delacourte A, Hof PR. (2000) Tau protein isoforms, phosphorylation and role in neurodegenerative disorders. *Brain Res Brain Res Rev* **33**, 95–130.
11. Lee VM. (1996) Regulation of tau phosphorylation in Alzheimer's disease. *Ann N Y Acad Sci* 777, 107–113.
12. Holzer M, Holzapfel HP, Zedlick D, Bruckner MK, Arendt T. (1994) Abnormally phosphorylated tau protein in Alzheimer's disease: Heterogeneity of individual regional distribution and relationship to clinical severity. *Neuroscience* **63**, 499–516.
13. Goedert M. (1993) Tau protein and the neurofibrillary pathology of Alzheimer's disease. *Trends Neurosci* **16**, 460–465.
14. Hampel H, Burger K, Pruessner JC, et al. (2005) Correlation of cerebrospinal fluid levels of tau protein phosphorylated at threonine 231 with rates of hippocampal atrophy in Alzheimer disease. *Arch Neurol* **62**, 770–773.
15. Augustinack JC, Schneider A, Mandelkow EM, Hyman BT. (2002) Specific tau phosphorylation sites correlate with severity of neuronal cytopathology in Alzheimer's disease. *Acta Neuropathol (Berl)* **103**, 26–35.
16. de Vrij FM, Fischer DF, van Leeuwen FW, Hol EM. (2004) Protein quality control in Alzheimer's disease by the ubiquitin proteasome system. *Prog Neurobiol* **74**, 249–270.
17. Zhang JY, Liu SJ, Li HL, Wang JZ. (2005) Microtubule-associated protein tau is a substrate of ATP/Mg(2+)-dependent proteasome protease system. *J Neural Transm* **112**, 547–555.
18. Shimura H, Schwartz D, Gygi SP, Kosik KS. (2004) CHIP-Hsc70 complex ubiquitinates phosphorylated tau and enhances cell survival. *J Biol Chem* **279**, 4869–4876.
19. Iqbal K, Grundke-Iqbal I. (1991) Ubiquitination and abnormal phosphorylation of paired helical filaments in Alzheimer's disease. *Mol Neurobiol* **5**, 399–410.
20. Kohnken R, Buerger K, Zinkowski R, et al. (2000) Detection of tau phosphorylated at threonine 231 in cerebrospinal fluid of Alzheimer's disease patients. *Neurosci Lett* **287**, 187–190.
21. Hampel H, Buerger K, Kohnken R, et al. (2001) Tracking of Alzheimer's disease progression with cerebrospinal fluid tau protein phosphorylated at threonine 231. *Ann Neurol* **49**, 545–546.
22. Buerger K, Teipel SJ, Zinkowski R, et al. (2002) CSF tau protein phosphorylated at threonine 231 correlates with cognitive decline in MCI subjects. *Neurology* **59**, 627–629
23. Buerger K, Zinkowski R, Teipel SJ, et al. (2002) Differential diagnosis of Alzheimer disease with cerebrospinal fluid levels of tau protein phosphorylated at threonine 231. *Arch Neurol* **59**, 1267–1272.
24. Cripps D, Thomas SN, Jeng Y, et al. (2006) Alzheimer disease-specific conformation of hyperphosphorylated paired helical filament-Tau is polyubiquitinated through Lys-48, Lys-11, and Lys-6 ubiquitin conjugation. *J Biol Chem* **281**, 10825–10838.
25. Weaver CL, Espinoza M, Kress Y, Davies P. (2000) Conformational change as one of the earliest alterations of tau in Alzheimer's disease. *Neurobiol Aging* **21**, 719–727.
26. Vincent I, Zheng JH, Dickson DW, Kress Y, Davies P. (1998) Mitotic phosphoepitopes precede paired helical filaments in Alzheimer's disease. *Neurobiol Aging* **19**, 287–296.
27. Mayya V, Rezaul K, Cong YS, Han D. (2005) Systematic comparison of a two-dimensional ion trap and a three-dimensional ion trap mass spectrometer in proteomics. *Mol Cell Proteomics* **4**, 214–223.
28. Gerber SA, Rush J, Stemman O, Kirschner MW, Gygi SP. (2003) Absolute quantification of proteins and phosphoproteins from cell lysates by tandem MS. *Proc Natl Acad Sci USA* **100**, 6940–6945.
29. Chen EI, Cociorva D, Norris JL, Yates JR, III. (2007) Optimization of mass spectrometry-compatible surfactants for shotgun proteomics. *J Proteome Res* **6**, 2529–2538.

Chapter 9

Proteomics Identification of Carbonylated and HNE-Bound Brain Proteins in Alzheimer's Disease

Rukhsana Sultana and D. Allan Butterfield

Summary

Free radicals and oxidative stress play a crucial role in the pathophysiology of a wide variety of diseases including cancer and neurodegenerative disorders. Reactive oxygen and reactive nitrogen species can react with biomolecules such as proteins, lipids, nucleic acid, etc. resulting in the formation of protein carbonyls, 3-nitrotyroine, HNE-bound proteins, etc. Such modifications in proteins often lead to functional impairment, and the identification of such oxidatively modified proteins may help in delineating the mechanism of disease 1pathogenesis or progression.

In this chapter, we described the protocol for the identification of oxidatively modified proteins, i.e., protein carbonyls and HNE-bound proteins in a given biological sample using three important techniques, i.e., proteomics coupled with mass spectrometry and immunochemical detection. These methods are placed in the context of our studies on Alzheimer's disease.

Key words: Proteomics, Oxidative stress, Protein carbonyls, 4-Hydroxy-2-nonenal, Isoelectric focusing, 2,4-Dinitrophenylhydrazine, Mass spectrometry, Alzheimer's disease

1. Introduction

Free radicals and oxidative stress play a crucial role in a wide variety of disease pathophysiology, including cancer and neurodegeneration *(1–6)*. Free radicals can damage virtually all biological molecules including DNA, RNA, a cholesterol, lipids, carbohydrates, proteins, and antioxidants. Oxidative modification of proteins can be induced in vitro by a wide array of pro-oxidant agents and occurs in vivo during aging and in certain disease conditions.

Proteins constitute one of the major targets of ROS, and the oxidation of proteins may lead to a loss of protein function as well as conversion of proteins to forms that are more susceptible to

Andrew K. Ottens and Kevin K.W. Wang (eds.), *Neuroproteomics*, Methods in Molecular Biology, vol. 566
DOI 10.1007/978-1-59745-562-6_9, © Humana Press, a part of Springer Science+Business Media, LLC 2009

protein degradation by proteasomes *(7, 8)*. There are numerous types of protein oxidative modifications, indexed by increased levels of protein carbonyls and 3-nitrotyrosine *(9–11)*. Protein carbonyls are formed by oxidation of the amino acid residues cysteine, methionine, tryptophan, arginine, lysine, proline, histidine, and others *(1, 12, 13)*. Protein carbonyls are also formed by reactions of lysine, cysteine, or histidine amino acids with α- and β-unsaturated aldehydes formed during the peroxidation of polyunsaturated fatty acids *(14, 15)*. A number of previous studies showed that carbonylation of proteins leads to impaired protein function, which suggests that oxidative modifications have physiological and pathological significance *(16–18)*. Hence, the identification of carbonylated proteins together with their functional study may help to identify metabolic or structural defects caused by oxidative modification.

Protein carbonyls are usually detected by derivatization of the carbonyl group with 2,4-dinitrophenylhydrazine (DNPH) with the formation of hydrazones *(19)*. These hydrazones can be detected spectrophotometrically at 375 nm; however, sample homogeneity or uniformity is a major concern. Another way to detect protein carbonyls is the immunochemical detection of hydrazones using an anti-DNP antibody that can give a clear indication of the amount of total protein carbonyls in a given sample. The latter method has been widely used to detect protein carbonyl formation **(Fig. 1)**.

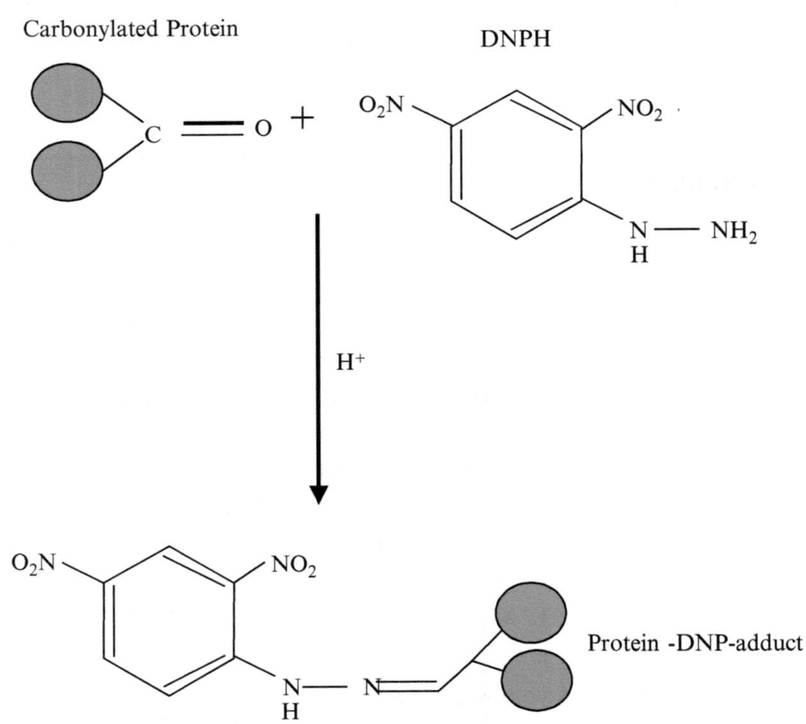

Fig. 1. Reaction of protein carbonyls with DNPH.

Hydroxynonenal (HNE) is an α-, β-unsaturated alkenal product of omega-6 polyunsaturated fatty acid peroxidation, and is a major cytotoxic end-product of lipid peroxidation *(14, 20, 21)*. HNE can form Michael adducts by covalently binding to cysteine, lysine, or histidine residues *(14)*, and can mediate oxidative stress-induced cell death in many cell types *(22, 23)*. HNE accumulates in cellular membranes at concentrations of 10 μM to 5 mM in response to oxidative insults *(14)*, and invokes a wide range of biological activities, including inhibition of protein and DNA synthesis *(24, 25)*, stimulation of C and D phospholipases *(26, 27)*, and activation of stress signaling pathways *(28–31)*.

In a number of neurodegenerative diseases, including Alzheimer's disease (AD), oxidative stress is evident in the brain *(31)*. For example, in brain from subjects diagnosed with AD and mild cognitive impairment (arguably an early form of AD) the levels of protein carbonyls *(18, 32)*, 3-NT *(33, 34)*, and free *(35)* and protein-bound *(21)* HNE were found to be significantly increased when compared with control subjects. Protein-bound HNE is normally detected in our laboratory using an immunochemical approach with an anti-HNE antibody (**Fig. 2**).

Two-dimensional gel electrophoresis (2D PAGE) involves the separation of proteins based on two physicochemical properties, i.e., isoelectric point and relative mobility or approximate

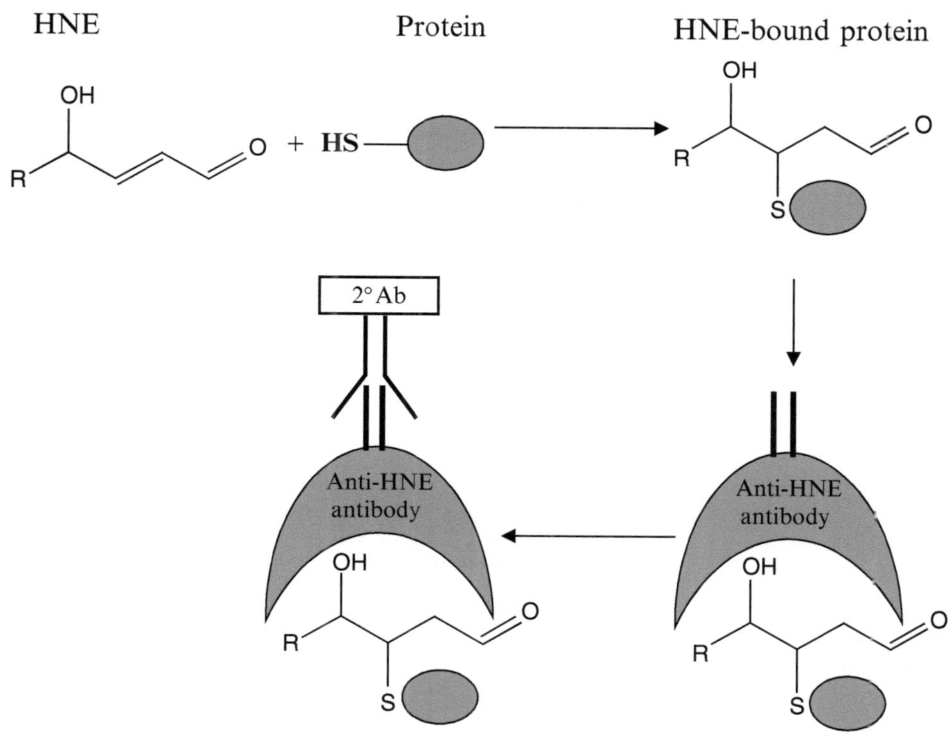

Fig. 2. Immunochemical detection of 4-hydroxy-2-nonenal (HNE) bound to proteins.

molecular mass *(36)*. In the first dimension, proteins are focused according to their isoelectric points (pIs) or the point at which the net charge on the protein is zero, using isoelectric focusing (IEF) electrophoresis. In the second dimension, proteins are separated based on their migration within an applied electric field according to their approximate molecular mass using sodium dodecyl sulfate poly-(acrylamide)-electrophoresis (SDS-PAGE). This 2D technique can provide the approximate pI and molecular mass (±10%) of most proteins, with some dramatic exceptions. The advantages of 2D PAGE involve the ability to separate a large number of proteins in a given sample where there exists a high probability that each individual spot represents an individual protein. Screening thousands of proteins at once, sometimes coupled to 2D Western blots, can provide information on post-translational modifications, in addition to the expression profile of proteins *(37)*.

Our laboratory first used proteomics to identify oxidatively modified protein in brain from subjects with AD and MCI, which extended our proteomic studies to early forms of AD and animal models of this disorder in addition to other neurodegenerative diseases *(18, 34, 38–44)*. In our proteomic studies, we carried out a comparative analysis between 2D PAGE and 2D Western blotting (for 3-nitrotyrosine or HNE-bound proteins) to detect oxidatively modified proteins (**Fig.** 3). Image analysis between the 2D PAGE and the 2D blot allows for the determination of protein spots that have an increased carbonyl or HNE reactivity

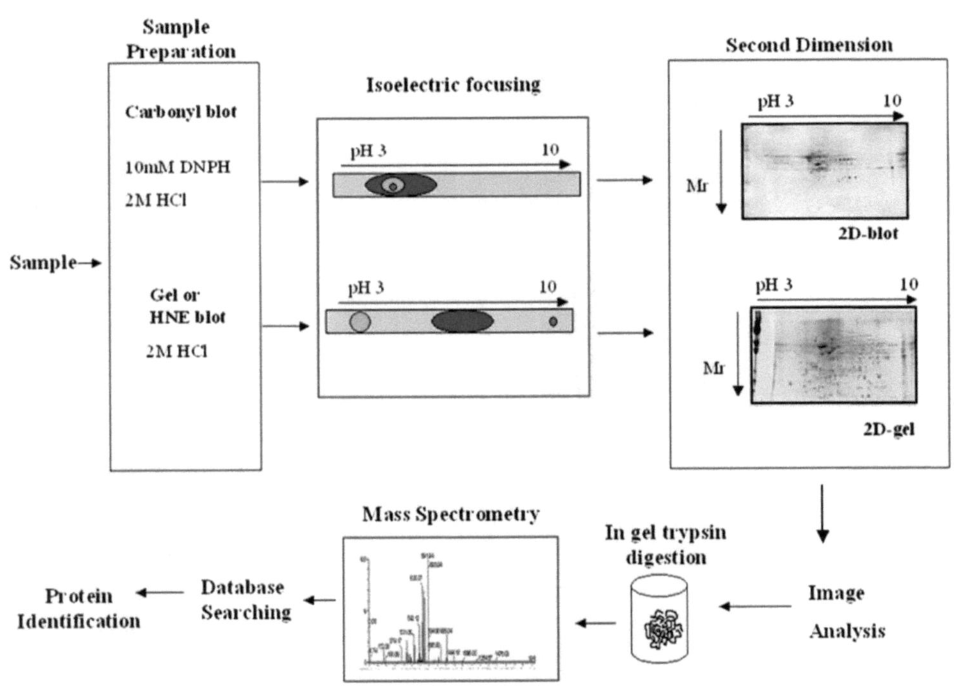

Fig. 3. Proteomics analysis.

compared to control. Oxidatively modified and dysfunctional proteins identified in AD brain from our laboratory are consistent with the pathology and altered biochemistry of AD and include brain proteins related to energy metabolism; proteasome function; cholinergic processes; Ca^{2+} homeostasis and excitotoxicity; regulation of cellular pH; structural proteins; mitochondrial function (including apoptosis); formation of neurofibrillary tangles, senile (neuritic) plaques, and cell-cycle activation; synaptic functioning, including long-term potentiation; and antioxidant function *(31)*. Each of these proteins and their associated pathways can be plausibly invoked in mechanisms of neurodegeneration, which ultimately underlie the memory and cognitive loss in this dementing disorder.

In our studies, we benefit from the Rapid Autopsy Program of the University of Kentucky Alzheimer's Disease Clinical Center. No specimen used in our laboratory has more than a 4-h postmortem interval (PMI). Longer PMIs make interpretation of results using brain highly suspect; the effects of PMI itself rather than disease play a significant role.

In this chapter, we described the protocols for the identification of oxidatively modified proteins, i.e., protein carbonyls and HNE-bound proteins, employing immunochemical detection of protein carbonyls and HNE-bound proteins coupled with neuroproteomic methods to take maximum advantage of these three techniques to detect the oxidatively modified proteins in a given brain sample.

2. Materials

2.1. Sample Preparation for Protein Carbonyl and HNE-bound Proteins

1. Sample homogenization buffer (pH 7.4): 10 mM HEPES, 137 mM NaCl, 4.6 mM KCl, 1.1 mM KH_2PO_4, 0.6 mM $MgSO_4$, 0.5 μg/mL leupeptin (stored as an aliquot at −20°C), 0.7 μg/mL pepstatin (stored as an aliquot at −20°C), 0.5 μg/mL type II S soybean trypsin inhibitor, 40 μg/mL PMSF dissolved in de-ionized (DI) water stored at 4°C.
2. DNPH solution: 10 mM 2,4-dinitrophenylhydrazine dissolved in 2 M HCl solution stored at room temperature.
3. Laemelli buffer stock: 0.125 M Tris–HCl (pH: 6.8), 4% SDS, and 20% glycerol added to make a final volume of 10 mL (store at room temperature).
4. Protein precipitation: Trichloroacetic acetic acid (100%, stored at 4°C) added to make a final 15% of the total volume.
5. Wash buffer I: 1:1 (v/v) ethanol/ethyl acetate (make fresh before use).

2.2. Isoelectric Focusing

1. Rehydration buffer: 8 M urea, 2 M thiourea, 2% CHAPS, 0.2% biolytes, 50 mM dithiothreitol (DTT), bromophenol blue dissolved in DI water (Make fresh before use).

2.3. Two-Dimensional SDS-Polyacrylamide Gel Electrophoresis

1. DTT equilibrium buffer (pH 6.8): 50 mM Tris–HCl, 6 M urea, 1% (m/v) SDS, 30% (v/v) glycerol, 0.5% DTT dissolved in DI water (make fresh before use).
2. IA equilibrium buffer (pH 6.8): 50 mM Tris–HCl, 6 M urea, 1% (w/v) SDS, 30% (v/v) glycerol, 4.5% iodoacetamide dissolved in DI water (this solution is light sensitive, and needs to be freshly prepared).
3. Running buffer (10×): 20% (w/v) glycine, 5% (w/v) Tris base, 1% (w/v) SDS in DI water (store at room temperature).
4. Fixing solution: 10% (v/v) methanol, 7% (v/v) acetic acid dissolved in DI water (store at room temperature).
5. SYPRO Ruby stain (store at room temperature).
6. Agarose solution: 0.5% low melting agarose is dissolved in 1× running buffer at 37°C.

2.4. Oxyblot Immunochemical Detection

1. Transfer buffer: 1% (v/v) 10× running buffer, 10% (v/v) methanol diluted with DI water and stored at 4°C.
2. Wash buffer-II: 0.01% (w/v) sodium azide and 0.2% (v/v) Tween 20 dissolved in phosphate buffered saline (PBS) stored at room temperature.
3. Blocking buffer: 2% bovine serum albumin (BSA) in wash buffer-II made fresh before use.
4. Primary antibody solution: Dilute anti-dinitrophenyl hydrazone antibody (1:100) (Chemicon International, Temecula, CA) or anti-HNE antibody (1:5,000) (Alpha Diagnostic, San Antonio, TX) in 20 mL of blocking buffer.
5. Secondary antibody solution: Dilute secondary antibody (anti-rabbit conjugated to alkaline phoshatase antibody, Sigma Aldrich, St Louis, MO) in 20 mL of wash buffer-II (1:3,000) (make fresh).
6. Developing solution: Dissolve one Sigma Fast tablet [5-bromo-4-chloro-3-indolyl phosphate/nitro blue tetrazolium (BCIP/NBT)] in 10-mL DI water (this solution is light sensitive – prepare fresh each time).

3. Methods

3.1. Brain Sample Preparation

1. Homogenize brain tissue in sample homogenization buffer (10% w/v). Centrifuge the samples at 2,500 × *g* and use the supernatant for proteomics (*see* **Note 1**).

2. Determine the sample protein concentration by a bicinchoninic acid (BCA) protein assay.
3. Sample derivatization for protein carbonyls: To 100–150 μg of the protein add four times the sample volume of DNPH, vortex and incubate the samples at room temperature for 20 min without shaking (*see* **Note 2**).
4. Sample amount of HNE-bound protein: No derivatization is needed. 150–250 μg of sample is used for the detection of HNE-bound proteins; the same amount of protein is used for an estimation of protein levels.
5. Add ice-cold TCA to the protein sample for a final volume of 30% and incubate on ice for 10 min (*see* **Note 3**).
6. Centrifuge the samples at 10,000 × *g* for 5 min at 4°C. Discard the supernatant and wash the pellet four times with ice-cold wash buffer-I at 10,000 × *g* for 5 min at 4°C.
7. Dry the final pellet and resuspend it in 200 μL of rehydration buffer. Vortex the samples at room temperature for 1–2 h (*see* **Note 4**).
8. Sonicate the sample at 2 rpm for 10 s.

3.2. Isoelectric Focusing of Samples or First Dimension

1. Select an immobilized pH gradient (IPG) strip of the desired pH range for IEF separation. Carefully load 180 μL of the sample into the bottom of a well in an IEF tray by using a micropipette (*see* **Note 5**).
2. Place the IPG strip gel side down on top of the sample, cover the IEF tray and place it in the IEF machine (*see* **Note 6**).
3. Start active rehydration of the IPG strips at 50 V, 20°C, overnight. Pause the program after 1 h and add 2 mL of mineral oil in each lane, then carryout the active rehydration step for about 16 h (*see* **Note 7**).
4. Wet paper wicks with 8 μL of nanopure water and place it between the electrodes and the IPG strip (*see* **Note 8**).
5. Carryout IEF at 20°C as follows: linear ramp to 300 V over 2 h, linear ramp to 500 V over 2 h, linear ramp to 1,000 V over 2 h, linear ramp to 8,000 V for 8 h, and a rapid ramp to 8,000 V over 10 h.
6. After completion of IEF, the IPG strip can be loaded directly for second dimension separation or stored at –80°C freezer until use (*see* **Note 9**).

3.3. Two-Dimensional Polyacrylamide Gel Electrophoresis

1. If applicable, remove the IPG strip from the –80°C freezer and thaw at room temperature for 30 min. Meanwhile warm an agarose solution at 37°C.
2. Add 4 mL of DTT equilibrium buffer to the tray holding the IPG strip and incubate at room temperature in the dark for 10 min.

3. Open a commercial SDS-PAGE 2D gel (e.g., from Bio-Rad) and rinse with DI water (both inside and outside) while waiting for IPG equilibration (*see* **Note 10**). Prepare the running buffer by diluting 100 mL of 10× Running Buffer stock with 900 mL of DI water in a measuring cylinder.
4. After 10 min of incubation in DTT equilibrium buffer, transfer the IPG strip into the next well in the equilibration tray, add 4 mL of IA equilibrium buffer, and incubate the IPG strip for another 10 min in the dark.
5. Wash the IPG strip in 1× running buffer (*see* **Note 11**).
6. Load the IPG strip with gel side facing up into the 2D gel (*see* **Note 12**).
7. Load 2 μL of unstained molecular mass marker into standard well adjacent to the IPG strip for 2D PAGE and stained molecular mass marker on a gel that will be used for 2D blot analysis.
8. Add the warm agarose solution into the IPG well of the 2D gel and push the IPG strip on either end until a contact is established between the gel and the IPG strip (*see* **Note 13**).
9. Allow 10 min for the agarose to solidify, place the gel in a tank filled with running buffer, and then fill the upper tank with running buffer.
10. Run the gels at 200 V for 65 min at room temperature or until the dye front (bromophenol blue) runs off the gel into the lower tank.
11. After 65 min of running the gel, disconnect the power supply and disassemble the gel unit. Break open the gels plate and cut one end of the gels at an angle to allow its orientation to be tracked.

3.4. Protein Staining

1. To the gels containing nonderivatized proteins with unstained marker, add 50 mL of fixative solution and incubate at room temperature for 60 min.
2. Remove the fixative solution and incubate gels in 50 mL of SYPRO Ruby gel stain from 4 h to overnight at room temperature.
3. Wash gels in DI water for 1 h and then scan with a fluorescence imager at an excitation wavelength of 300 or 490 nm and an emission wavelength of 640 nm.

3.5. Western Blotting

1. Soak a precut nitrocellulose membrane and a sheet of filter paper in transfer buffer for 10 min.
2. Make a transfer sandwich in the following order: first place one soaked filter paper on the semi-dry transfer unit platform, followed by the nitrocellulose membrane, then the gel and one more sheet of filter paper, and carryout the transfer at 15 V for 2 h at room temperature (*see* **Note 14**).

3. Once the transfer is completed, remove the membrane and incubate it in 20 mL of blocking buffer for 1 h at room temperature on a rocking platform.
4. After 1 h, add a 1:100 solution of anti-DNPH or a 1:5,000 solution of anti-HNE in blocking buffer and incubate for 1–2 h at room temperature on a rocking platform.
5. Remove the primary antibody and wash the membrane three times for 5 min each with 50 mL of Wash Buffer-II.
6. Add the secondary antibody (1:3,000) in 20 mL of Wash Buffer-II and incubate the membrane on a rocking platform for 1 h.
7. Wash the membrane three times for 5 min each with Wash Buffer-II.
8. Develop the membrane using Developing Solution for 10–30 min until the desired appearance of spots.
9. Wash the membrane with DI water to stop the color development and dry the membrane between Kim-Wipes. Use a flatbed scanner to capture the image of the blot (see example in **Fig. 3**).

3.6. Image Analysis

1. Carry out computerized image analysis using an appropriate software package (for our laboratory we use PD Quest image analysis software from Bio-Rad) to determine the levels of specific protein carbonyls or specific HNE-bound proteins (*see* **Note 15**).
2. The protein spots showing significantly increased protein carbonyls or HNE-bound protein levels (Student's *t* test, $p < 0.05$) are excised from the gel and digested with trypsin. Generally, a minimum of six samples from each group is recommended for correct identification of modified proteins (*see* **Note 15**).
3. Submit the resulting tryptic peptides for mass spectrometry analysis. This mass information (peptide mass finger printing or tandem mass spectrometry) is used to interrogate appropriate databases (for example, ExPASY) to lead to the identification of the proteins.
4. In new sample sets, the protein that is identified by mass spectrometry should have its identity confirmed either by sequence analysis of each peptide (MS/MS) or by immunoprecipitation.

4. Notes

1. This will remove the nuclear fraction and unbroken cells from the samples.

2. This step is needed only for protein carbonyl detection; if one is analyzing HNE-bound protein, skip this step and load the same amount of protein described in **step 4** of **Subheading 3.1**.
3. This procedure will precipitate the protein
4. This step ensures proper solubilization of the proteins.
5. Our protocol was written for Bio-Rad (Hercules, CA, USA) IPG strips, but the methodology is generally applicable to other commercial products. Avoid air bubbles as these will interfere with current flow.
6. Make sure that the positive end of strip is toward the positive end of the IEF tray. This is important for correct connections and proper IEF.
7. Addition of oil will prevent the evaporation of solvent from the sample.
8. This will act like a shock absorber and prevent burning of the strip.
9. Keep the gel side facing up for the IPG strip in a disposable tray. Frozen IPG strips look milky white in appearance.
10. Remove extra water with Kim-Wipes.
11. This step will remove excess equilibration buffer from the IPG strip.
12. Do not push the IPG strip into the well at this point.
13. Avoid bubbles while adding agarose and also make sure that the IPG strip is in parallel contact with the gel.
14. Smoothly and gently roll a glass rod over the gel to remove bubbles that are trapped between the nitrocellulose membrane and gel. Once the sandwich is ready, roll the glass rod once again to eliminate any trapped air bubble in the transfer sandwich.
15. To determine the proteins showing significant oxidative modification, first a match set is created separately for gels or blots by selecting a gel or blot with the best resolution or separation among those obtained, followed by normalization and manual matching of a minimum of 25–30 protein spots for gels and 20–25 spots for blots. Following the creation and matching of the gels and blots, a high-match set is created for gels and blots using the first match sets. This high-match set is normalized to the actual protein content as measured by the intensity of a protein stain such as SYPRO ruby to determine the carbonyl or HNE-bound protein immunoreactivity on the blots. That is, the computer-assisted image analysis program is used to determine a specific oxidative modification by dividing the intensity of the blot spot by the intensity of the gel spot.

Acknowledgments

This work was supported in part by NIH grants to D.A.B. [AG-10836; AG-05119].

References

1. Butterfield, D. A., and Stadtman, E. R. (1997) Protein oxidation processes in aging brain. *Adv. Cell Aging Gerontol.* **2**, 161–191.
2. Butterfield, D. A., and Lauderback, C. M. (2002) Lipid peroxidation and protein oxidation in Alzheimer's disease brain: Potential causes and consequences involving amyloid beta-peptide-associated free radical oxidative stress. *Free Radic. Biol. Med.* **32**, 1050–1060.
3. Joshi, G., Sultana, R., Tangpong, J., Cole, M. P., St Clair, D. K., Vore, M., Estus, S., and Butterfield, D. A. (2005) Free radical mediated oxidative stress and toxic side effects in brain induced by the anti cancer drug adriamycin: Insight into chemobrain. *Free Radic. Res.* **39**, 1147–1154.
4. Behl, C. (2005) Oxidative stress in Alzheimer's disease: Implications for prevention and therapy. *Subcell. Biochem.* **38**, 65–78.
5. Gotz, M. E., Kunig, G., Riederer, P., and Youdim, M. B. (1994) Oxidative stress: Free radical production in neural degeneration. *Pharmacol. Ther.* **63**, 37–122.
6. Beal, M. F. (1996) Mitochondria, free radicals, and neurodegeneration. *Curr. Opin. Neurobiol.* **6**, 661–666.
7. Grune, T., and Davies, K. J. (2003) The proteasomal system and HNE-modified proteins. *Mol. Aspect. Med.* **24**, 195–204.
8. Grune, T., Reinheckel, T., and Davies, K. J. (1997) Degradation of oxidized proteins in mammalian cells. *FASEB J.* **11**, 526–534.
9. Aksenov, M. Y., Aksenova, M. V., Butterfield, D. A., Geddes, J. W., and Markesbery, W. R. (2001) Protein oxidation in the brain in Alzheimer's disease. *Neuroscience* **103**, 373–383.
10. Boyd-Kimball, D., Sultana, R., Abdul, H. M., and Butterfield, D. A. (2005) Gamma-glutamylcysteine ethyl ester-induced up-regulation of glutathione protects neurons against Abeta(1-42)-mediated oxidative stress and neurotoxicity: Implications for Alzheimer's disease. *J. Neurosci. Res.* **79**, 700–706.
11. Sultana, R., Newman, S., Mohmmad-Abdul, H., Keller, J. N., and Butterfield, D. A. (2004) Protective effect of the xanthate, D609, on Alzheimer's amyloid beta-peptide (1–42)-induced oxidative stress in primary neuronal cells. *Free Radic. Res.* **38**, 449–458.
12. Dalle-Donne, I., Scaloni, A., Giustarini, D., Cavarra, E., Tell, G., Lungarella, G., Colombo, R., Rossi, R., and Milzani, A. (2005) Proteins as biomarkers of oxidative/nitrosative stress in diseases: The contribution of redox proteomics. *Mass Spectrom. Rev.* **24**, 55–99.
13. Davies, K. J., Delsignore, M. E., and Lin, S. W. (1987) Protein damage and degradation by oxygen radicals. II. Modification of amino acids. *J. Biol. Chem.* **262**, 9902–9907.
14. Esterbauer, H., Schaur, R. J., and Zollner, H. (1991) Chemistry and biochemistry of 4-hydroxynonenal, malonaldehyde and related aldehydes. *Free Radic. Biol. Med.* **11**, 81–128.
15. Refsgaard, H. H., Tsai, L., and Stadtman, E. R. (2000) Modifications of proteins by polyunsaturated fatty acid peroxidation products. *Proc. Natl Acad. Sci. USA* **97**, 611–616.
16. Yin, D., and Chen, K. (2005) The essential mechanisms of aging: Irreparable damage accumulation of biochemical side-reactions. *Exp. Gerontol.* **40**, 455–465.
17. Sultana, R., Boyd-Kimball, D., Poon, H. F., Cai, J., Pierce, W. M., Klein, J. B., Merchant, M., Markesbery, W. R., and Butterfield, D. A. (2006) Redox proteomics identification of oxidized proteins in Alzheimer's disease hippocampus and cerebellum: An approach to understand pathological and biochemical alterations in AD. *Neurobiol. Aging* **27**, 1564–1576.
18. Butterfield, D. A., Poon, H. F., St Clair, D., Keller, J. N., Pierce, W. M., Klein, J. B., and Markesbery, W. R. (2006) Redox proteomics identification of oxidatively modified hippocampal proteins in mild cognitive impairment: Insights into the development of Alzheimer's disease. *Neurobiol. Dis.* **22**, 223–232.
19. Reznick, A. Z., and Packer, L. (1994) Oxidative damage to proteins: spectrophotometric method for carbonyl assay. *Methods Enzymol.* **233**, 357–363.
20. Mattson, M. P., Fu, W., Waeg, G., and Uchida, K. (1997) 4-Hydroxynonenal, a product of lipid peroxidation, inhibits dephosphorylation of the microtubule-associated protein tau. *Neuroreport* **8**, 2275–2281.

21. Lauderback, C. M., Hackett, J. M., Huang, F. F., Keller, J. N., Szweda, L. I., Markesbery, W. R., and Butterfield, D. A. (2001) The glial glutamate transporter, GLT-1, is oxidatively modified by 4-hydroxy-2-nonenal in the Alzheimer's disease brain: The role of Abeta1-42. *J. Neurochem.* **78**, 413–416.
22. Choudhary, S., Zhang, W., Zhou, F., Campbell, G. A., Chan, L. L., Thompson, E. B., and Ansari, N. H. (2002) Cellular lipid peroxidation end-products induce apoptosis in human lens epithelial cells. *Free Radic. Biol. Med.* **32**, 360–369.
23. Uchida, K. (2003) 4-Hydroxy-2-nonenal: A product and mediator of oxidative stress. *Prog. Lipid Res.* **42.**, 318–343.
24. Uchida, K., and Stadtman, E. R. (1992) Modification of histidine residues in proteins by reaction with 4-hydroxynonenal. *Proc. Natl Acad. Sci. USA* **89**, 4544–4548.
25. Wonisch, W., Kohlwein, S. D., Schaur, J., Tatzber, F., Guttenberger, H., Zarkovic, N., Winkler, R., and Esterbauer, H. (1998) Treatment of the budding yeast *Saccharomyces cerevisiae* with the lipid peroxidation product 4-HNE provokes a temporary cell cycle arrest in G1 phase. *Free Radic. Biol. Med.* **25**, 682–687.
26. Rossi, M. A., Di Mauro, C., and Dianzani, M. U. (2001) Experimental studies on the mechanism of phospholipase C activation by the lipid peroxidation products 4-hydroxynonenal and 2-nonenal. *Int. J. Tissue React.* **23**, 45–50.
27. Natarajan, V., Scribner, W. M., and Taher, M. M. (1993) 4-Hydroxynonenal, a metabolite of lipid peroxidation, activates phospholipase D in vascular endothelial cells. *Free Radic. Biol. Med.* **15**, 365–375.
28. Abdul, H. M., and Butterfield, D. A. (2007) Involvement of PI3K/PKG/ERK1/2 signaling pathways in cortical neurons to trigger protection by cotreatment of acetyl-L-carnitine and alpha-lipoic acid against HNE-mediated oxidative stress and neurotoxicity: Implications for Alzheimer's disease. *Free Radic. Biol. Med.* **42**, 371–384.
29. Tamagno, E., Robino, G., Obbili, A., Bardini, P., Aragno, M., Parola, M., and Danni, O. (2003) H2O2 and 4-hydroxynonenal mediate amyloid beta-induced neuronal apoptosis by activating JNKs and p38MAPK. *Exp. Neurol.* **180**, 144–155.
30. Zhang, H., Court, N., and Forman, H. J. (2007) Submicromolar concentrations of 4-hydroxynonenal induce glutamate cysteine ligase expression in HBE1 cells. *Redox Rep.* **12**, 101–106.
31. Butterfield, D. A., Reed, T., Newman, S., and Sultana, R. (2007) Roles of amyloid b-peptide-associated oxidative stress and brain protein modifications in the pathogenesis of Alzheimer's disease and mild cognitive impairment. *Free Radic. Biol. Med.* **43**, 658–677
32. Hensley, K., Hall, N., Subramaniam, R., Cole, P., Harris, M., Aksenov, M., Aksenova, M., Gabbita, S. P., Wu, J. F., Carney, J. M., Lovell, M. A., Markessbery, W. R., and Butterfield, D. A., (1995) Brain regional correspondence between Alzheimer's disease histopathology and biomarkers of protein oxidation. *J. Neurochem.* **65**, 2146–2156
33. Butterfield, D. A., Reed, T., Perluigi, M., De Marco, C., Coccia, R., Keller, J.N., Markesbery, W. R., and Sultana, R. (2007) Elevated levels of 3-nitrotyrosine in brain from subjects with amnestic mild cognitive impairment: Implications for the role of nitration in the progression of Alzheimer's disease. *Brain Res.* **1148**, 243–248
34. Sultana, R., Poon, H. F., Cai, J., Pierce, W. M., Merchant, M., Klein, J. B., Markesbery, W. R., and Butterfield, D. A. (2006) Identification of nitrated proteins in Alzheimer's disease brain using a redox proteomics approach. *Neurobiol. Dis.* **22**, 76–87.
35. Markesbery, W. R., and Lovell, M. A. (1998) Four-hydroxynonenal, a product of lipid peroxidation, is increased in the brain in Alzheimer's disease. *Neurobiol. Aging* **19**, 33–36.
36. Rabilloud, T. (2002) Two-dimensional gel electrophoresis in proteomics: Old, old fashioned, but it still climbs up the mountains. *Proteomics.* **2**, 3–10.
37. Anderson, N. L., Matheson, A. D., and Steiner, S. (2000) Proteomics: Applications in basic and applied biology. *Curr. Opin. Biotechnol.* **11**, 408–412.
38. Boyd-Kimball, D., Castegna, A., Sultana, R., Poon, H. F., Petroze, R., Lynn, B. C., Klein, J. B., and Butterfield, D. A. (2005) Proteomic identification of proteins oxidized by Abeta(1-42) in synaptosomes: Implications for Alzheimer's disease. *Brain. Res.* **1044**, 206–215.
39. Boyd-Kimball, D., Poon, H. F., Lynn, B. C., Cai, J., Pierce, W. M., Jr., Klein, J. B., Ferguson, J., Link, C. D., and Butterfield, D. A. (2006) Proteomic identification of proteins specifically oxidized in *Caenorhabditis elegans* expressing human Abeta(1-42): Implications for Alzheimer's disease. *Neurobiol. Aging* **27**, 1239–1249.
40. Castegna, A., Aksenov, M., Aksenova, M., Thongboonkerd, V., Klein, J. B., Pierce, W. M., Booze, R., Markesbery, W. R., and Butterfield, D. A. (2002) Proteomic identification

of oxidatively modified proteins in Alzheimer's disease brain. Part I: Creatine kinase BB, glutamine synthase, and ubiquitin carboxy-terminal hydrolase L-1. *Free Radic. Biol. Med.* **33**, 562–571.

41. Castegna, A., Aksenov, M., Thongboonkerd, V., Klein, J. B., Pierce, W. M., Booze, R., Markesbery, W. R., and Butterfield, D. A. (2002) Proteomic identification of oxidatively modified proteins in Alzheimer's disease brain. Part II: Dihydropyrimidinase-related protein 2, alpha-enolase and heat shock cognate 71. *J. Neurochem.* **82**, 1524–1532.
42. Perluigi, M., Fai Poon, H., Hensley, K., Pierce, W. M., Klein, J. B., Calabrese, V., De Marco, C., and Butterfield, D. A. (2005) Proteomic analysis of 4-hydroxy-2-nonenal-modified proteins in G93A-SOD1 transgenic mice--a model of familial amyotrophic lateral sclerosis. *Free Radic. Biol. Med.* **38**, 960–968.
43. Poon, H. F., Castegna, A., Farr, S. A., Thongboonkerd, V., Lynn, B. C., Banks, W. A., Morley, J. E., Klein, J. B., and Butterfield, D. A. (2004) Quantitative proteomics analysis of specific protein expression and oxidative modification in aged senescence-accelerated-prone 8 mice brain. *Neuroscience* **126**, 915–926.
44. Sultana, R., Perluigi, M., and Butterfield, D. A. (2006) Redox proteomics identification of oxidatively modified proteins in Alzheimer's disease brain and in vivo and in vitro models of AD centered around Abeta(1-42). *J. Chromatogr. B Analyt. Technol. Biomed. Life. Sci.* **833**, 3–11.

Chapter 10

Mass Spectrometric Identification of In Vivo Nitrotyrosine Sites in the Human Pituitary Tumor Proteome

Xianquan Zhan and Dominic M. Desiderio

Summary

The chemically stable tyrosine nitration of a protein involves the addition of a nitro group ($-NO_2$) to the phenolic ring of a tyrosine residue, which may be associated with nervous system physiological and pathological processes. Identification of nitrotyrosine sites on a protein could clarify the functional significance of the modification. Due to the rarity of nitrotyrosine sites in a proteome, tandem mass spectrometry, coupled with different techniques that isolate and enrich nitrotyrosine-containing proteins from a pituitary proteome, is currently the most effective method for site identification. Commercially available nitrotyrosine polyclonal/monoclonal antibodies enable one to detect nitrotyrosine-containing proteins in a two-dimensional gel electrophoresis (2DGE) map, and to preferentially enrich nitrotyrosine-containing proteins with immunoprecipitation. Our present protocols have integrated different isolation/enrichment techniques (2DGE; Western blots; nitrotyrosine immunoaffinity precipitation) and two different tandem mass spectrometry methods (MALDI-MS/MS; ESI-MS/MS) to determine the amino acid sequence of nitrotyrosine-containing peptides that derive from nitrated proteins. Bioinformatics tools are then used to correlate nitrotyrosine sites with a functional domain/motif in order to understand the relationship between tyrosine nitration and the structural/functions of proteins.

Key words: Nitrotyrosine, Nitroproteomics, Two-dimensional gel electrophoresis, Nitrotyrosine immunoaffinity enrichment, Tandem mass spectrometry, Bioinformatics

1. Introduction

Protein tyrosine nitration (NO_2-Tyr-Prot) is a potential marker of oxidative/nitrosative injuries *(1, 2)* and results from, not only the main in vivo peroxynitrite pathway, but also myeloperoxidase and other metalloperoxidase reaction pathways *(3, 4)*. The nitration (addition of a $-NO_2$ group) of a tyrosine residue in a protein decreases the electron density of the phenolic ring of tyrosine

Andrew K. Ottens and Kevin K.W. Wang (eds.), *Neuroproteomics*, Methods in Molecular Biology, vol. 566
DOI 10.1007/978-1-59745-562-6_10, © Humana Press, a part of Springer Science+Business Media, LLC 2009

(1, 2), affects the chemical properties of the tyrosine residue, and competes with the phosphorylation of a tyrosine residue. Nitration involves the redox signaling system, occurs under physiological conditions, is enhanced under pathological conditions, and can be reversed by enzymatic or nonenzymatic mechanisms *(3)*. Therefore, tyrosine nitration can either increase or decrease a protein's function, and is associated with many physiological/ pathological processes such as neurodegenerative diseases, tumor, and inflammation diseases *(2)*.

Extensive data demonstrate that reactive oxygen species (ROS) and reactive nitrogen species (RNS) are involved in the multiple hypothalamic–pituitary–target organ axes systems, and are elevated in pituitary tumors. Nitric oxide synthases are expressed in the human and rat pituitary, and have an elevated activity in pituitary adenomas *(5–8)*. Nitric oxide participates in activating the release of luteinizing hormone-releasing hormone (LHRH) and follicle-stimulating hormone-releasing hormone (FSHRH) from the hypothalamus, and LH and FSH from the pituitary *(9–12)*. Nitric oxide might also stimulate or inhibit the secretion of prolactin *(13–15)*, regulate the secretion of growth hormone in the normal human pituitary and in acromegaly *(5, 16–18)*, and can play an important role in hypothalamic–pituitary–adrenocortical axis inhibition of the release of ACTH *(19)*. Our studies show that nitrotyrosine-containing proteins are present in normal human pituitary postmortem *(2, 20)* and nonfunctional pituitary adenomas postsurgical resection *(1)*. Therefore, the role of ROS/ RNS may be important in normal pituitary function and relevant to dysfunction in pituitary adenoma. Elucidation of nitrotyrosine sites could improve our understanding of the role of tyrosine nitration in pituitary physiological and pathological processes.

The identification of nitrotyrosine sites is challenging due to its rarity in a proteome. The combination of soft ionization and tandem mass spectrometry offers promise for identifying nitrotyrosine site on a protein *(1, 2, 20)*. However, mass spectrometry is limited in its sensitivity (generally high femtomole to low picomole), which is an issue since nitroproteins are at low abundance in the in vivo pituitary proteome. Therefore, isolation and enrichment of nitrotyrosine-containing proteins or peptides is needed prior to mass spectrometry analysis. In our studies, two methods were employed to isolate and enrich nitrotyrosine-containing proteins from a pituitary proteome prior to mass spectrometry: two-dimensional gel electrophoresis (2DGE) plus nitrotyrosine Western blotting analysis *(2, 20)*, and nitrotyrosine immunoaffinity enrichment *(1)*. The nitrotyrosine-containing proteins were enzymatically digested, and tandem mass spectrometry was used to obtain the amino acid sequence. Nitrotyrosine sites were then located to the structural/functional domain of a nitrated protein in order to clarify the role of tyrosine nitration. Ultimately, the presented methods can be readily adapted to the study of other neurological diseases.

2. Materials

2.1. Two-Dimensional Gel Electrophoresis

1. Prepare a 0.9% sodium chloride solution with deionized distilled water (ddH_2O) (*see* **Note 1**).
2. Homogenizing buffer (10 mL): 2 M acetic acid and 0.1% (v/v) mercaptoethanol. Store at 4°C.
3. Bicinchoninic acid (BCA) protein assay reagent kit (Pierce, Rockford, IL).
4. Protein extracting buffer (1 mL): 7 M urea, 2 M thiourea, 4% (w/v) CHAPS, 100 mM DTT (add prior to use), 0.5% v/v pharmalyte (add prior to use), and a trace of bromophenol blue (*see* **Note 2**).
5. Immobiline pH-gradient DryStryp (GE Life Sciences, Piscataway, NJ) (*see* **Note 3**).
6. Rehydration buffer (1 mL): 7 M urea, 2 M thiourea, 4% (w/v) CHAPS, 60 mM DTT (add prior to use), 0.5% v/v Pharmacia IPG buffer (add prior to use), and a trace of bromophenol blue.
7. Resolving-gel buffer stock (4×): 1.5 M Tris–HCl (pH 8.8). Filter solution through a 0.45-μm filter. Store at 4°C.
8. Reducing equilibration buffer (50 mL): 375 mM Tris–HCl (pH 8.8), 6 M urea, 2% (w/v) SDS, 20% (v/v) glycerol, 2% w/v DTT (add prior to use), and a trace of bromophenol blue.
9. Alkylation equilibration buffer (50 mL): 375 mM Tris–HCl (pH 8.8), 6 M urea, 2% (w/v) SDS, 20% (v/v) glycerol, 2.5% w/v iodoacetamide (add prior to use), and a trace of bromophenol blue.
10. 40% acrylamide/bisacrylamide stock solution (29:1): 40% w/v acrylamide, 1.38% w/v *N,N'*-methylenebisacrylamide (*see* **Note 4**) (Bio-Rad, Hercules, CA).
11. 10% ammonium persulfate (3 mL) is prepared prior to use.
12. SDS electrophoresis buffer (1×; 25 L): 25 mM Tris, 192 mM glycerine, and 0.1% SDS. Store at room temperature (*see* **Note 5**).
13. 1% Agarose sealing solution (100 mL) is prepared with the SDS electrophoresis buffer and kept at ca. 80°C prior to use (*see* **Note 6**).

2.2. Visualization of 2DGE-Separated Proteins

1. Fixing solution (250 mL): 50% (v/v) methanol and 5% (v/v) acetic acid.
2. Sensitizing solution (250 mL): 0.02% (w/v) sodium thiosulfate.
3. Silver reaction solution (250 mL): 0.1% (w/v) silver nitrate, with 200 μL 37% (v/v) formaldehyde added prior to use.
4. Developing solution (250 mL): 3% (w/v) sodium carbonate, with 100 μL 37% (v/v) formaldehyde added prior to use.
5. Stopping solution (250 mL): 5% (v/v) acetic acid.
6. Storing solution (250 mL): 8.8% glycerol.
7. Destaining solution (200 mL): 100 mL of 7.5 mM potassium ferricyanide is mixed with 100 mL of 25 mM sodium thiosulfate prior to use.

2.3. 2DGE-Based Western Blotting for Nitrotyrosine

1. Polyvinylidene fluoride (PVDF) membrane: Immobilon-P transfer membrane (Millipore, Bedford, MA), or Hybond-P 20 × 20 cm (Amersham) (*see* **Note 7**).
2. PVDF membrane equilibration buffer (1 L): 25 mM Tris, 192 mM Glycine, and 20% (v/v) methanol. Store at room temperature.
3. Gel equilibration buffer (1 L): 25 mM Tris, 192 mM glycine, 10% (v/v) methanol. Store at room temperature.
4. Anode transfer buffer R stock solution (10×, 1 L): 36.3% (w/v) Tris–base (pH 10.4).
5. Anode transfer buffer R (1 L): 100 mL anode transfer buffer R stock solution (10×), 200 mL methanol, and 700 mL ddH_2O.
6. Anode transfer buffer S stock solution (10×) (1 L): 3.03% w/v g Tris–base (pH 10.4).
7. Anode transfer buffer S (1 L): 100 mL anode transfer buffer S stock solution (10×), 200 mL methanol, and 700 mL ddH_2O.
8. Cathode transfer buffer T stock solution (10×) (1 L): 5.2% (w/v) 6-amino-*n*-hexanoic acid (pH 7.6).
9. Cathode transfer buffer T (1 L): 100 mL cathode transfer buffer T stock solution (10×), 200 mL methanol, and 700 mL ddH_2O.
10. NovaBlot Electrode Paper (Amersham).
11. 10 mM phosphate buffered saline (PBS) (1 L).
12. PBST (1 L): 10 mM PBS, 0.2% (v/v) Tween-20, and 0.01% sodium azide.
13. 0.3% BSA/PBST (1 L): 0.3% (w/v) BSA, 10 mM PBS, 0.2% (v/v) Tween-20, and 0.01% (w/v) sodium azide.

14. Primary antibody: #N0409 rabbit anti-human nitrotyrosine antibody (Sigma, St. Louis, MO). Dilute (1:1,000 = v:v) in a 0.3% BSA/PBST solution (1 μg Ab/mL) (*see* **Note 8**).
15. Secondary antibody: #31340 goat anti-rabbit alkaline phosphase-conjugated IgG (Pierce). Dilute (1:5,000 = v:v) in a 0.3% BSA/PBST solution.
16. Development reagent: 5-bromo-4-chloro-3-indolyl phosphate/nitro blue tetrazolium (BCIP/NBT) (1-Step™ NBT/BCIP; Pierce).

2.4. Immunoaffinity Enrichment of Nitrotyrosine-Containing Proteins

1. Handee™ Spin Cup Columns (Pierce).
2. Handee™ Microcentrifuge tubes (Pierce).
3. ImmunoPure® Immobilized Protein G plus (Pierce) (*see* **Note 9**).
4. Binding/washing buffer (BupH™ Modified Dulbecco's PBS) (Pierce): 140 mM NaCl, 8 mM sodium phosphate, 2 mM potassium phosphate, and 10 mM KCl; pH 7.4 when reconstituted.
5. Disuccinimidyl suberate (DSS), 2 mg/tube (Pierce) (*see* **Note 10**).
6. ImmunoPure® IgG Elution Buffer (Pierce), pH 2.8, contains a primary amine.
7. #AB5411 rabbit anti-nitrotyrosine polyclonal antibody (Millipore, Bedford, MA), or MAB5404 mouse anti-nitrotyrosine monoclonal antibody (Millipore) (*see* **Note 8**).
8. M-PER® mammalian protein extraction reagent (Pierce Catalog No. 78501).
9. pH-neutralized solution: 1 M Tris (pH 9.5).

2.5. Trypsin Digestion and Mass Spectrometric Characterization of Isolated Nitroproteins

1. Sequencing grade modified trypsin (5 × 20 μg aliquots; Promega, Madison, WI). Store at –20°C (*see* **Note 11**).
2. Trypsin resuspension buffer (Promega): 50 mM acetic acid (pH 2.8).
3. Trypsin-dissolving solution (100 μL, pH 8.2): mix 5 μL of trypsin resuspension buffer (pH 2.8) and 95 μL of 200 mM NH_4HCO_3 (pH 8.2).
4. Buffer X: a solution that contained 1 M Tris (pH 9.5), 100 mM dithiothreitol, and 100 mM iodoacetamide
5. 30 mM potassium ferricyanide (50 mL): 495 mg potassium ferricyanide is dissolved in 50-mL ddH_2O and vortexed. Store at 4°C.
6. 100 mM sodium thiosulfate (50 mL): 1.24 g sodium thiosulfate ($Na_2S_2O_3 \cdot 5H_2O$) is dissolved in 50-mL ddH_2O and is vortexed. Store at 4°C.

7. Silver destaining solution: Mix 30 mM of potassium ferricyanide with 100 mM of sodium thiosulfate (1:1 v/v) prior to use.
8. 200 mM ammonium bicarbonate (50 mL): 0.792 g ammonium bicarbonate is dissolved in 50-mL ddH_2O. Store at 4°C.
9. 100 mM ammonium bicarbonate (50 mL): 0.396 g ammonium bicarbonate is dissolved in 50-mL ddH_2O. Store at 4°C.
10. 50 mM ammonium bicarbonate (50 mL): 0.198 g ammonium bicarbonate is dissolved in 50-mL ddH_2O. Store at 4°C.
11. 1% trifluoroacetic acid (TFA): 0.5 mL TFA is diluted with 49.5-mL ddH_2O. Store at 4°C.
12. 0.1% TFA (1 mL): 0.1 mL of 1% TFA is diluted with 0.9-mL ddH_2O prior to use.
13. ZipTipC18 microcolumn (Millipore).
14. 50% acetonitrile/0.1% TFA (1 mL): 0.5 mL of acetonitrile, 0.1 mL of 1% TFA, and 0.4 mL of ddH_2O are mixed prior to use.
15. 10 mg/mL α-cyano-4-hydroxycinnamic acid (CHCA) stock solution (1 mL): 10 mg CHCA in 1 mL of 50% (v/v) acetonitrile/0.1% (v/v) TFA (make prior to use).
16. 2.5 mg/mL α-cyano-4-hydroxycinnamic acid (CHCA) solution: 25 μL of 10 mg/mL CHCA stock solution is mixed with 75 μL of 50% (v/v) acetonitrile/0.1% (v/v) TFA prior to use.
17. 85% v/v acetonitrile/0.1% v/v TFA (1 mL): 0.85 mL of acetonitrile, 0.1 mL of 1% TFA, and 0.05 mL of ddH_2O are mixed prior to use.
18. 2% actonitrile/0.5% acetic acid (1 mL): 20 μL of acetonitrile and 5 μL of acetic acid are mixed with 975 μL of ddH_2O prior to use.
19. Mobile phase A (100 mL): 0.1% v/v formic acid in ddH_2O.
20. Mobile phase B (100 mL): 90% v/v acetonitrile and 0.1% v/v formic acid in ddH_2O.
21. Capillary column (8-cm long): a New Objective PicoFrit 360-μm (OD), 75-μm (ID), and 15-μm tip pores (ID) packing Magic C18AQ material (5-μm beads, 200 Å pores) (Michrom Bioresources, Auburn, CA).

3. Methods

In order to identify low-abundance nitrotyrosine sites in human pituitary tumor, the proteome is separated with 2DGE, followed by a nitrotyrosine Western blotting assay to locate the nitrotyrosine

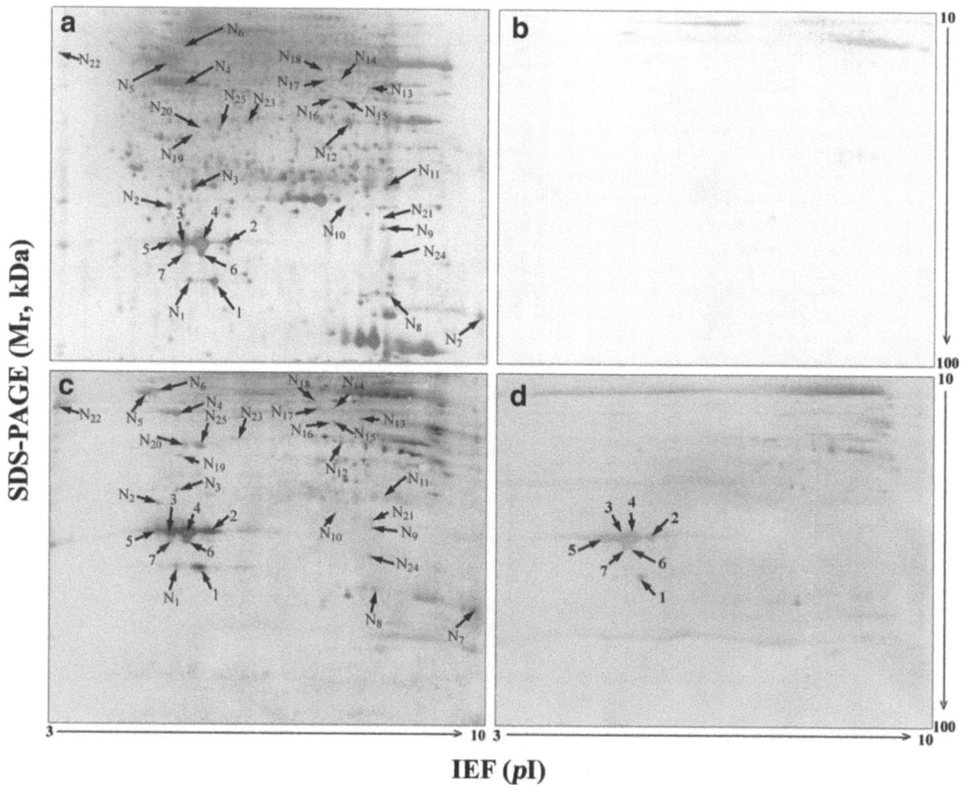

Fig. 1. Two-dimensional Western blot analysis of anti-3-nitrotyrosine-positive proteins from human pituitary tissue (70 μg protein per 2D gel). (**a**) A silver-stained 2D gel of human pituitary proteins. (**b**) A silver-stained 2D gel after the transfer of proteins onto a PVDF membrane. (**c**) A Western blot image of anti-3-nitrotyrosine-positive proteins (anti-3-nitrotyrosine antibodies + secondary antibody). (**d**) A negative control Western blot to show the cross-reaction of the secondary antibody (only the secondary antibody; no anti-3-nitrotyrosine antibody) [reproduced from ***(20)*** with permission from Elsevier Science].

immunopositive proteins (**Fig. 1**) *(2, 20)*. Alternatively, the nitrotyrosine-containing proteins are first enriched with nitrotyrosine immunoaffinity precipitation *(1)*. Commercially available anti-nitrotyrosine monoclonal and polyclonal antibodies are used for the 2DGE-based Western blot and the immunoprecipitation of nitrotyrosine-containing protein. The discerned nitroproteins are subjected to trypsin digestion and are analyzed with tandem mass spectrometry to identify each nitrotyrsoine site. Nitrotyrosine-containing peptides present different tandem mass spectrum features between MALDI-MS/MS (**Fig. 2**) *(20)* and ESI-MS/MS (**Fig. 3**) *(2)*; therefore, the two ionization methods provide complementary information to identify nitrotyrosine sites. The sites can then be located to functional domains/motifs with bioinformatics (**Fig. 4**) *(1)*.

3.1. Preparation of Samples

1. Obtain normal human pituitary tissue (or other CNS tissues) from autopsy or neurosurgical resections of pituitary adenoma

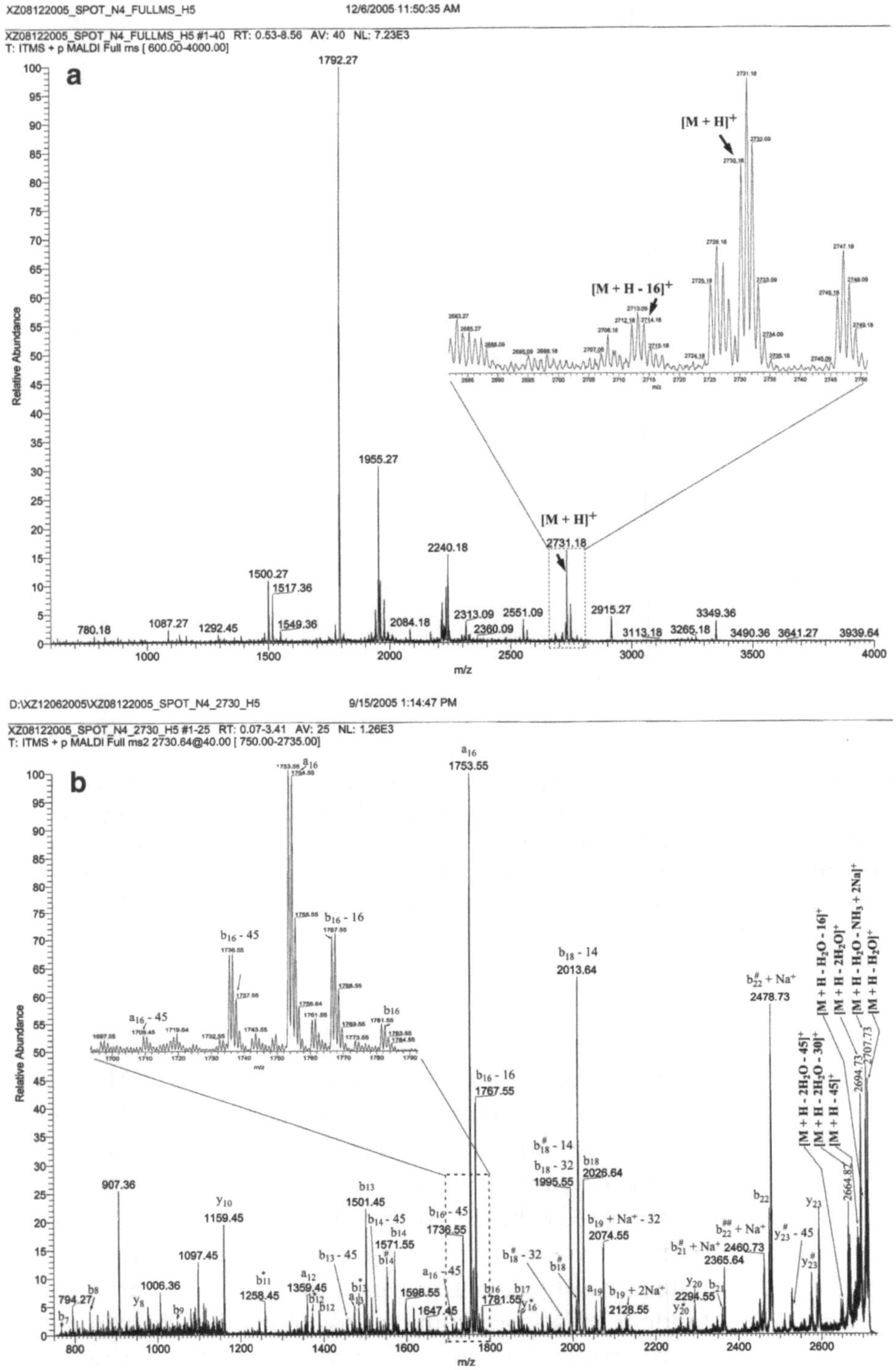

Fig. 2. Mass spectra of the nitropeptide, ^{120}HSTLVNDAY*KTLLAPLSRGLYLLK^{143} from human mitochondrial co-chaperone protein HscB (SWISS-PROT number = Q8IWL3). (**a**) MS spectrum containing the precursor ion at an *m/z* of 2,731.2. (**b**) MS^2 spectrum of the peptide with the Y^{128} nitration site. *Asterisk* indicates the loss of NH_3, *hash* the loss of H_2O. M–14 = M + 2H–O; M–16 = M–O; M–32 = M–2O; M–45 = M + H–NO_2 [reproduced from *(20)* with permission from Elsevier Science].

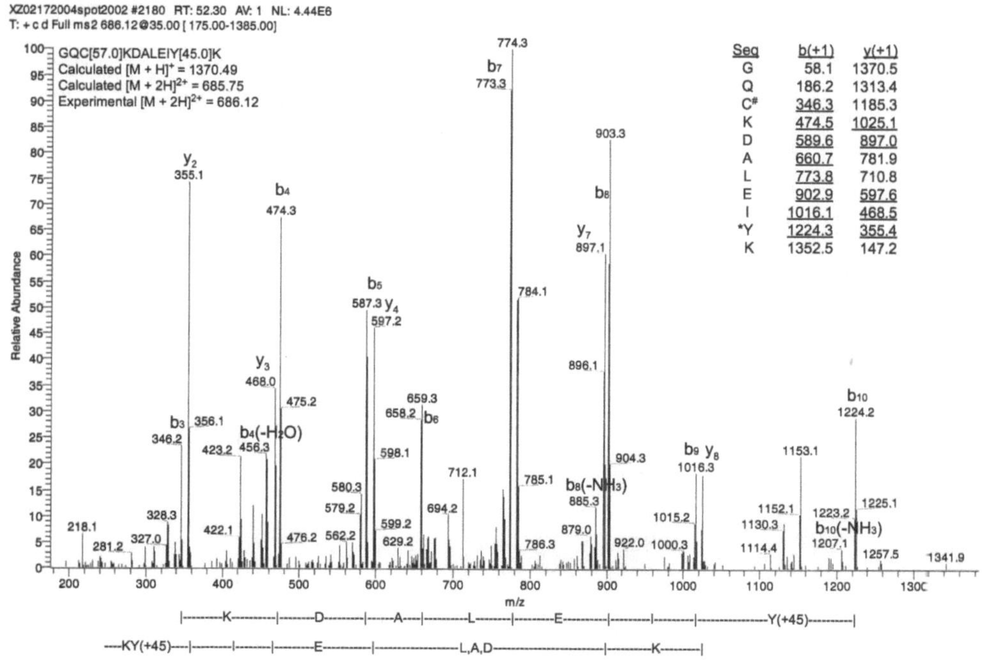

Fig. 3. SEQUEST (*top-right*) and de novo (*bottom*) sequence correlation for an MS² spectrum of the precursor $[M+2H]^{2+}$ ion at an *m/z* of 686.12 (RT = 52.30 min, scan number 2180) for the nitrotyrosyl peptide ²²⁸GQC#KDALEIYK²³⁸ (Tyr-237) derived from synaptosomal-associated protein (spot 1) [reproduced from *(2)* with permission from Elsevier Science].

tissue, freeze immediately in liquid nitrogen, and store at −80°C until ready to process.

2. Add 2 mL of 0.9% sodium chloride and lightly shake to remove blood from the surface of the tissue. Repeat two more times (*see* **Note 12**).
3. Add 10 mL of homogenization buffer for every ca. 0.5–0.6 g of tissue and homogenize (1 min; repeat ten times) with a tissue homogenizer at 13,000 rpm and 4°C (e.g., Polytron Model P710/35, Brinkmann Instruments, Westbury, NY). Then sonicate the homogenate for 20 s.
4. Lyophilize 1-mL aliquots of the homogenate and store at −80°C.
5. Determine the lyophilized protein content with a bicinchoninic acid (BCA) protein assay kit (*see* **Note 13**).
   (a) 280 μg of a lyophilized pituitary sample is added to 264 μL of a solution that contained 8 M urea and 4% CHAPS. Let it stand for 2 h, sonicate for 5 min, rotate for 1 h, sonicate for 5 min, rotate for 1 h, and centrifuge at 15,000 × *g* for 20 min.
   (b) Preparation of BSA standard solutions: Dilute the BSA standard (2 mg/mL) with ddH_2O to generate the following series: 2,000, 1,500, 1,000, 750, 500, 250, 125, and 25 μg/mL; also ddH_2O (0 μg/mL).

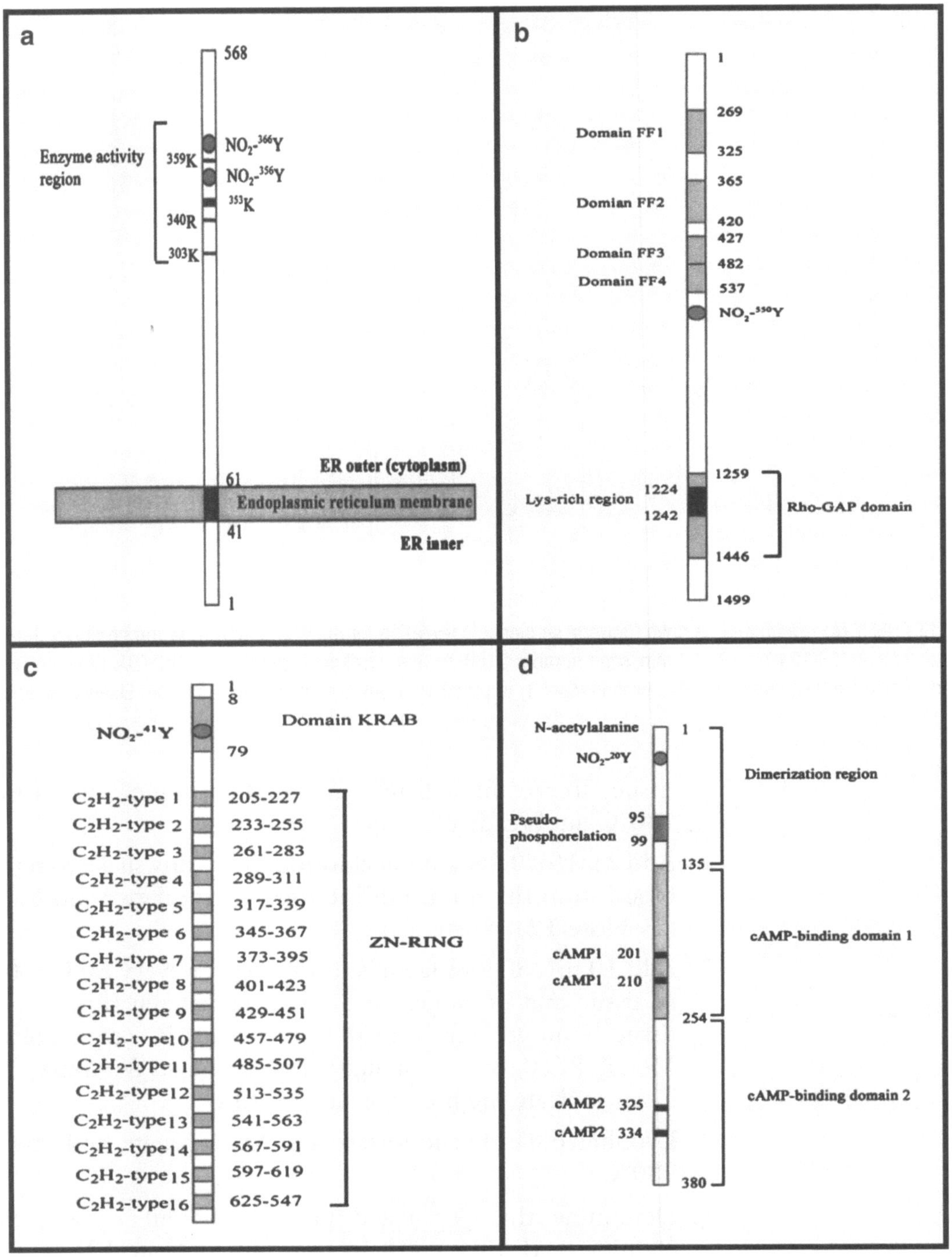

Fig. 4. Nitration site and functional domains of four nitroproteins. (**a**) Sphingosine-1-phosphate lyase 1. The site [353] K is a pyridoxal phosphate binding motif. (**b**) Rho-GTPase-activating protein 5. (**c**) Zinc finger protein 432. The KRAB domain is a transcriptional suppressor. The ZN-RING is a DNA-binding region. (**d**) cAMP-dependent protein kinase type I-beta regulatory subunit [reproduced from *(1)* with permission from Elsevier Science].

(c) BCA working solution: Mix 50 parts of reagent A and 1 part of reagent B (50:1) prior to use.

(d) 0.1 mL of sample or standard solution is mixed with 2 mL of BCA working solution (1:20). Incubate at 37°C for 30 min. Cool to room temperature (ca. 10 min). Measure the O.D. value ($A_{562\ nm}$) on a spectrophotometer.

(e) Use linear regression to calculate the standard linear line (BSA concentration vs. $A_{562\ nm}$) and obtain a regression equation that uses $A_{562\ nm}$ to calculate the protein concentration of a lyophilized pituitary sample(s).

6. Protein extraction for 2DGE: Weigh enough lyophilized pituitary sample (70 μg of protein for an 18-cm immobilized pH gradient (IPG) strip pH 3–10) to mix with 250 μL of the protein-extracting buffer. The mixture is vortexed for 5 min, sonicated for 5 min, and rotated for 50 min. Add 110 μL of rehydration buffer. The mixture is sonicated for 5 min, rotated for 50 min, vortexed for 5 min, and centrifuged for 20 min at 15,000 × *g*. Collect the supernatant, now referred to as the "protein sample solution" *(21)*.

3.2. Two-Dimensional Gel Electrophoresis

3.2.1. Rehydration of IPG Dry Strip

1. 350 μL of the "protein sample solution" is pipeted into the slot in the rehydration tray for an 18-cm Dry IPG strip.
2. Remove the plastic cover from the dry IPG strip (do not touch the gel-side) (*see* **Note 3**). Place the IPG strip gel-side-down onto the "protein sample solution" and distribute the "protein sample solution" evenly along the whole IPG strip length (avoid bubbles).
3. Overlay the IPG strip with 3–4 mL of mineral oil to prevent evaporation. Leave the strip to be rehydrated overnight (ca. 18 h) at room temperature.
4. Blot away the mineral oil on the plastic side of the rehydrated IPG strip with a paper towel. Rinse the IPG strip in water for 5 s. Dry the plastic side with a paper towel; blot the gel side to remove excess water with a lint-free tissue (i.e., KimWipe; *see* **Note 14**).

3.2.2. First-Dimension, Isoelectric Focusing

1. Isoelectric focusing (IEF) is assumed to be performed on the Amersham IPGphor™ Isoelectric Focusing System.
2. Wet two small pieces of paper wick (sample application paper) with 10-μL ddH_2O, and remove excess water with a KimWipe. Clean the IPG strip holder: add a few strip holder cleaner (GE-Life Sciences) into the IPG strip holder, rub with a KimWipe, rinse with ddH_2O for five times, and dry.
3. Place a wet paper wick on the electrode wire at both ends of the IPG strip holder (*see* **Note 15**).

4. Place the rinsed IPG strip(s) with the gel-side down into the IPG strip holder. The acidic (pointed) end of the IPG strip is positioned on the pointed-end of the IPG strip holder. Both ends of IPG strips contact the wet paper wick; avoid bubbles.
5. Overlay the IPG strip with 3–4 mL of mineral oil. Place the lid on the IPG strip holder. Avoid any bubbles.
6. Load the assembled strip holder into the IPGphor unit, with the pointed end to the back (+) plate, and square end over the front (–) plate; align straight. Close the IPGphor cover to hold the gel onto the holder electrodes and press the holder to the platform electrodes.
7. IEF parameters: Set the maximum current per strip to 30 μA and temperature to 20°C. Use the following program: (a) 250 V, 1 h, 125 Vh, step and hold; (b) 1,000 V, 1 h, 500 Vh, gradient; (c) 8,000 V, 1 h, 4,000 Vh, gradient; (d) 8,000 V, 4 h, 32,000 Vh, step and hold; (e) 500 V, 0.5 h, 250 Vh, step and hold. The total time is 7.5 h, 36,875 Vh. Input the number of IPG strips. After 2 h, change the anode paper wick.
8. When finished, remove the IPG strip and lay on its plastic back to blot off the mineral oil. Wrap in a sheet of plastic wrap, and store at –80°C (*see* **Note 16**).

3.2.3. Cast SDS-PAGE Gel(s)

Cast 12 PAGE resolving gels (gel concentration = 12%) with a Bio-Rad PROTEN-plus multicasting chamber (*see* **Note 17**).

1. Mix 180 mL of 40% (w/v) acrylamide/bisacrylamide stock solution (29:1), 150 mL of 1.5 M Tris–HCl (pH 8.8), and 270 mL of ddH_2O; de-gas with a vacuum pump for 10 min.
2. Add 3 mL of 10% ammonium persulfate and 150 μL of TEMED to the mixture solution; mix gently; and avoid bubbles.
3. Gently pour the solution into the holding chamber (1 L plastic bottle with attached tubing at its bottom).
4. Connect the tubing from the holding chamber to the inlet port on the multicasting chamber. Place a gel comb in the first gel cassette.
5. Elevate the holding chamber above the level of the multicasting chamber to fill the gel cassette up to the level of the comb.
6. Remove the comb and overlay the gels immediately with ddH_2O. Allow the gels to polymerize for >1 h.

3.2.4. Second-Dimension, SDS-PAGE

SDS-PAGE is assumed to be performed on a Bio-Rad PROTEAN plus® Dodeca™ vertical cell electrophoresis system.

1. Connect the inlet tubing from the Dodeca cell tank to a carboy that holds 25 L of electrophoresis buffer. Fill the Dodeca buffer tank with electrophoresis buffer. Set the water bath temperature to 15°C and turn the circulator on for >1 h prior to use.

2. Remove the focused IPG strips from the freezer, and place them (one in each slot) in an equilibrium tray with the gel-side facing up. Cover each strip with ca. 4 mL of reducing equilibrium buffer. Rock the tray gently for 10 min. Pour out the reducing equilibrium buffer, and immediately pour in ca. 4 mL of alkylation equilibrium buffer for each slot; rock the tray gently for 10 min.
3. During equilibrium, disassemble the multicasting chamber. Remove a gel cassette, pour out the water that was used to overlay the gel, rinse three times the revealed gel well with ddH_2O, and blot away excess water with a KimWipe. Position the gel cassette in a gel-stander.
4. Remove the equilibrated IPG strips, rinse with electrophoresis buffer in a 20-cm long cylinder, and blot away the liquid on the IPG strip surface with a KimWipe. Position the IPG strip onto the longer glass plate and over the well of the SDS-PAGE gel (gel-side facing front with the plastic back contacting the longer glass plate and the pointed end of IPG strip to the left).
5. Add quickly ca. 3 mL of hot 1% agarose solution (ca. 80°C) into the well of the SDS-PAGE gel and push the IPG strip quickly into the unpolymerized agarose solution. Make sure that the top-side of the IPG strip aligns with the top of the shorter glass plate. Let the agarose polymerize for 10 min.
6. Lift the lid of the Dodeca tank. Using two hands, insert vertically the gel cassette between plastic gaskets (gasket should be flared out toward the electrode card), hinged-side down (PROTEIN plus hinged spacer plate). The top of the gel (with the IPG strip) is positioned next to the cathode (black electrode card; –) such that the sample migrates horizontally toward the anode (red electrode card; +).
7. Adjust the level of electrophoresis buffer up to the middle of the top spacer (*see* **Note 18**). Place the lid on the tank, and make sure that the pump tubing is connected to the top of the lid via the quick-connect fittings. Set the buffer recirculation pump (Bio-Rad) to maximal scale (100), and turn it on.
8. Connect the Dodeca Cell to a PowerPac 200 power supply; and set it to the constant voltage mode. Run at 200 V for 370 min.
9. Afterward, dissemble the electrophoresis system, remove and place the PROTEAN plus hinged spacer plate on the benchtop – short plate facing upward and the hinge to the left. Insert the gel releaser between the short plate and the long plate at the top right corner. Pull the gel releaser up until the gel cassette is opened completely (180°). Gently remove the gel from the plate, taking care to avoid tearing the gel (*see* **Note 19**).

10. Then the 2DGE-separated proteins are visualized with silver stain (*see* **Subheading 3.3**) or are transferred to a PVDF membrane for Western blot analysis (*see* **Subheading 3.4**).

3.3. Visualization of 2DGE-Separated Proteins

3.3.1. Silver Staining

1. Place the gel into a clean flat tray.
2. Add 250 mL of fixing solution and shake for 20 min. Replace fixative with 250 mL of 50% (v/v) methanol and shake for 10 min. Replace methanol with 250 mL of ddH_2O and shake slowly for 10 min. Discard the liquid.
3. Add 250 mL of sensitizing solution and shake slowly for 1 min. Replace with 250 mL of ddH_2O and shake slowly for 1 min; repeat one more time. Discard the liquid.
4. Add 250 mL of silver reaction solution and shake slowly for 20 min. Replace with 250 mL of ddH_2O, and shake slowly for 1 min; repeat one more time. Discard the liquid.
5. Add 250 mL of developing solution and shake until the desired intensity of staining is reached (usually ca. 3 min). Discard the liquid.
6. Add 250 mL of stopping solution and shake slowly for 10 min. Replace with 250 mL of ddH_2O and shake slowly for 5 min. Discard the liquid.
7. Add 250 mL of storing solution and keep at 4°C.

3.3.2. Modified Silver Staining

If the first silver staining does not produce the desired result, then use the following modified silver staining procedure:

1. Wash the gel with ca. 250 mL ddH_2O for ca. 10 min. Discard the liquid.
2. Pour 200 mL of destaining solution onto the gel, and shake slowly until all staining is removed (ca. 1–5 min). Discard the liquid.
3. Wash the destained gel with ca. 250-mL ddH_2O, shaking slowly for 3 min. Repeat six more times.
4. Silver restaining: Exactly follow **steps 2–7** in **Subheading 3.3.1**.

3.4. 2D Western Blot for Nitrotyrosine Immunoreactivity

1. After electrophoresis, remove the 2D gel, cut a notch in the upper left-hand corner (acidic end) to help orientate the 2D gel. Soak the 2D gel in gel equilibration buffer for at least 10 min.
2. Preparation of the PVDF membrane: Cut a sheet of PVDF membrane to match the gel size (20 × 17.2 cm), and place it into 100% methanol for 10 min, wash with ddH_2O for 5 min, and label its hydrophilic side (The hydrophilic side can easily be found by watching the water moving slower on the hydrophilic side than the other side) using a pencil to write the date on the margin of the hydrophilic side of

the PVDF membrane (*see* **Note 20**). Next, equilibrate the PVDF membrane in equilibration buffer for at least 10 min (always keep the PVDF membrane wet).

3. The transfer is assumed to be performed with an Amersham Multiphor II semidry electrotransfer system. Assemble the electroblotting cassette according to the following procedure (*see* **Note 21**):
   (a) Saturate the anode electrode plate with ddH_2O, and remove excess water with paper. Put the plate onto the buffer tank.
   (b) Immerse six sheets of filter paper in anode transfer buffer R, and place them carefully atop one another onto the anode plate.
   (c) Soak three sheets of filter paper in anode transfer buffer S, and place them carefully on top of the first six filter sheets.
   (d) Wet the PVDF membrane in anode transfer buffer S for 30 s, and place the PVDF (hydrophilic side up) on top of the filter paper stack.
   (e) Put the 2D gel onto the PVDF membrane.
   (f) Immerse nine sheets of filter paper in cathode transfer buffer T, and place them carefully on top of the gel.
   (g) Saturate the cathode electrode plate with ddH_2O, and remove excess water with filter paper; place the plate on top of the stack.
   (h) Connect all units of the transfer system.
4. Connect to a power supply (e.g., Amersham EPS 3501XL). Electrotransfer is performed at a constant current of 0.8 mA/cm^2 for 100 min. After the transfer, remove the top filter papers and draw a line with a pencil around the 2D gel to define the borders on the margin of PVDF membrane, which also orients the gel.
5. Place the PVDF membrane (protein-side-up) on a flat clean dish. Add 100 mL of 0.3% BSA/PBST and block for 60 min at room temperature with gentle shaking. After blocking, rinse the PVDF membrane twice with ddH_2O.
6. Add 100 mL of diluted primary antibody (100 μL rabbit anti-nitrotyrosine antibody is diluted with 100 mL of 0.3% BSA/PBST). Incubate for 1 h at room temperature with gentle shaking. Pour out the primary antibody solution. Wash with 200 mL of PBST and shake for 15 min; repeat three more times.
7. Rinse the blot twice with ddH_2O.
8. Add 100 mL of diluted secondary antibody (20-μL goat anti-rabbit alkaline phosphase-conjugated IgG diluted in

100 mL of 0.3% BSA/PBST). Incubate for 1 h at room temperature with gentle shaking. Pour out the secondary antibody solution. Add 200 mL of PBST and wash with gentle shaking for 15 min; repeat two more times. Wash with 200 mL of PBS with gentle shaking for 15 min; repeat two more times. Rinse for four times with ddH_2O (*see* **Note 22**).

9. Add enough 1-Step™ NBT/BCIP substrate to cover the PVDF membrane (protein-side up). Gently shake until the desired color appears (ca. 20 min); if the amount of protein is very low, you may need a longer developing time. Wash for 10 min with ddH_2O.
10. Dry the PVDF membrane between two sheets of filter paper. Store the membrane between two transparent plastic sheets.

3.5. Image Analysis of 2DGE and 2D Western Blots

1. Capture a digitized image of the PVDF membrane and the corresponding silver-stained 2D gel (e.g., with a flatbed scanner).
2. Import the digitized image into 2D gel image analysis software (e.g., Bio-Rad PDQuest) to automatically define the boundaries of the imaged spots.
3. Match the immunopositive Western blotting spots to the corresponding silver-stained 2D gel spots by software; then check manually each pair of matched spots (**Fig. 1**) (*see* **Note 23**).

3.6. Nitrotyrosine Immunoaffinity Precipitation

1. Protein extraction for immunoprecipitation: (a) weigh a portion of pituitary tissue (e.g., ca. 62 mg wet weight) into a 1.5-mL Eppendorf tube; (b) rinse three times with binding/washing buffer to remove blood from the tissue surface; (c) add 600 μL of Pierce M-PER® mammalian protein extraction buffer that is compatible with immunoprecipitation (10:1 = buffer:tissue), vortex for 5 min, homogenize for 5 min, sonicate for 20 s, rotate for 2 h, sonicate for 20 s, and centrifuge at 15,000 × *g* for 30 min; (d) transfer the supernatant to a new tube, which is referred as the extracted protein sample.
2. Measure the protein content with the modified procedure of Bradford Protein Assay (Bio-Rad): (a) Prepare 0.1 N HCl (10 mL): 0.833-mL HCl is mixed with 9.167-mL ddH_2O. (b) Prepare 10 mg/mL ovalbumin standard in ddH_2O. (c) Dilute 0.1 N HCl with ddH_2O (1:9): 100 μL of 0.1 N HCl is mixed with 800 μL of ddH_2O. (d) Dilute 10 mg/mL ovalbumin standard with the protein extraction buffer (i.e., Pierce M-PER® mammalian protein extraction buffer) (1:10): 10 μL of 10 mg/mL ovalbumin standard is mixed with 90 μL of protein extraction buffer. (e) Dilute Bio-Rad Dye with ddH_2O (1:4): 2-mL dye is mixed with 6-mL

ddH$_2$O. (f) Make the ovalbumin standard series (duplicate): 0 μg/mL [0 μL (d) + 90 μL (c) + 910 μL (e)], 1 μg/mL [1 μL (d) + 89 μL (c) + 910 μL (e)], 2 μg/mL [2 μL (d) + 88 μL (c) + 910 μL (e)], 4 μg/mL [4 μL (d) + 86 μL (c) + 910 μL (e)], 6 μg/mL [6 μL (d) + 84 μL (c) + 910 μL (e)], 8 μg/mL [8 μL (d) + 82 μL (c) + 910 μL (e)], 10 μg/mL [10 μL (d) + 80 μL (c) + 910 μL (e)], and 12 μg/mL [12 μL (d) + 78 μL (c) + 910 μL (e)]. At the same time, make the sample reaction system (duplicate): 1 μL protein sample + 89 μL (c) + 910 μL (e). (g) Mix and leave it to stand at room temperature for 5 min. Measure the O.D. value ($A_{595\ nm}$) on a spectrophotometer. (h) Use linear regression to calculate the standard linear line (Ovalbumin concentration vs. $A_{595\ nm}$) and obtain a regression equation that uses $A_{595\ nm}$ to calculate the protein concentration of the extracted protein sample.

3. Immunoprecipitation of nitrotyrosine-containing proteins is assumed to be carried out mainly with a Pierce Seize X mammalian immunoprecipitation kit. Equilibrate the immobilized protein G, anti-nitrotyrosine antibody, and binding/washing buffer to room temperature.
4. Gently swirl the bottle of ImmunoPure immobilized protein G beads to resuspend fully the beads, and add 400 μL of beads into a 0.5-mL Handee spin-cup column that is placed inside a Handee microcentrifuge tube. Centrifuge for 1 min at 3,000 × *g*. Discard the flow-through. Wash the beads twice with 400 μL of binding/washing buffer (invert the tube ten times and centrifuge at 3,000 × *g* for 1 min). Put the spin-cup into a new microcentrifuge tube.
5. Add 300 μL of binding/washing buffer and 100 μL (100 μg) of anti-nitrotyrosine antibody, and invert the spin-cup quickly ten times to mix the antibody and beads. Incubate at room temperature for 1 h with gentle rotation to allow the antibody to bind to the protein G. Centrifuge for 1 min at 3,000 × *g*, and discard the flow-through. Wash three times with 500 μL of binding/washing buffer (invert ten times, centrifuge at 3,000 × *g* for 1 min). Put the spin-cup into a new microcentrifuge tube.
6. Dissolve 2 mg of disuccinimidyl suberate (DSS) in 80 μL of dimethyl sulfoxide (DMSO). Dilute 25 μL of the DSS solution with 400 μL of binding/washing buffer, and add 425 μL of diluted DSS solution to the washed beads with immobilized protein G antibodies. Invert the tube quickly ten times, and incubate for 30 min with gentle rotation to allow the antibodies to be crosslinked with the immobilized protein G (*see* **Note 10**). Centrifuge for 1 min at 3,000 × *g*, and discard the flow-through.

7. Wash the antibody–protein G-coupled beads with 500 μL of elution buffer (pH 2.8) (invert ten times, centrifuge for 1 min at 3,000 × *g*); repeat four more times. Put the spin-cup into a new microcentrifuge tube.
8. Wash the antibody–protein G-beads with 500 μL of binding/washing buffer (invert tube ten times, centrifuge 1 min at 3,000 × *g*); repeat three more times.
9. Dilute the extracted pituitary protein samples with binding/washing buffer (at least a 1:1 (v/v) dilution). Add 500 μL of diluted sample to the antibody–protein G-beads spin-cup column, and mix quickly by inverting the tube ten times. Incubate overnight (gentle rocking at 4°C) to bind nitrotyrosine-containing proteins with anti-nitrotyrosine antibodies. Centrifuge for 1 min at 3,000 × *g* and discard the flow-through.
10. Wash the nitroprotein–antibody–protein G-beads with 400 μL of binding/washing buffer (invert tube ten times, centrifuge for 1 min at 3,000 × *g*) to remove any nonbound proteins; repeat two more times.
11. Elute the bound nitroproteins with 200 μL of elution buffer (pH 2.8) containing primary amines (gently mix, centrifuge for 1 min at 3,000 × *g*); repeat two more times. Collect the eluants containing the nitroproteins and add 10 μL of the pH-neutralizing solution per 200 μL of eluant. If necessary, the eluant may be partially dried in a vacuum centrifuge to increase the concentration of nitroprotein (*see* **Note 24**).

3.7. Digestion of Nitrotyrosine-Containing Protein with Trypsin

3.7.1. Trypsin Digestion of Immunoprecipitated Nitroproteins (See Note 25) (1)

1. Remove a vial of sequencing grade-modified trypsin (20 μg) and equilibrate to room temperature.
2. Add 100 μL of trypsin-dissolving solution to the 20 μg of trypsin and mix.
3. Add 20 μL of each nitroprotein eluant into a 0.5-mL siliconized tube (*see* **Note 26**). Add 1 μL of buffer X and mix.
4. Add to the 21-μL neutralized sample, 25 μL of trypsin solution and 54 μL of ddH_2O to prepare the enzyme digestion reaction system (final concentration of NH_4HCO_3 = 50 mM, pH 8.1).
5. Incubate at 37°C overnight.

3.7.2. In-Gel Trypsin Digestion of Nitroproteins (See Note 25) (21)

1. Excise the silver-stained gel spots and place them into a 1.5-mL siliconized tube (*see* **Note 26**). Wash six times with 500 μL of ddH_2O.
2. Transfer the gel into another new 1.5-mL siliconized tube and mince it into several pieces (ca. 0.5–1 mm^3) with a pipet tip.
3. Add 20 μL of fresh silver destaining solution to the gel pieces until the brown color is reduced to a pale yellow (ca. 1–2 min).

4. Remove the destaining solution and wash the gel pieces 5–6 times with 20 μL of ddH_2O until the yellow color disappears.
5. Incubate the gel pieces for 20 min in 20 μL of 200 mM ammonium bicarbonate. Discard the buffer, and wash the gel pieces with 20 μL of ddH_2O.
6. Dehydrate the gel pieces with 30 μL of acetonitrile, repeating application until the gel pieces turn an opaque white. Dry for ca. 30 min in a vacuum centrifuge at 30°C.
7. Dissolve a vial of lyophilized trypsin powder (20 μg) in 100 μL of trypsin resuspension buffer. Dilute 20 μL of the trypsin solution (200 ng/μL) to 16 ng/μL with 230 μL of 50 mM ammonium bicarbonate.
8. Add 20–30 μL (0.32–0.48 μg) of the diluted trypsin solution to each tube until the gel pieces are fully covered. Incubate for ca. 18–20 h at 37°C and then cool for 30 min at 4°C.
9. Gently centrifuge for 10 s, sonicate in a water bath (30°C) for 5–6 min, and centrifuge at 12,000 × *g* for 2 min. Carefully transfer the supernatant containing the tryptic peptide mixture into a 0.5-mL siliconized tube.
10. Add 10–15 μL of 50 mM ammonium bicarbonate onto the gel pieces and incubate for 10 min. Follow **step 9** to further extract tryptic peptides. Combine this peptide extract into the same tube as the initial extract. Repeat this step.

3.8. Preparation of Tryptic Digests for Mass Spectrometry Analysis

Extract and clean tryptic peptides with a C18 ZipTip:

1. Prepare the ZipTip by washing with 10 μL of acetonitrile five times and then with 10 μL of 50% acetonitrile five times. Equilibrate the ZipTip with 10 μL of 0.1% TFA five times. Bind the tryptic peptides by pipeting the sample up and down 15 times. Wash twice with 10 μL of 0.1% TFA.
2. For MALDI-MS/MS analysis, add 2 μL of 2.5-mg/mL α-cyano-4-hydroxycinnamic acid (CHCA) solution in a clean 0.5-mL siliconized tube, gently and slowly pipet the 2-μL solution up and down through peptide-C18 beads for six times; at the seventh time, directly pipet down the purified tryptic peptide mixture onto a MALDI plate and air-dry.
3. For LC-ESI-MS/MS analysis, add 6 μL of 85% v/v acetonitrile/0.1% v/v TFA in a clean 0.5-mL siliconized tube, gently and slowly pipet this 6-μL solution up and down through peptide-C18 beads for ten times; at the last time, pipet down the tryptic peptide mixture in the tube, air-dry the eluate, and store at –20°C. Prior to MS analysis, add 6 μL of 2% acetonitrile/0.5% acetic acid to redissolve the dried tryptic peptide mixture.

3.9. Mass Spectrometric Identification of Nitration Sites

3.9.1. MALDI-MS/MS

1. The tryptic peptide mixture is assumed to be analyzed with a vMALDI-LTQ mass spectrometer (ThermoElectron) in the "*N*th order double play" data-dependent experiment mode to obtain tandem mass spectra for tryptic peptides.
2. The instrument parameters are as follows:
   (a) For the vMALDI source, enable the crystal-positioning system (CPS) and auto spectrum filter (ASF). Set up the ASF threshold with 500 counts for an MS scan and 250 counts for an MS^2 scan.
   (b) Enable automatic gain control (AGC) to allow the vMALDI software to automatically adjust the number of laser shots to maintain the quality of the vMALDI spectra.
   (c) For an MS scan, use the high-mass range (m/z 600–4,000) of the LTQ with a normal scan rate, full scan, positive polarity, profile data type, and five microscans.
   (d) Collect MS^2 scans for the 50 most intense peaks in each full MS spectrum with the settings: high-mass range (m/z 50–4,000), normal scan rate, positive polarity, profile data type, an isolation width of 3.0 Th, a normalized collision energy of 40, a default charge state of 1, a minimal signal threshold of 100 counts, an activation q_z value of 0.25, an activation time of 30 ms, and five microscans.
3. Program an experimental sequence with the Xcalibur software to obtain MS/MS spectra for tryptic peptides in each MALDI spot analyzed.
4. Input the MS/MS data into protein database correlation software (e.g., Bioworks from ThermoElectron running the SEQUEST algorithm) to identify the probable amino acid sequence of each tryptic peptide and the corresponding protein by searching against publicly available nonredundant protein databases (e.g., Swiss-Prot and NCBInr databases). Set the software to consider mass modifications of +45 Da ($+NO_2-H$) at Tyr and +57 Da ($+NH_2COCH_2-H$) at Cys. Each positive search result is confirmed with a manual interpretation of the MS and MS^2 data (K or R at the C terminus; K or R preceding the N terminus; 0 or 1 missed trypsin cleavage site(s); singly charged b-, y-, and a-ions; Homo sapiens; and high-quality MS and MS^2 spectra).
5. For each matched tandem mass spectrum, return to the corresponding MALDI spot and manually acquire the summation of n = 20–200 MS^2 spectra using the vMALDI-LTQ Tune Page to improve the signal-to-noise (S/N) ratio.
6. Use the new MS^2 spectra to determine the nitration site(s) of the detected nitroproteins. MALDI-MS spectra of nitrotyrosine-containing peptides demonstrate a unique characteristic – the loss of one or two oxygen atoms from the nitro group (**Fig. 2**) (*see* **Note 27**).

3.9.2. LC-ESI-MS/MS

1. LC-ESI-MS/MS analysis of purified tryptic peptide mixtures is assumed to be carried out with an online capillary liquid chromatography (LC) – electrospray ionization (ESI) –quadrupole-ion trap mass spectrometer (Q-IT) (LCQ^{Deca}, ThermoElectron) which is also managed with Xcalibur software.
2. The LCQ^{Deca} ESI instrument parameters are an ESI voltage of 2.0 kV and an ion transfer capillary temperature of 110°C; the "triple-play" data-dependent scan mode that is used to acquire the MS/MS data – a full-range MS scan (*m/z* 200–2,000) followed by three MS/MS scans of three most intense peaks from the MS scan with an isolation width value of *m/z* 2, an activation *q* value of 0.25, an activation time of 100 ms, a normalized collision energy of 35%, a default charge state of 2, minimum precursor ion signal of 5×10^5, and minimum daughter ion signal of 1×10^5 counts. Use a 1 pmol/μL standard solution of synthetic des-arg bradykinin to determine the instrument sensitivity and mass accuracy. In the MS mode, the $[M + 2H]^{2+}$ ion of des-arg bradykinin at *m/z* 452.7 should have a signal intensity of 1×10^7 (arbitrary units) at a direct-infusion flow-rate of 0.5 μL/min. In the MS/MS mode, the precursor ion at *m/z* 452.7 produces fragment ions at *m/z* 404.2, 710.4, and 807.4.
3. Inject (manually or by autosampler) 6 μL of a tryptic peptide mixture onto the 8-cm long capillary column.
4. Elute peptides with the following gradient method at a flow rate of 35 μL/min: (a) 100% mobile phase (mp) A for 5 min; (b) a linear gradient to 65% mp B within 30 min; (c) maintain at 65% mp B for 15 min; (d) a linear gradient back to 100% mp A within 5 min; and (e) maintain at 100% mp A until the mass spectrometry analysis is complete.
5. Input the MS/MS data into protein database correlation software as described in **step 4** of **Subheading 3.9.1** Each positive search result – nitration of a Tyr residue – is confirmed with a manual check of the original LC, MS, and MS/MS data to determine each nitration site. During the analysis of those nitrated proteins, the following experimental criteria are applied: K or R at the C terminus; K, R, or D preceding the N terminus; 0 or 1 missed trypsin cleavage sites; singly charged product b- and y-ions, and a match to a homo sapiens protein sequence. De novo sequencing independently interprets the MS/MS data to accurately obtain the amino acid sequence. The amino acid sequence from de novo sequencing is used to search the human SWISS-PROT protein database with the SIB BLAST search engine (http://us.expasy.org/tools/blast/). Unlike MALDI-MS spectra, an ESI-MS spectrum of a nitrotyrosine-containing peptide does not demonstrate any unique characteristics **(Fig. 3)** (*see* **Note 27**).

3.10. Bioinformatics Determination of the Domain/Motif Where Nitration Occurred

1. Input the Swiss-Prot access number or the full amino acid sequence of each nitrated protein into the ScanProsite software (http://us.expasy.org/tools/scanprosite) to scan the functional/structure domains of each nitroprotein.
2. Input the Swiss-Prot access number or the full amino acid sequence of each nitrated protein into the MotifScan software (http://myhits.isb-sib.ch/cgi-bin/motif_scan) to scan the motifs of each nitroprotein.
3. Locate the tyrosine nitration-site within functional/structural domain or motifs. An example of a domain/motif where nitration occurs is shown in **Fig. 4** (*see* **Note 28**).

4. Notes

1. The water that is used to make a buffer/solution and to wash a gel should be deionized with distilled water that has a resistance of 18.2 MΩ cm. We often use Millipore water.
2. Urea decomposes at temperatures above 30°C; therefore, all urea solution should be kept a temperature below 30°C.
3. The dimension of an Amersham IPG dry strip is 0.5-mm thick and 3-mm wide with different lengths (7, 11, 13, 18, and 24 cm). Strips with different pH ranges are available (e.g., 3–10, 4–7, 6–9, 4–5, etc.) with either a linear or nonlinear pH gradient. Importantly, it is critical that the IPG buffer used match with the strip, otherwise IEF will not work properly *(22)*.
4. Acrylamide and bisacrylamide in the monomeric form are neurotoxic. Avoid inhaling or skin exposure. Polymerize any unused monomer with an excess of ammonium persulfate for ecological disposal *(22, 23)*.
5. Avoid inhaling SDS powder (be particularly cautious when weighing) as SDS is a respiratory irritant.
6. Pay attention not to use ddH_2O instead of SDS electrophoresis buffer when making the agarose-sealing solution. Be sure not to boil the agarose solution when heating; if the solution turns a light yellow color, then replace it with a fresh solution.
7. An Immobilon-P transfer membrane has two pore sizes: 0.45 and 0.2 μm. The 0.45-μm PVDF membrane is mostly used to transfer proteins, whereas the 0.2-μm PVDF membrane is preferred for transfer of peptides or small molecular weight proteins.
8. Polycolonal or monocolonal anti-nitrotyrosine antibodies can be used, which are commercially available from Upstate Biotechnology, International Chemicon, and Sigma.

9. ImmunoPure® immobilized protein G plus is a cross-linked 6% beaded agarose, supplied as a 50% slurry (for example, 200 μL of settled gel is equivalent to 400 μL of 50% slurry) that contains 0.02% sodium azide. The binding capacity of this product is greater than 20 mg human IgG per mL of settled gel. Store gel at 4°C.
10. Disuccinimidyl suberate (DSS) is a water-insoluble, strong polypeptide cross linker. It should be first dissolved in an organic solvent such as DMSO, and added to the aqueous reaction mixture. The moisture-sensitive DSS should be stored at 4–8°C in a desiccator. The DSS vial must be equilibrated to room temperature before opening to avoid moisture condensing onto the compound. Prepare the DSS solution just prior to use, because DSS readily hydrolyzes to become non-reactive. The reaction rate of completing DSS hydrolysis increases at greater pH, thus a lower pH is optimal for these experiments. Hydrolysis occurs more readily in dilute protein or peptide solutions. In concentrated protein solutions, the acylation reaction is favored. Over-crosslinking can result in the loss of biological activity of enzymes, antibodies, etc. due to a conformational change or DSS modification of lysine groups involved in binding a substrate or antigen. You can adjust the molar ratio of the reagent to the target to minimize activity loss.
11. Promega's Sequencing-Grade Modified Trypsin is porcine trypsin modified by reductive methylation. Its resistance to autolysis is two times greater than unmodified trypsin. The trypsin is further treated by L-1-tosylamido-2-phenylethyl chloromethyl ketone (TPCK) that inhibits chymotrypsin activity without effect on trypsin, followed by affinity purification, which increases the activity and stability of the enzyme. The modified trypsin has a maximal activity at pH 7–9, is reversibly inactivated at pH 4, is resistant to mild denaturing conditions (0.1% SDS, 1 M urea, or 10% acetonitrile), and retains 48% activity in 2 M guanidine HCl.
12. In washing off blood from a tissue's surface, some of that tissue might be lost.
13. The BCA reagent is not compatible with thiourea, DTT, IPG buffer, and bromophenol blue. Thus, a certain amount of each lyophilized pituitary sample must be reconstituted in a solution of 8 M urea and 4% CHAPS for BCA protein concentration analysis. For different experiments, use the BCA assay to determine protein concentration relative to a fixed concentration sample standard.
14. It is necessary for a rehydrated IPG strip to be rinsed thoroughly with ddH_2O to remove all mineral oil and undissolved components.

15. The wet paper wick not only conducts current, but also removes salts from the sample. Tissue samples all contain salts that can interfere with IEF. If the sample salt concentration is too high, then the paper wick should be replaced once with a new wet paper wick during the IEF separation.
16. After IEF, clean the IPG strip holder using the Ettan IPGphor strip holder cleaning solution (Amersham Biosciences, Cat# 80–6452–78) and distilled water, and allow the holder to air dry. Wipe off mineral oil from lids with paper towel.
17. The Bio-Rad PROTEN-plus Multicasting chamber may cast 1–12 gels at a time. Gels cast as described here can be used within a week when they are covered with a moist filter paper and stored at 4°C. Our studies with 2D gels determined that the reproducibility of a single-concentration gel is better than that of a gradient gel *(24, 25)*.
18. Ensure that the running buffer level is appropriate. If the gel cassette is completely immersed in buffer, then current will leak around the gel, which negatively affects the electrophoresis results. If the buffer level is not high enough to cover the entire gel area, then separation may not occur or the gel could overheat.
19. After electrophoresis, use only water to wash the Dodeca electrophoresis tank, gaskets, electrode cards, and lid. For long-term storage, flush all parts thoroughly with water to completely remove any residual buffer.
20. One side of a PVDF membrane is hydrophobic and the other hydrophilic. Be sure to place the hydrophilic side atop the gel to allow passage of proteins into the membrane. The hydrophobic side will stop the migration of the proteins to keep them from coming off of the membrane.
21. Pay careful attention to avoid forming and introducing air bubbles in **steps (b)–(g)** of **Subheading 3.4**; bubbles severely affect the semidry electrotransfer.
22. The wash steps after blocking, incubation with the primary antibody, and incubation with the secondary antibody must be sufficient to remove excess salt and Tween-20, which could precipitate on the PVDF membrane and increase the background when visualized.
23. Our studies *(24, 25)* demonstrated the high levels of reproducibility and resolution of 2DGE. For the IEF first dimension, a high reproducibility can be obtained with commercially available IPG strips. For the second-dimensional SDS-PAGE, different types of gel system are available: vertical vs. horizontal gel systems; constant-percentage vs. gradient gels. Compared to a horizontal Multiphor gel system (pre-cast

180 × 245 × 0.5 mm gradient SDS gels; analyzed one gel at a time), the vertical Dodeca gel system (single-concentration 190 × 205 × 1 mm gels; analyzed up to 12 gels at a time) provided better spatial and quantitative reproducibility *(24, 25)*. Moreover, for between-gel comparison, the same experiment conditions (including the loading amount of protein, sample-dissolving solution, IEF, IPG strip equilibration solution, SDS-PAGE, and gel image process) should be used to conduct 2DGE experiments *(26)*.

24. If the immunoprecipitated nitroproteins are used for 1D SDS-PAGE and Western blot assay, the bound nitroproteins may be eluted with 100 μL of 1% SDS/62.5 mM Tris–HCl (pH 7.0) (incubate at 60°C for 20 min; centrifuge for 1 min at 3,000 × *g*); repeat two more times. Collect the eluants containing the nitroproteins, and partially dry it in a vacuum centrifuge to increase the concentration of nitroprotein. This is a stronger elution method than the acidic elution buffer (pH 2.8). After this elution, the cross-linked antibody–protein G beads cannot be reused.
25. Gloves and a head cap should be used to avoid keratin contamination from skin and hair *(21)*.
26. All tubes and pipet tips that contact samples should be siliconized, or designed for low-retention to avoid the loss of proteins or peptides.
27. The ultraviolet laser photodecomposes nitro groups ($-NO_2$) to form unique ions in a MALDI-MS spectrum of nitrotyrosine-containing peptides; that decomposition event decreases the intensity of precursor ions for MS/MS analysis *(27, 28)*. In turn, if this type of decomposition is observed, then the presence of a nitrotyrosine is confirmed. Nitro group decomposition does not occur with ESI. However, the neutral loss of a nitro group (–45 Da) does occur with MALDI-MS/MS and ESI-MS/MS. The presence of an immonium ion at an *m/z* of 181.06 can also help to identify the existence of nitrotyrosine *(20)*. However, the immonium ion often is not detected when using an ion-trap mass spectrometer (LCQ and LTQ), because the low-mass cut-off is automatically set near or above this *m/z* value relative to the size of the peptide. The issue is avoided when using a quadrupole-time-of-flight tandem mass spectrometer.
28. The function or role of a nitrotyrosine modification can be explored by locating the site to a functional/structural domain/motif. The functions of most nitrated proteins identified in our studies were obtained from exploring the literature with bioinformatic tools. ScanProsite and Motifscan software mine the literature for information on domains/motifs of

proteins. Once a nitropeptide's sequence is determined, it can then be localized to a known domain/motif. The SwissProt annotation page of a protein provides functional/structural information that is linked to the corresponding original literature.

Acknowledgments

The authors acknowledge financial support from NIH (NS 42843 to DMD; RR 16679 to DMD).

References

1. Zhan, X., and Desiderio, D. M. (2006) Nitroproteins from a human pituitary adenoma tissue discovered with a nitrotyrosine affinity column and tandem mass spectrometry. *Anal. Biochem.* **354**, 279–289.
2. Zhan, X., and Desiderio, D. M. (2004) The human pituitary nitroproteome: Detection of nitrotyrosyl-proteins with two-dimensional Western blotting, and amino acid sequence determination with mass spectrometry. *Biochem. Biophys. Res. Commun.* **325**, 1180–1186.
3. Scaloni, A. (2006) Mass spectrometry approaches for the molecular characterization of oxidatively/nitrosatively modified proteins, in *Redox Proteomics: From Protein Modification to Cellular Dysfunction and Diseases* (Dalle-Donne, I., Scaloni, A., and Butterfield, D. A., eds.), Wiley, Hoboken, NJ, pp. 59–100.
4. Khan, J., Brennan, D. M., Bradley, N., Gao, B., Bruckdorfeer, R., and Jacobs, M. (1998) 3-nitrotyrosine in the proteins of human plasma determined by an ELISA method. *Biochem. J.* **330**, 795–801.
5. Lloyd, R. V., Jin, L., Qian, X., Zhang, S., and Scheithauer, B. W. (1995) Nitric oxide synthase in the human pituitary gland. *Am. J. Pathol.* **146**, 86–94.
6. Pawlikowshi, M., Winczyk, K., and Jaranowska, M. (2003) Immunohistochemical demonstration of nitric oxide synthase (NOS) in the normal rat pituitary gland, estrogen-induced rat pituitary tumor and human pituitary adenomas. *Folia Histochem. Cytobiol.* **41**, 87–90.
7. Ueta, Y., Levy, A., Powell, M. P., Lightaman, S. L., Kinoshita, Y., Yokota, A., Shibuya, I., and Yamashita, H. (1998) Neuronal nitric oxide synthase gene expression in human pituitary tumours: a possible association with somatotroph adenomas and growth hormone-releasing hormone gene expression. *Clin. Endocrinol. (Oxford)* **49**, 29–38.
8. Kruse, A., Broholm, H., Rubin, I., Schmidt, K., and Lauritzen, M. (2002) Nitric oxide synthase activity in human pituitary adenomas. *Acta Neurol. Scand.* **106**, 361–366.
9. McCann, S. M., Karanth, S., Mastronardi, C. A., Dees, W. L., Childs, G., Miller, B., Sower, S., and Yu, W. H. (2001) Control of gonadotropin secretion by follicle-stimulating hormone-releasing factor, luteinizing hormone-releasing hormone, and leptin. *Arch. Med. Res.* **32**, 476–485.
10. McCann, S. M., Haens, G., Mastronardi, C., Walczewska, A., Karanth, S., Rettori, V., and Yu, W. H. (2003) The role of nitric oxide (NO) in control of LHRH release that mediates gonadotropin release and sexual behavior. *Curr. Pharm. Des.* **9**, 381–390.
11. Ceccatelli, S., Hulting, A. L., Zhang, X., Gustafsson, L., Villar, M., and Hokfelt, T. (1993) Nitric oxide synthase in the rat anterior pituitary gland and the role of nitric oxide in regulation of LH secretion. *Proc. Natl Acad. Sci. USA* **90**, 11292–11296.
12. Pinilla, L., Gonzalez, L. C., Tena-Sempere, M., Bellido, C., and Aguilar, E. (2001) Effects of systemic blockade of nitric oxide synthases on pulsatile LH, prolactin, and GH secretion in adult male rats. *Horm. Res.* **55**, 229–235.
13. Duvilanski, B. H., Zambruno, C., Seilicovich, A., Pisera, D., Lasaga, M., Diaz, M. C., Belova, N., Rettori, V., and McCann, S. M. (1995) Role of nitric oxide in control of prolactin release by the adenohypophysis. *Proc. Natl Acad. Sci. USA* **92**, 170–174.

14. Schwartz, J. (2000) Intercellular communication in the anterior pituitary. *Endocr. Rev.* **21**, 488–513.
15. Brunetti, L., Ragazzoni, E., Preziosi, P., and Vacca, M. (1995) A possible role for nitric oxide but not for prostaglandin E2 in basal and interleukin-1-B-induced prolactin release in vitro. *Pharmacol. Lett.* **56**, 277–283.
16. Bocca, L., Valenti, S., Cuttica, C. M., Spaziante, R., Giordano, G., and Giusti, M. (2000) Nitric oxide biphasically modulates GH secretion in cultured cells of GH-secreting human pituitary adenomas. *Minerva Endocrinol.* **25**, 55–59.
17. Cuttica, C. M., Giusti, M., Bocca, L., Sessarego, P., De Martini, D., Valenti, S., Spaziante, R., and Giordano, G. (1997) Nitric oxide modulates in vivo and in vitro growth hormone release in acromegaly. *Neuroendocrinology* **6**, 426–431.
18. Pinilla, L., Tena-Sempere, M., and Aguilar, E. (1999) Nitric oxide stimulates growth hormone secretion in vitro through a calcium- and cyclic guanosine monophosphate-independent mechanism. *Horm. Res.* **51**, 242–247.
19. Riedel, W. (2002) Role of nitric oxide in the control of the hypothalamic–pituitary–adrenocortical axis. *Z. Rheumatol.* **59**, II/36–II/42.
20. Zhan, X., and Desiderio, D. M. (2007) Linear ion-trap mass spectrometric characterization of human pituitary nitrotyrosine-containing proteins. *Int. J. Mass Spectrom.* **259**, 96–104.
21. Zhan, X., and Desiderio, D. M. (2003) A reference map of a human pituitary adenoma proteome. *Proteomics* **3**, 699–713.
22. Zhan, X. (2005). Two-dimensional electrophoresis, in *Experimental Protocols for Medical Molecular Biology in Chinese and English* (Zheng, W., ed.), Peking Union Medical College Press, Beijing, pp. 93–108.
23. Westermeier, R. (ed.) (1997) *Electrophoresis in Practice: A guide to Methods and Applications of DNA and Protein Separations.* VCH Wiley, Weinheim.
24. Zhan, X., and Desiderio, D. M. (2003) Differences in the spatial and quantitative reproducibility between two second-dimensional gel electrophoresis. *Electrophoresis* **24**, 1834–1846.
25. Zhan, X., and Desiderio, D. M. (2003) Spot volume vs. amount of protein loaded onto a gel. A detailed, statistical comparison of two gel electrophoresis systems. *Electrophoresis* **24**, 1818–1833.
26. Zhan, X., and Desiderio, D. M. (2003) Heterogeneity analysis of the human pituitary proteome. *Clin. Chem.* **49**, 1740–1751.
27. Petersson, A. S., Steen, H., Kalume, D. E., Caidahl, K., and Roepstorff, P. (2001) Investigation of tyrosine nitration in proteins by mass spectrometry. *J. Mass Spectrom.* **36**, 616–625.
28. Sarver, A., Scheffler, N. K., Shetlar, M. D., and Gibson, B. W. (2001) Analysis of peptides and proteins containing nitrotyrosine by matrix-assisted laser desorption/ionization mass spectrometry. *J. Am. Soc. Mass Spectrom.* **12**, 439–448.

Chapter 11

Improved Enrichment and Proteomic Analysis of Brain Proteins with Signaling Function by Heparin Chromatography

Kurt Krapfenbauer and Michael Fountoulakis

Summary

Detection of low-abundance proteins with signaling function is essential for the identification of biomarkers and novel drug targets. We present a protocol for specific enrichment of secreted proteins with signaling function by combining subcellular fractionation with heparin chromatography. The subcellular fractionation includes the preparation of a fraction enriched in cytosolic proteins. A further enrichment was achieved by heparin chromatography. The proteins eluted from the heparin column were analyzed by MudPIT tandem mass spectrometry and identified with the use of an in silico algorithm. Forty-eight percent of the identified proteins (188 out of 391) bound to the heparin matrix. Fifty-four percent of them (101) are secreted proteins with signaling function and 23% (44) of the enriched signaling proteins had not been detected by 2D PAGE without application of the heparin enrichment step. The heparin chromatography method can be combined with other proteomics enrichment approaches, such as ion exchange or reversed phase chromatography.

Key words: Brain proteome, Subcellular fractionation, Heparin chromatography, Protein enrichment, Secreted proteins, Signaling function, LC-MS/MS analysis, MudPIT analysis, In silico analysis

1. Introduction

Visualization and identification of the low-abundance brain proteins may facilitate the identification of novel drug targets and diagnostic markers. However, not all proteins of an organism are expressed in amounts sufficient for detection by two-dimensional polyacrylamide gel electrophoresis (2D PAGE). In order to visualize and identify low-copy number gene products, we

Andrew K. Ottens and Kevin K.W. Wang (eds.), *Neuroproteomics,* Methods in Molecular Biology, vol. 566
DOI 10.1007/978-1-59745-562-6_11, © Humana Press, a part of Springer Science+Business Media, LLC 2009

enriched low-abundance cytosolic proteins by applying heparin chromatography. This step was chosen because heparin has a high protein binding capacity and can discriminate and enrich proteins with minor differences in their pI values and glycosylation patterns *(1, 2)*. As most secreted proteins are glycosylated *(3)*, heparin sepharose and/or lectin columns (*see* **Note 1**) are versatile tools for the enrichment of many classes of glycosylated proteins such as proteins with signaling functions, growth factors, coagulation factors, and steroid receptors *(4, 5)*. The ligand in a heparin sepharose column is a naturally occurring sulfated glycosaminoglycan, which is extracted from the native proteoglycan of porcine intestinal mucosa. Heparin consists of alternating units of uronic acid and d-glucosamine, most of which are substituted with one or two sulfate groups. Immobilized heparin has two main modes of interaction with proteins: It can operate as an affinity ligand, e.g., in its interaction with coagulation factors and it can also function as a high capacity cation exchanger because of its anionic sulfate groups, leading thus to an additional enrichment of positively charged proteins.

We applied heparin chromatography to enrich secreted rat brain proteins with signaling function prior to proteomic analysis. The column was operated with a syringe instead of a liquid chromatography pump. Elution was performed by increasing the ionic strength with 2 M NaCl. Separation of the eluted proteins was carried out by one-dimensional (1D) polyacrylamide gel electrophoresis (PAGE) and the proteins were identified by multidimensional protein identification technology (MudPIT) tandem mass spectrometry *(6, 7)* in combination with in silico analysis (*see* **Note 2**) **(Fig. 1)**. The analysis resulted in the identification of 391 different gene products. One hundred and eighty eight proteins bound to the heparin matrix, 101 of which were secreted proteins with signaling functions. Forty-four of the enriched proteins had not been detected by 2D PAGE without previous enrichment by combination of subcellular prefractionation and heparin chromatography. Heparin chromatography specifically enriched several enzymes that had not been identified before *(8)*, like peptidyl-glycine alpha-amidating monooxygenase, sodium/hydrogen exchanger 5, etc. For these proteins no spots/bands had been detected without enrichment by heparin chromatography.

Moreover, heparin chromatography may be useful in the depletion of albumin from body fluids, such as plasma and cerebrospinal fluid, in which it represents more than 50% of total proteins *(1, 9)*. For example, serum albumin, which is represented by a strong band in the starting material **(Fig. 2)**, was completely recovered in the flow-through fraction. Specific removal of albumin, as well as of other high abundance proteins allowed the visualization and identification of minor components of the samples,

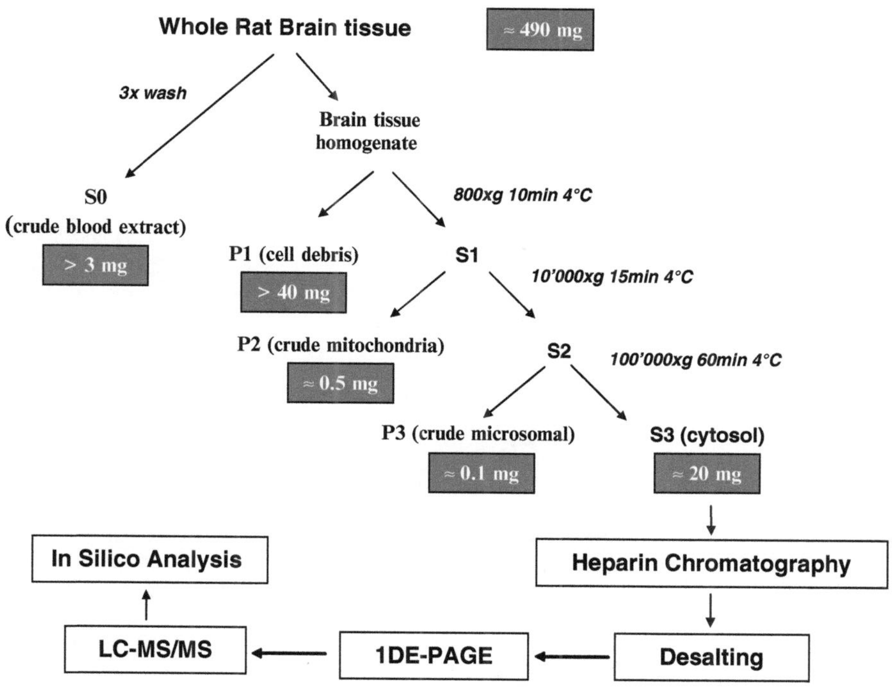

Fig. 1. Schematic for centrifugal prefractionation of rat brain proteins. Different centrifugal forces lead to enrichment of cellular components such as mitochondria, microsomal, and cytosolic proteins. The cytosolic fraction was subjected to further fractionation by heparin chromatography followed by separation on a 10% homogenous polyacrylamide gel and protein identification by LC-MS/MS.

whose levels may change in certain disorders *(10)*. In addition to the easier design of protein purification steps, use of selected chromatography steps prior to the LC-MS/MS can significantly facilitate the analysis of complex protein mixtures.

2. Materials

2.1. Sample Preparation

1. Homogenization buffer: 20 mM HEPES, 320 mM sucrose, 1 mM EDTA, 5 mM dithiothreitol, 1 tablet of protease inhibitor cocktail (Roche Diagnostics, Indianapolis, IN), 0.2 mM sodium metavanadate, 1 mM sodium fluoride. Combine all ingredients in a Falcon tube and bring to 50 mL with ultrapure water; agitate at room temperature (avoid heat) on an end-over-end rotator until completely dissolved. Must be prepared fresh before use.

2.2. Heparin Chromatography

1. Heparin column: HiTrap Heparin HP (Cat. no 17-0406-01, GE Healthcare, Munich, Germany). It is made of polypropylene, which is biocompatible and does not interact with most

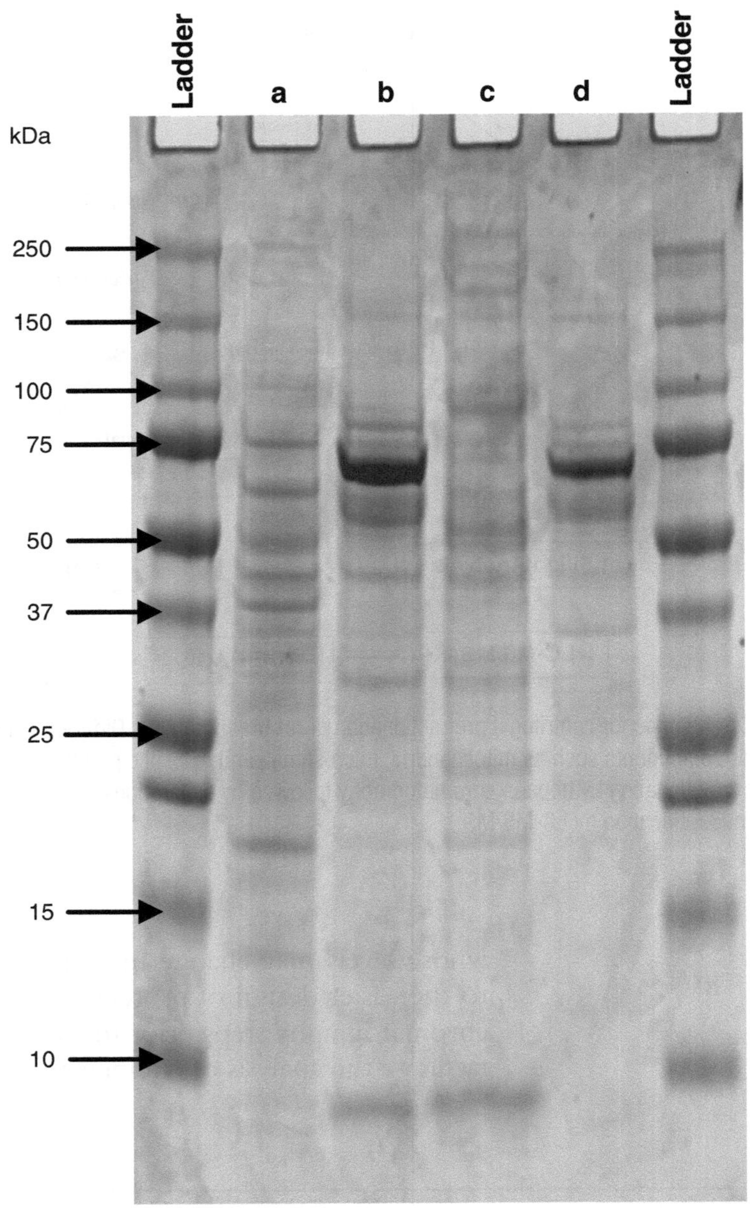

a...... Brain Tissue Homogenate

b...... Supernatant (S3)

c...... Heparin

d...... Heparin Flow Through

Fig. 2. 1DE gel analysis of fractions eluted from heparin chromatography. Fraction enriched with cytosolic rat brain proteins was prepared as stated in **Subheading 3.1**, further separated as stated in **Subheading 3.4**, and analyzed by 1DE as stated in **Subheading 3.5**. Gels were stained with colloidal Coomassie blue with protein bands identified by LC-MS/MS.

nonglycosylated biomolecules. The prefilled column can be easily operated using a syringe, a peristaltic pump, or in a chromatography system such as FPLC. The column volume is 1 mL, ligand 10 mg heparin/mL medium, mean particle size is 34 μm to highly cross-linked spherical agarose as bead structure. Flow rates should be 0.2–1.0 mL/min.

2. 1 M Na_2HPO_4 stock buffer: Dissolve 178.0 g of disodium hydrogen phosphate dihydrate in 500 mL of distilled water and adjust it to 1 L with distilled water. Dispense the solution into 100-mL aliquots and sterilize them by autoclaving for 20 min at 15 psi on the liquid cycle. Store at 25°C.
3. 1 M NaH_2PO_4 stock buffer: Dissolve 120.0 g of sodium dihydrogenphosphate monohydrate in 500 mL of distilled water and dilute it to 1 L with distilled water. Dispense the solution into 100-mL aliquots and sterilize them by autoclaving for 20 min at 15 psi on the liquid cycle. Store at 25°C.
4. Binding buffer (10 mM sodium phosphate, pH = 7.0): Add 5.8 mL of 1 M Na_2HPO_4 stock buffer and 4.2 mL of the 1 M NaH_2PO_4 stock buffer and adjust the volume to 1 L with distilled water in a volumetric flask. Pass the solution through a 0.22-μm filter, and store in aliquots (100 mL) at 25°C.
5. Elution buffer (10 mM sodium phosphate, 2 M NaCl, pH = 7): Dissolve 116.9 g sodium chloride in 100 mL of distilled water, add 5.8 mL of 1 M Na_2HPO_4 stock buffer, and add 4.2 mL of 1 M NaH_2PO_4 stock buffer. Dilute to 1 L with distilled water in a volumetric flask. Pass the solution through a 0.22-μm filter, and store in 100-mL aliquots at 25°C.

2.3. Chromatographic Desalting with Poros Column

1. Stock-slurry: Wet 1.0 g Poros 20R2 chromatographic media (Applied Biosystems, Foster City, CA) with 3 mL of ethanol, followed immediately by 7 mL of distilled water, stir gently, and store at 4°C.
2. Binding mobile phase (0.1% TFA): Dissolve 1 mL TFA (trifluoroacetic acid) in 1,000 mL Milli-Q water, stir, and store at 25°C. Prepare fresh.
3. Elution phase (70% acetonitrile, containing 0.1% TFA): Mix 700 mL of acetonitrile with ultrapure water containing 1 ml of TFA and bring to 1,000 mL.

2.4. Bradford Assay

1. Mix 10 mL of dye reagent (Protein assay solution, Art. No. 500–0205, Bio-Rad, Hercules, CA) with 50 mL ultrapure water and bring to 100 mL. Prepare fresh, protect the solution from light, and store at 4°C.
2. Standard bovine serum albumin (BSA) stock solution: Dilute 1:20 the contents of one BSA standard ampule (Pierce, Cat.

No. 23209, Rockford, IL) into clean Eppendorf tubes, preferably using the same diluents as the samples. Each 1 mL of 2.0 mg/mL albumin standard is sufficient to prepare a set of diluted standards for either working range. Prepare fresh or store 1-mL aliquots at –20°C.

2.5. 1D PAGE

1. LDS sample buffer: 2.5 μL (4×) NuPAGE LDS sample buffer (Invitrogen, Carlsbad, CA), 1.0 μL NuPAGE 10× reducing agent (Invitrogen), 6.5 μL distilled water. Combine all ingredients in a 1.5-mL Eppendorf tube and agitate at room temperature. Must be prepared fresh before use.
2. Running buffer: Prepare 1× SDS running buffer by adding 50 mL of 20× NuPAGE MES running buffer (Invitrogen) to 950 mL of deionized water in a graduated cylinder. Prepare fresh, cover with Para-Film, and invert to mix.
3. NuPAGE Novex Bis–Tris Mini Gel system (10% BT linear gradient, NuPAGE, Cat. No. NP0315BOX, Invitrogen).
4. NuPAGE antioxidant solution (Cat. No. NP0005, Invitrogen).
5. NuPAGE sample reducing agent (10×), (Cat. No. NP0004, Invitrogen).
6. Prestained protein standard (SeeBlue Pre-stained, Cat. No. LC5625, Invitrogen).
7. Digestion solution: 5 mM ammonium bicarbonate (pH 8.8) containing 50 ng of mass spectrometry grade trypsin (Promega, Madison, WI, USA).

2.6. MudPIT Analysis

2.6.1. Microcapillary Column

1. 4 cm × 50 μm i.d. × 5 μm C18 microSPE precolumn (Polymicro Technologies, Phoenix, AZ).
2. Model P-2000 Laser Puller (Sutter Instrument Co., Novato, CA).
3. Stainless steel pressurization device (Brechbuehler, Inc., Houston, TX, or MTA for blueprints kindly provided by John Yates, Scripps Research Institute, La Jolla, CA).
4. 85 cm × 15 μm i.d. × 3 μm C18 (packed capillary column, Polymicro Technologies).
5. 10 μm i.d. × 150 μm o.d. fused silica capillary (Polymicro Technologies).
6. M-520 Inline Micro Filter Assembly and F-185 Microtight 0.0155 × 0.025 Sleeves (UpChurch Scientific, Oak Harbor, WA).

2.6.2. Multidimensional Chromatography and Tandem Mass Spectrometry

1. Buffer A: 5% acetonitrile, 0.1% formic acid, use HPLC-grade water.
2. Buffer B: 80% acetonitrile, 0.1% formic acid, use HPLC-grade water.

3. P-775 MicroTee Assemblies and F-185 Microtight 0.0155 × 0.025 Sleeves (UpChurch Scientific).
4. Gold wire (Scientific Instrument Services, Inc., Ringoes, NJ).
5. Agilent 1100 series G1379A degasser, G1311A quaternary pump, G1329A autosampler, G1330B autosampler thermostat, and G1323B controller (Agilent Technologies, Palo Alto, CA).
6. LCQ DECA-XPplus tandem mass spectrometer (Thermo Electron, San Jose, CA).
7. Thermo Electron Nanospray II ion source or PicoView Sources from New Objective (Woburn, MA).

3. Methods

3.1. Sample Preparation

1. Whole rat brain tissue (≈490 mg) was washed three times with 10 mL of sucrose homogenization buffer to remove excess of blood (*see* **Note 3**).
2. For protein extraction, whole brain tissue is suspended in sucrose homogenization buffer in a glass-teflon potter homogenizer (Elvehjem Potter) (*see* **Notes 4** and **5**). After ten strokes at 500 rpm and 4°C, centrifuge at 800 × *g* for 10 min at 4°C to sediment undissolved material [cell debris (P1), *see* **Fig.** 1].
3. The supernatant (S1) is centrifuged at 10,000 × *g* for 10 min at 4°C to remove the crude mitochondrial fraction (P2).
4. The supernatant (S2) is further centrifuged at 100,000 × *g* for 1 h to sediment undissolved material (crude microsomal fraction P3) leaving supernatant (S3) enriched with cytosolic proteins (*see* **Note 6**).

3.2. Determination of Protein Concentration

Protein quantification is performed with the Coomassie brilliant blue method known as the Bradford assay *(11)*. Following protocol is applied:

1. Label cuvettes as necessary for the standard curve and the protein samples.
2. Pipette 0, 5, 10, 15, 20, 25, 50, 100 μL of standard BSA stock solution into the cuvettes containing 2 mL of the Bradford reagent solution. Balance to a total volume of 2.1 mL with distilled water.
3. Protein samples to be measured should be appropriately diluted. Pipette 40 μL of LDS sample buffer into a 1.5-mL Eppendorf tube and add 5 μL of the protein sample. Add an

appropriate volume of distilled water to each tube to give a total volume of 1 mL.

4. Pipette 1 mL of the Bradford reagent into a cuvette and add 100 μL of the diluted sample and 100 μL of the BSA-standard solution, respectively (*see* **Notes 7–9**).
5. Measure absorbance at a wavelength of 595 nm in a UV–Vis spectrometer following the manufacturer's instructions.
6. Generate a standard curve with the BSA data and calculate the concentration of the diluted sample. Multiply by the total dilution factor to calculate the initial sample concentration.

3.3. Heparin Chromatography

The sample solution is first adjusted to the composition of the binding buffer. This is done by diluting the sample (5 mg total cytosolic brain protein) with 25 mL of the binding buffer. The sample is centrifuged before applying to the column (*see* **Notes 6** and **10**). The flow rate of the heparin column is 1 mL/min with the use of a hand-driven syringe.

1. A 25-mL syringe is filled with binding buffer. In addition to this, the stopper is removed and the column is connected to the syringe with the provided adapter to avoid introducing air into the column.
2. The twist-off outlet cap is removed and the heparin sepharose is washed with ten column volumes of binding buffer to equilibrate the column.
3. The sample (prepared as described above) is then applied onto the column using a syringe fitted to the luer adaptor.
4. The column is washed with at least five volumes of binding buffer or until no protein appears in the effluent. Effluent is checked by the Bradford reaction.
5. To elute the proteins, the column is washed with five volumes of elution buffer in a single 10-mL Falcon tube without fractionation (*see* **Notes 11–13**).

3.4. Desalting with Poros 20R2 Columns

The protein fraction eluted from the heparin column is desalted by using reversed phase chromatography and dried using a speed vac apparatus.

1. Pipette 1 mL of the Poros stock slurry into a spin column (load capacity: ~4–5 mg of protein, check with the measurement of OD at 280 nm) and wash with five column volumes (5 mL) of 0.1% TFA (v/v).
2. Apply the protein fraction from the heparin elution and wash with 10 mL of binding solution (0.1% TFA).
3. Elute the proteins with 1 mL of 70% acetonitrile in 0.1% TFA (v/v).

4. Dry in a speed vac apparatus, and reconstitute in 25 μL of LDS sample buffer.
5. Determine the protein concentration by the Bradford assay (**Subheading 3.2**). The protein concentration should be approximately 8–12 mg/mL.

3.5. 1D PAGE: Sample Load and Running Conditions

1. These instructions assume the use of the NuPAGE Novex Bis–Tris Mini Gel system (*see* **Note 14**), though they are easily adapted to other formats such as Hoeffer SE-400 or SE-600 gel system. Carefully remove the comb from the Novex ready mini gel and wash the wells with running buffer with syringe fitted with a 22-gauge needle.
2. Adjust 15 μg of protein sample to a volume of 20 μL with LDS sample buffer and heat at 70°C for 10 min (*see* **Note 15**). Afterward, chill the samples on ice, and before applying to the gel spin down at 3,000 × *g* for 30 s.
3. Fill the upper buffer chamber with 200 mL of running buffer containing 500 μL of antioxidant solution.
4. Fill the lower buffer chamber with at least 600 mL of running buffer.
5. Load 20 μL of each sample into the corresponding gel well. Reserve one well for prestained molecular weight markers.
6. Assemble the gel unit and connect to a power supply. If cooling is available for the whole gel unit, electrophoresis can be performed with constant voltage until the dye front (bromophenol blue) runs off the gel.

Running condition	200 V constant
Running time	35 min
Expected	40–55 mA/gel
Current	100–125 mA/gel

7. After approximately of 35 min running, disconnect the gel unit and remove the gel.
8. Shake the gel in 50-mL Coomassie Brilliant Blue R-250 staining solution for at least 12 h. Decant staining solution and replace with a minimum of 100 mL of Coomassie R-250 destaining solution per gel. Shake gel for at least 2 h. Change destaining solution until a clear background is observed (*see* **Notes 16** and **17**).
9. Scan the gels in a Molecular Dynamics Personal densitometer. Images can be processed using Photoshop (Adope) and PowerPoint (Microsoft) software. Protein bands can be

quantified using the Image Master 2D Elite software (Amersham Pharmacia Biotechnology) or similar imaging system (BioSpectromAC Imaging System, UVP).

10. Excise all observed bands from the gel (about 20 bands) and place them in Eppendorf tubes. Destain the gel bands with 30% (v/v) acetonitrile in 0.1 M ammonium bicarbonate and dry in a speed vac apparatus. Rehydrate the dried gel pieces with 5 μL of trypsin digestion solution, centrifuge for 1 min, and leave at room temperature for about 12 h. After digestion, extract the peptides by adding 5 μL of water, followed by 10 μL of 75% (v/v) acetonitrile 10 min later, containing 0.3% (v/v) trifluoroacetic acid. Complete the extraction by vortexing at 100 rpm, for 5 min, 25°C and centrifuge the tubes for 1 min, at 1,000 × *g*, 25°C.

3.6. Tandem Mass Spectrometry Analysis: LC-MS/MS

Analysis of tryptic digests is performed by the LC-MS/MS approach termed multidimensional protein identification technology (MudPIT) as described in *(6, 7)* with minor modifications. It combines multidimensional liquid chromatography with electrospray ionization tandem mass spectrometry, integrating strong cation-exchange (SCX) resin and reversed-phase resin in a biphasic column. Analysis is usually performed in duplicate and the separation should be reproducible to within 0.5% between two analyses. Analysis of replicates demonstrated that the fraction of identifications shared between replicates [expressed as: (number of proteins identified in the replicate run identifying fewer proteins)/(number of proteins identified in the replicate run identifying more proteins)] was approximately 0.8. For values less than 0.8 more replicates are recommended.

3.6.1. Setup of the LC-MS/MS

1. The microcapillary column is attached to the P-775 Micro-Tee Assemblies.
2. One connection point of the cross contains the transfer line from the HPLC pump. This consists of a 4 cm piece of 50 μm i.d. capillary precolumn packed with 5 μm C18 micro SPE particules.
3. A second connection point ahead of the packed capillary column is used as a split/waste line. This split line allows the majority of the flow to exit through the split; therefore, very low flow rates can be achieved through the packed capillary microcolumn. The size and the length of this section of capillary depend on the flow rate from the pump and the length of the microcolumn.
4. Another connection contains a section of gold wire to apply 2,400 V for electrospray ionization to occur.

5. Place the packed, loaded, and washed column into a MicroTee on a stage, which in this case is designed for the ThermoFinnigan DECA-XPplus series mass spectrometer. This stage performs a threefold purpose: to support the MicroTee and hold it in place along with the connections, to electrically insulate the MicroTee from contact with its surroundings when it is held at high voltage potential, and to allow for fine position adjustments of the microcolumn with respect to the entrance of the mass spectrometer (heated capillary) by using an XYZ manipulator.
6. Measure the flow from the tip of the capillary microcolumn, using graduated glass capillaries. To do this, set the flow rate of the Agilent1100 to 0.1 mL/min from the controller. The target flow rate at the tip should be approximately 200–300 nL/min and a back pressure on the Agilent1100 of between 30 and 50 bars. If the flow rate is too high, cut off a portion of the split line capillary. This will cause more of the flow to exit out of the split and cause less flow through the microcolumn. If the flow is too low, a longer piece of 50-μm capillary or a section with a smaller inner diameter can be used to force more flow through the microcolumn. Measuring the flow rate and adjusting the split line may have to be repeated a number of times until the target flow rate is reached.
7. Prior to initiating a run, position the microcolumn using the XYZ manipulator so that the needle tip is within 5 mm from the orifice of the mass spectrometer's heated capillary.
8. Load the sample onto the micro-SPE nanoLC column at approximately 8 μL/min, which requires <2 min for a 10-μL solution with a sample loss of <5% (due to the syringe and valve adapters). The 85-cm long 3-μm particle packed 15-μm-i.d. capillary provides a separation peak capacity of approximately 10^3 at a constant pressure of 10,000 psi. Gradient elution (**Table 1**) resolves peptides in time, which are electrosprayed, via a replaceable nanoESI emitter made from a 10-μm-i.d. × 150-μm-o.d. fused silica capillary with a 2-μm-i.d orifice for efficient ionization connecting via a zero dead volume stainless steel union fitting. The ESI source is interfaced to an ion trap MS/MS for peptide detection.

3.6.2. Instrument Method Used in the MS/MS Analysis

The following settings are similar for a typical data-dependent MS/MS acquisition as already described in *(6)*. The method consists of a continual cycle beginning with one MS scan, which records all of the *m/z* values of the ions present at that moment in the gradient, followed by rounds of MS/MS on the three most intense ions. Full MS spectra are recorded on the peptides over a 400–1,600*m/z* range. Dynamic exclusion is activated to improve the protein identification capacity during the analysis. In the

Table 1
Gradient profile

Time (min)	Flow rate (mL/min)	% Buffer A	% Buffer B
0	0.15	100	0
16	0.15	70	30
17	0.15	30	70
20	0.15	0	100

"Timed Events" window of the instrument setup, the following sequence should be inserted:

Time (min) = 0.00, Settings = Contact 1, Value = Open
Time (min) = 0.05, Settings = Contact 1, Value = Closed
Time (min) = 0.10, Settings = Contact 1, Value = Open

3.7. Identification of Putative Secreted Proteins by In Silico Analysis

Database searching is an important component in the analysis of a complex peptide mixture, which requires significant computing power. In addition, researchers/bioinformaticians who have the necessary expertise can design their own algorithm to identify specific proteins/peptides with signaling function. Secreted proteins are normally characterized by a signal peptide sequence that helps in threading the N terminus of the protein through a membrane before it is cleaved off by a signal peptidase. Signal peptides are only loosely defined, often with a positively charged, polar section, followed by a hydrophobic stretch, and a short pattern around the cleavage site.

1. We developed an in-house, web-based software package to identify or predict signaling peptides. However, similar software platforms are commercially available, (e.g., Ingenuity software systems, Hampton, UK). The software is based on a set of specialized, manually created Hidden Markov Models (HMMs) that attempt to recognize the (sparse) sequence features common to signal peptides or anchors, respectively. As these sequence signals cannot be reliably predicted, the "signal" and "anchor" scores that any input sequence is assigned are fed into a support vector machine (SVM) in a second analysis step. The SVM is trained on a set of bona fide examples for both classes. On this training set, the SVM obtained the following results on three training sets (signal – anchor – neither):

(a) *Signal.* 96% sensitivity and 96% specificity

(b) *Anchor.* 87% sensitivity and 98% specificity

(c) *Neither.* 96% sensitivity and 97% specificity

2. These results are as good as the best claims in the literature (note, however, that most tools only look at the signal peptides and do not attempt to predict membrane anchors).
3. For the default setting of this Web interface, a list is produced that simply classifies each input sequence in either of the three categories "Signal," "Anchor," or "Normal" (the latter indicating that neither signal peptide nor membrane anchor are predicted).

4. Notes

1. Immobilized lectins can also be used as an alternative, since the affinity of glycosylated proteins for lectins is stronger than for heparin. However, binding capacity and the affinity of proteins to lectins is strongly dependent on the pH of the binding buffer, flow rate, and temperature leading to reproducibility problems when compared to heparin chromatography
2. Although many different instrumental setups for MudPIT have been designed, the components are similar, and include an injector, two isocratic or gradient LC pumps, a single column for the first dimension, one or more columns for the second dimension, one or more computer-controlled switching valves, and an appropriate detection system. However, a variety of other combination modes can be used like column-switching 2D LC separations methods, which include ion exchange-reversed phase-, size exclusion-reversed phase-, reversed phase-size exclusion-, ion exchange-size exclusion-, and/or normal phase-reversed phase chromatography as already described in *(6)*.
3. The present protocol can be used for the analysis of brain from other species as well as for other organs such as kidney, liver, etc. In bloody tissues, several high-abundance plasma proteins such as albumin, immunoglobulin, etc. are dominant and suppress the detection of lower-abundance proteins by 1D- and 2D PAGE or mass spectrometry. Therefore, removal of excess blood proteins prior to homogenization is a prerequisite for the detection of the low-abundance components derived from the original tissues *(9)*. To remove the excess of blood from the tissues, gently wash the tissue sample under iso-osmotic conditions (homogenization buffer), which has

no influence on the protein content of the solid tissue. Keep the wash buffer ice-cold and immediately continue with the homogenization procedure.

4. Brain tissue should be homogenized in a way to minimize proteolysis and other modes of protein degradation. Homogenization should be done at as low a temperature as possible under iso-osmotic conditions. This step should be carried out as quickly as possible, using ice-cold homogenization buffer containing protease inhibitors.

5. After homogenization, the sample will become highly viscous as a consequence of the release of high-molecular-weight chromosomal DNA. Nucleic acids increase background, cause smearing, and can clog gel pores when performing gel electrophoresis. Shear the chromosomal DNA by pipetting the sample up and down or by sonicating with sonicator tip immersed in a chilled sample cup. Depending on the power output of the sonicator, between 30 s and 1 min of sonication at half power should be sufficient to reduce the viscosity to a manageable level. As an alternative, chromosomal DNA can be sheared by repeated passage through a 23-gauge hypodermic needle, but this can have a negative effect on the integrity of some proteins or by the addition of 0.1 U of DNase I to the sample and incubation at 37°C for 15 min prior to heparin chromatography. This will decrease the viscosity of the mixture and prevent column clogging during sample loading, washing, or elution.

6. Remove all particulate materials by ultracentrifugation. Solid particles, lipids, and DNA must be removed because they will block the heparin column.

7. Selection of the protein standard is potentially the greatest source of error in a protein assay. The best choice for a standard is a purified, known concentration of the predominant protein found in the sample. This is not always possible or necessary; in some cases, a rough estimate of the total protein concentration in the sample is sufficient. If a highly purified version of the protein of interest is not available or if it is too expensive to use as the standard, the alternative is to choose a protein that will produce a similar color response curve with the selected protein assay method. For general protein determination, BSA works well as a protein standard because it is widely available in high purity and relatively inexpensive.

8. Protein assays are used to determine protein concentration by comparing sample assay response to that of a standard whose concentration is known. Protein samples and protein standards are processed in the same manner by mixing them with assay reagent and using a spectrophotometer to measure the absorbance.

9. When mixing solution in the cuvette, ensure that liquid does not splash out. Mix carefully by pipetting up and down. The volume of a cuvette is 50–2,000 μL.
10. Preserve sample quality by preparing the sample just prior to the chromatography step or storing samples in aliquots at −80°C. Do not expose samples to repeated thawing or freezing.
11. Forty-eight percent of brain proteins were eluted with 1–2 M NaCl. About 62% of the input (5 mg) was recovered in the flow-through and wash fractions.
12. We used a single salt step to elute bound proteins, thus avoiding the use of a lengthy linear gradient that results in the formation of large numbers of fractions and does not deliver efficient protein separation, as we had previously observed with heparin chromatography *(1)*.
13. Presence of interfering substances, like salt, residual buffers, detergents, and other charged small molecules, in the sample can cause insufficient focusing of protein bands, and should be removed or maintained at a low concentration. Removal of interfering substances can be performed by dialysis, gel filtration, or precipitation/resuspension. However, reversed-phase chromatography and ultrafiltration are the most efficient methods resulting in minimal sample loss. Dialysis is time consuming, requires a large volume of solution, and leads to the dilution of the sample. Gel filtration may be acceptable, but often results in appreciable protein loss. Precipitation/resuspension is quicker, but less effective for removing salts, and can also result in loss of proteins (especially peptides).
14. Electrophoresis is usually carried out using Bis–Tris polyacrylamide gels with a SDS/MES buffer system. However, many other types of gels are used successfully, including urea-polyacrylamide gels and nondenaturing polyacrylamide gels.
15. To avoid modification of proteins, never store samples for a long time after adding sample buffer. If samples are stored in a buffer containing urea for a longer time, urea can be hydrolyzed to isocyanate, which modifies proteins by carbamylation. Also do not heat proteins in the presence of urea, since this increases the reaction rate of carbamylation.
16. Gels can be left in destaining solution for up to 3 days without significant change in band intensity and background clarity. For long-term storage (over 3 days), keep the gel in a 0.01% sodium azide solution.

17. Comparison of the 2D PAGE results of analysis of proteins applied to, and eluted from, heparin columns shows that not all proteins can be visualized by Coomassie blue staining, leading thus to a lower number of identified proteins in comparison with those acquired with 1D PAGE and systematic identification by LC-MS/MS *(1)*.

References

1. Karlsson, K., Cairns, N., Lubec, G., and Fountoulakis, M. (1999) Enrichment of human brain proteins by heparin chromatography. *Electrophoresis* **20**, 2970–2976.
2. Shin, J. H., Krapfenbauer, K., and Lubec, G. (2006) Large-scale identification of cytosolic mouse brain proteins by chromatographic prefractionation. *Electrophoresis* **27**, 2799–2813.
3. Conrad, E. H. (ed.) (1998) *Heparin-binding proteins.* Academic, San Diego, CA.
4. Shin, J. H., Krapfenbauer, K., and Lubec, G. (2005) Column chromatographic prefractionation leads to the detection of 543 different gene products in human fetal brain. *Electrophoresis* **26**, 2759–2778.
5. Pollak, D., Krapfenbauer, K., Fountoulakis, M., Peyrl, A., and Lubec, G. (2004) Expressional pattern of known and predicted signaling proteins in seven human cell lines. *J. Chromatogr. B* **808**, 185–208.
6. Florens, L., and Washburn, M. P. (2006) Proteomic analysis by multidimensional protein identification technology. *Methods Mol. Biol.* **328**, 159–175.
7. Cagney, G., Park, S., Chung, C., Tong, B., O'Dushlaine, C., Shields, D. S., and Emili, A. (2005) Human tissue profiling with multidimensional protein identification technology. *J. Proteome Res.* **4**, 1757–1767.
8. Krapfenbauer, K., Fountoulakis, M., and Lubec, G. (2003) A rat brain protein expression map including cytosolic and enriched mitochondrial and microsomal fractions. *Electrophoresis* **24**, 1847–1870.
9. Fountoulakis, M., Juranville, J. F., Jiang, L., Avila, D., Roeder, D., Jakob, P., Berndt, P., Evers, S., and Langen, H. (2004) Depletion of the high-abundance plasma proteins. *Amino Acids* **27**, 249–259.
10. Lubec, G., Krapfenbauer, K., and Fountoulakis, M. (2003) Proteomics in brain research: potentials and limitations. *Prog. Neurobiol.* **69**, 193–211.
11. Bradford, M. M. (1976) A rapid and sensitive method for the quantitation of microgram quantities of protein utilizing the principle of protein-dye binding. *Anal. Biochem.* **72**, 248–254.

Chapter 12

Calmodulin-Binding Proteome in the Brain

Zhiqun Zhang, Firas H. Kobeissy, Andrew K. Ottens, Juan A. Martínez, and Kevin K.W. Wang

Summary

Calcium dyshomeostasis is involved in neuropathological conditions such as traumatic brain injury (TBI), stroke, and neurodegenerative diseases. Under such conditions in the brain, calmodulin (CaM), a Ca^{2+} sensor, mediates critical signaling functions through binding and regulating a diverse population of downstream targets referred to as calmodulin-binding proteins (CaMBPs). We developed a CaM-affinity capture method followed by reversed-phase liquid chromatography tandem mass spectrometry (RPLC-MSMS) to study the calcium-dependent CaM-binding proteome in rat brain. A total of 69 potential CaMBPs were identified by this proteomic technique, of which 26 were known CaMBPs and 43 were putative novel CaMBPs. This study shows that the CaM-affinity capture when coupled with tandem mass spectrometry may serve as an effective tool toward constructing a brain CaM-binding proteomic network. The general methods described here can be applied to study possible alternations of calmodulin-binding proteome in neurological, neurodegenerative, and psychiatric disorders.

Key words: Calmodulin, Calmodulin-binding protein, Mass spectrometry, Affinity chromatography, Protein interactions

1. Introduction

Calmodulin (CaM) is one of the major calcium (Ca^{2+}) sensors that are central in regulating multiple intracellular events. CaM exerts its regulatory functions through the modulation of a diverse number of CaM-binding proteins (CaMBPs) in a Ca^{2+}-dependent manner *(1–3)*. With the extraordinarily high concentration of CaM (10–100 μM) found in neurons of the central nervous system (CNS) *(4)*, there is significant interest in the downstream pathways involving CaM regulation. Several CaMBPs have

Andrew K. Ottens and Kevin K.W. Wang (eds.), *Neuroproteomics,* Methods in Molecular Biology, vol. 566
DOI 10.1007/978-1-59745-562-6_12,

already been characterized including adenyl cyclases (AC1 and AC8), calcineurin A, CaM-dependent protein kinase I, II, IV, neuronal nitric oxide synthase, and various calcium-ion channels. Known CaMBPs are involved in synaptic plasticity, learning, and memory *(4, 5)*; however, it is necessary to identify all unknown CaMBPs in order to fully construct a calcium–CaM-mediated protein network in the brain.

In the past, traditional molecular biology techniques, such as the calmodulin overlay technique (CaMBOT), have been used to isolate and identify individual CaMBPs *(6)*. A genetic and proteomic approach to screen protein–protein interaction-based expression libraries with CaM has proven to be a powerful way of identifying novel CaMBPs in *Arabidopsis, yeast, and humans (7–9)*. Those techniques have provided remarkable results; however, there are notable limitations. The laborious procedures of radioisotope labeling, isolation, and sequencing of cDNA encoding for CaMBPs and the critical requirement for efficiency and specificity of many steps (transcription, translation, fusion, etc.) limit the effective coverage of the CaM-binding proteome.

Recent developments in mass spectrometry technology, coupled to advances in the field of bioinformatics, have allowed us to investigate CaMBPs in a high-throughput manner. In this study we have exploited Ca^{2+}-dependent CaMBP interaction with CaM sepharose and the subsequent contrary calcium-chelating dissociation of CaMBPs from the sepharose to purify a CaM-binding subproteome. Following resolution by one-dimensional sodium dodecyl sulfate-polyacrylamide gel electrophoresis (1D-SDS-PAGE), the purified CaMBPs were profiled by RPLC-MSMS analysis *(10)*.

2. Materials

2.1. Brain Tissue Collection and Protein Extraction

1. Adult male Sprague–Dawley rats, average weight around 280–300 g (Harlan, Indianapolis, IN).
2. 1× Phosphate buffered saline (PBS) buffer, 3.2 mM Na_2HPO_4, 0.5 mM KH_2PO_4, 1.3 mM KCl, 135 mM NaCl, pH 7.4 (Cell Signaling Technology, Danvers, MA).
3. Small mortar and pestle set (Fisher Scientific).
4. Triton lysis buffer: 50 mM Tris–HCl (pH 7.4), 5 mM ethylenediamine tetraacetic acid (EDTA), 5 mM ethylene glycol tetraacetic acid (EGTA), 1% Triton X-100, and 1 mM dithiothreitol (DTT); store at 4°C (*see* **Note 1**).
5. Complete, Mini; Protease Inhibitor Cocktail Tablet (Roche Biochemicals, Indianapolis, IN) (*see* **Note 2**).

6. Bio-Rad DC Protein Assay Reagent Package contains reagents A, B, and S (Bio-Rad #500-0116).
7. 2 mg/ml Albumin standard (Pierce #23210 or equivalent).
8. Dry ice.

2.2. CaM Affinity Capture and Elution

1. Sepharose CL4B (Sigma, St. Louis, MO, CL4B200); store at 4°C (*see* **Note 3**).
2. CaM-agarose (Sigma, St. Louis, MO, P4385); store at 4°C (*see* **Note 4**).
3. Equilibration buffer (Buffer A): 20 mM Tris–HCl (pH 7.4); 150 mM NaCl and 1 mM DTT; store at 4°C.
4. 100 mM $CaCl_2$.
5. Binding buffer: Buffer A with 10 mM $CaCl_2$; store at 4°C.
6. Washing buffer: Buffer A with 1 mM $CaCl_2$; store at 4°C.
7. Elution buffer: Buffer A with 5 mM EDTA; store at 4°C.
8. Millipore YM-10 filter unit (Millipore, Bellerica, MA).
9. Novex 2× tricine–SDS sample buffer (LC1676, Invitrogen, Carlsbad, CA); store at 4°C.
10. Hamilton syringe (100 μL)

2.3. SDS-Polyacrylamide Gel Electrophoresis (SDS-PAGE)

1. 10× Tris/tricine/SDS: 100 mM Tris, pH 8.3, 100 mM tricine, 0.1% SDS (Bio-Rad).
2. Novex® 10–20% tricine gel 1.0 mm, 10 wells (Invitrogen).
3. Prestained molecular weight markers: Precision Plus Protein All Blue Standards (Bio-Rad); store at –20°C.

2.4. Coomassie Blue Staining

1. Staining solution: 40% methanol, 50% deionized water, 10% acetic acid (HOAc), and 0.25% (w/v) Coomassie Brilliant Blue R-250 (Bio-Rad).
2. Destaining solution: 40% methanol, 50% deionized water, and 10% HOAc.

2.5. In-Gel Digestion

1. 100 mM ammonium bicarbonate (NH_4HCO_3, 1.98 mg/mL) with 50% (v/v) acetonitrile.
2. 10 mM DTT (1.54 mg/mL) in 100 mM NH_4HCO_3.
3. 55 mM iodoacetamide (10.2 mg/mL) in 50 mM NH_4HCO_3.
4. 12.5 μg/mL trypsin in 50 mM NH_4HCO_3, pH 8 (*see* **Note 5**).
5. Promega Trypsin Gold (V5280); one vial contains 100-μg trypsin. Add 900 μL 50 mM HOAc to ten nonstick 1.5-mL tubes and chill to 4°C. Add 1 mL 50 mM HOAc to trypsin vial. Aliquot 100-μL trypsin solution to each chilled tube. Store at –70°C until use. Once a vial is taken from –70°C storage, aliquot 20 μL into 50 nonstick 600-μL tubes (on ice), and store at –20°C.

6. 50% Water and 50% acetonitrile (V/V).
7. Dry incubator set at 37 and 56°C.
8. SpeedVac™ concentrator or equivalent.

3. Methods

To profile the rat brain calmodulin-binding proteome, we utilized a CaM-affinity capture method combined with RPLC-MSMS-based proteomics. The specificity of the proteome methodology is derived from the Ca^{2+}-dependent CaM–CaMBPs interaction. Most of CaMBPs have reversible association with CaM in the presence of Ca^{2+}. When incubated with the Ca^{2+}chelator (EDTA), the majority of CaMBPs bound to the CaM-agarose will be specifically released producing an enriched CaMBP. Following separation by SDS-PAGE and Coomassie blue staining, the purified CaMBP proteome can be analyzed by RPLC-MSMS. The flowchart of the methods was summarized in **Fig. 1**.

3.1. Preparation of Brain Samples

1. Animal surgical procedures were conducted in compliance with the Animal Welfare Act and the University of Florida Institutional Animal Care and Use Committee (IACUC) and the National Institutes of Health guidelines detailed in the Guide for the Care and Use of Laboratory Animals. Adult male (280–300 g) Sprague–Dawley rats were initially anesthetized with 4% isoflurane in a carrier gas of O_2 (4 min) followed by maintenance anesthesia of 2.5% isoflurane in the same carrier gas until the animals were euthanized by decapitation.
2. Cortex tissue was removed, rinsed with ice-cold PBS, and halved. Brain tissues were rapidly dissected, rinsed in ice-cold PBS, and snap-frozen in liquid nitrogen.

3.2. Protein Exaction with Triton-Lysis Buffer

1. Chill mortar, pestle, and spatula in bucket with ample dry ice for at least 20 min. Remove brain sample from Eppendorf tube and grind with a pestle over dry ice to a fine powder.
2. The pulverized brain tissue powder was then lysed for 90 min at 4°C in Triton lysis buffer with complete protease inhibitor cocktail (*see* **Note 6**). Vortex the brain cortex lysate every 15 min and then centrifuge at 15,000 × *g* for 10 min at 4°C.
3. Transfer the supernatant to a new tube and perform a DC protein assay (Bio-Rad, Hercules, CA) to determine protein concentration (*see* **Note 7**).

3.3. CaM-Agarose Affinity Purification

1. Before CaM-agarose purification, the CaM-agarose and sepharose CL4B need to be washed thoroughly with ten volumes

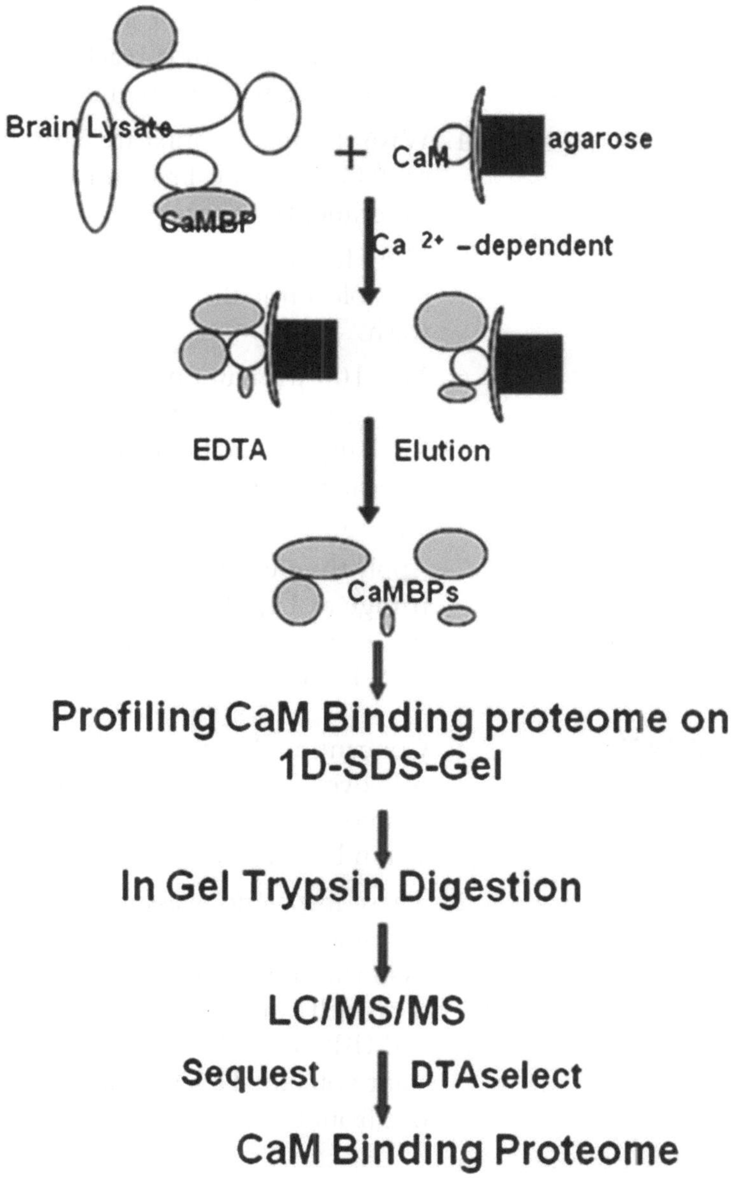

Fig. 1. Schematic of calmodulin-affinity purification RPLC-MSMS methodology to study the rat brain calmodulin-binding proteome. Reprinted in part from *(10)* with permission from Landes Bioscience, copyright 2006.

of buffer A (*see* **Note 8**). Here, sepharose CL4B and CaM-agarose with plain lysis buffer are used as quality controls.

2. Carefully pour 600 µL (50% slurry) of CaM-agarose or sepharose CL4B along with a small stir bar into a flat 2-mL tube on a stirring plate. Transfer 400 µL (2.25 mg/mL) of rat brain lysate to the tube with the slurry once the stir bar begins to stir at low speed.

3. Slowly add 110-μL $CaCl_2$ stock (100 mM) to the mixture drop by drop until it reaches 10 mM (*see* **Note 9**). Incubate the mixture with gentle mixing at 4°C for 4 h.
4. Wash the agarose mixture with the washing buffer eight times at 4°C; keep the first and last flow through for the control experiments.
5. Use Hamilton syringe to aspirate the supernatant as much as possible until the agarose becomes semidry during the last wash.
6. Mix 100-μL elution buffer that contains 5 mM EDTA with the agarose gently for 5 min and then leave it on the bench for 1 min. Use Hamilton syringe to aspirate the supernatant. Repeat this step once. Collect all the eluate and dilute with equal volume of Buffer A.
7. Apply the elution on the surface of Millipore YM-10, then centrifuge at 14,000 × *g* for 1 h at 4°C until the volume reduces to 20 μL. The concentrated eluate can now be analyzed by SDS-PAGE.

3.4. SDS-PAGE

1. Concentrated samples are mixed with an equivalent volume of Novex 2× tricine–SDS sample buffer. The samples are then heated for 90 s in a boiling water bath and centrifuged for 1 min before loading.
2. These instructions assume the use of Invitrogen Novex Mini gel system. Fill the buffer chamber with 1× Tris/tricine/SDS running buffer. Wash the wells before loading the samples with running buffer. Carefully load 30 μL concentrated CaMBPs into a 10–20% tricine gel sample well, and run at constant voltage (120 V) until the tracking dye has just migrated out from the lower slot (*see* **Note 10**).

3.5. Coomassie Blue Staining

1. After the electrophoresis is finished, carefully move the gel to a new plastic container and cover it to exclude airborne contaminants. Rinse the gel twice with 100-mL distilled water (5 min each wash) (*see* **Note 11**).
2. Fixing and staining of gel is accomplished by soaking the gel in a staining solution for 1 h on a tabletop shaker. The gel is then destained by soaking in the destaining solution on the shaker with periodical replacing of the destaining solution until desired contrast is achieved.
3. Store the gel in 10% acetic acid at 4°C until the next step (purified CaM-binding proteome is shown in **Fig. 2**).

3.6. In-Gel Digestion for RPLC-MSMS Analysisa

1. Wash the gel with water three times (5 min each wash) before excising the gel. Excise each visible band from the stained gel with a razor, cut it into four 1-mm cubes and place into a 0.5-mL microcentrifuge tube. Wash the gel cubes three times (5 min per wash) with 200-μL water. Use gel loading pipette tips to remove the solution and discard.
2. Add 200 μL of 50% 100 mM ammonium bicarbonate/50% acetonitrile and vortex 1 h or place overnight on a shaker. Remove the solution and discard. Repeat this wash one additional time (1 min). Resulting gel particles should be clear. Dehydrate gel pieces by adding 100 μL of 100% acetonitrile. At this point the gel pieces should shrink and become an opaque-white color. Remove acetonitrile and speed vacuum for 10–20 min.
3. Rehydrate the cubes with 30 μL of freshly prepared 10 mM DTT solution and incubate for 30 min at 56°C.
4. Replace the DTT solution with roughly the same volume of freshly prepared 55 mM iodoacetamide in 50 mM ammonium bicarbonate, and incubate for 30 min in the dark at room temperature.
5. Remove the iodoacetamide solution and wash the gel pieces with water. Remove water and dehydrate with 100% acetonitrile. The gel pieces should shrink and become an opaque-white color. If they do not, remove the acetonitrile and repeat the washing–dehydration cycle until they do.
6. Remove the acetonitrile and air-dry the gel pieces for 5–10 min.
7. Rehydrate gel particles in 15 μL of a 12.5 ng/μL trypsin solution (or volume required to cover the expanding gel pieces) and place on ice for 30 min.
8. Remove excess trypsin solution and overlay the rehydrated gel particles with 20 μL of 50 mM ammonium bicarbonate to keep them immersed throughout digestion. Incubate the gel cubes overnight at 37°C.
9. Peptides are extracted in water and 50% water/50% acetonitrile sequentially as follows: Add 100 μL of water, vortex (time), and spin the tube (time–speed). Transfer the digest solution to a clean siliconized tube. Repeat the process with 50% water/50% acetonitrile.
10. Dry peptide-containing solution by speed vacuum and resuspend in mobile phase solution for capillary RPLC-MSMS.

3.7. Capillary RPLC-MSMS-Based Protein Identification

Capillary RPLC-MSMS-based protein identification was performed as described in the previous chapter (*see* Chapter 13) and our previous work *(11)*. Briefly, sample digests (2 μL) were

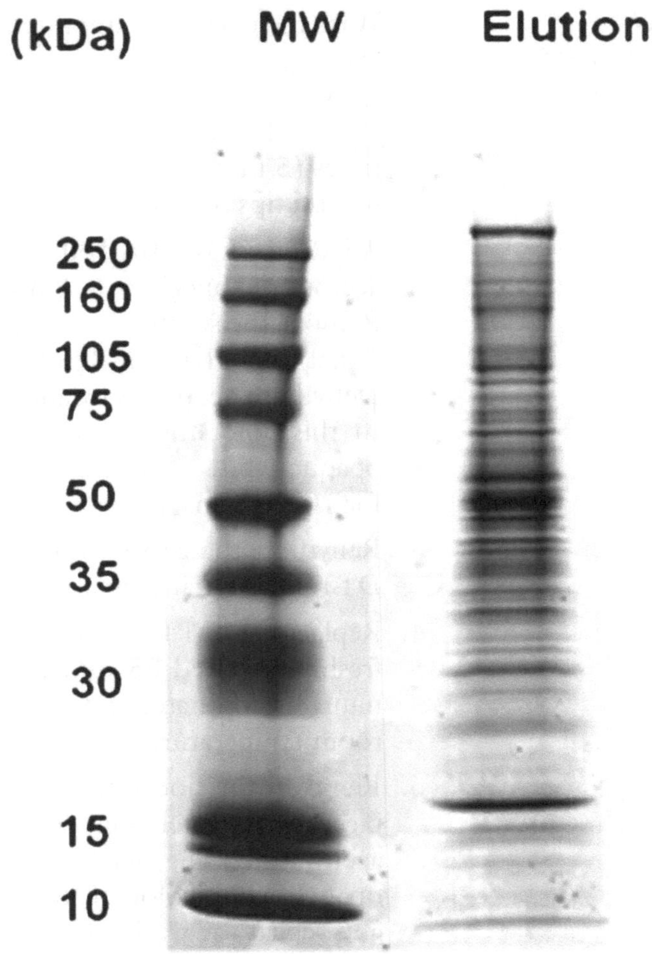

Fig. 2. Coomassie-stained gel of calmodulin-binding rat brain proteome captured by affinity chromatography as indicated in the **Subheading 3**. *Left lane* is molecular weight markers; *right lane* is rat brain calmodulin-binding proteome profile. Reprinted in part from *(10)* with permission from Landes Bioscience, copyright 2006.

loaded via an autosampler onto a 100-μm × 5-cm C-18 reverse phase capillary column at 1.5 μL/min. Peptide elution was performed by linear gradient: 5–60% methanol in 0.4% acetic acid over 30 min at 200 nL/min. Tandem mass spectra were collected in data-dependant mode (three most intense peaks) on a Thermo Electron LCQ Deca XP plus ion trap mass spectrometer. Tandem mass spectra were searched against an NCBI rat indexed RefSeq protein database using Sequest. Filtering and sorting was performed with DTAselect software by peptide number and Sequest cross-correlation values (X_{corr} values of 1.8, 2.5, 3.5 for +1, +2, +3 charge states, respectively) *(12)*. Peptides filtered and sorted by DTAselect were assigned to specific protein accession numbers [National Center for Biotechnology Information (NCBI)]. A total of 69 potential CaMBPs were identified by this proteomic technique, of which 26 were known CaMBPs and 43 were putative novel CaMBPs *(10)*. The general method can be applied to neurodisorders in which there might be a defect in the calmodulin-binding proteome, such as traumatic brain injury (TBI) or Alzheimer's disease *(10, 13, 14)*.

4. Notes

1. EGTA does not dissolve quickly; vortex and add NaOH to adjust pH 7.5–8.0. Add fresh DTT when preparing the lysis buffer.
2. One protease inhibitor cocktail tablet per 10-mL lysis buffer.
3. Sepharose CL4B served as control, non-CaMBP.
4. The calmodulin agarose needs to be washed with ten volumes of Buffer A thoroughly before use. Calmodulin-binding capacity of agarose will vary with batch. Make sure that the manufacturer provides specific information to determine protein-to-agarose ratio for optimum binding.
5. Trypsin quality is very important. We recommend aliquoting Promega stock trypsin for single use whenever trypsin bicarbonate solution is prepared.
6. The amount of lysis buffer required depends on amount of brain tissue obtained. We recommend a lysis buffer volume equivalent to 4× the brain powder obtained.
7. For DC protein assay we recommend a 1:5–1:10 dilution of lysate to fall within the linear range of assay (0.0125–2 mg/mL).
8. To minimize the damage to the agarose use wide-bore pipette tips during wash. Simple nick of pipette tip with scissors to expand bore size will do fine.
9. Because the endogenous CaM in samples reduces the efficiency of affinity capture of brain CaMBDPs to CaM agarose, it is necessary to dissociate CaM–CaMBPs complexes in the presence of EDTA and EGTA. The lysate should be mixed with excess amount of CaM-agarose followed by additional excess $CaCl_2$ (10 mM) to allow competitive association of CaMBPs to CaM agarose. The optimal excess amount of CaM-agarose for CaM affinity capture appears to be 100-µg CaM-immobilized agarose per 300-µg rat brain protein lysate.

10. Whenever possible, lanes 1 and 10 should be left empty. It is also helpful to load the samples in an intermittent manner, which makes it easier to cut the gel and avoid contamination for the mass spectrometry study.
11. Always wear gloves throughout all steps. Unless stated otherwise, type I water should be used throughout the procedure.

Acknowledgments

We wish to thank Stephen F. Larner for insightful discussion and editing assistance. This work was supported by the Department of Defense grant DAMMED-03-1-0066, and NIH grants R01 NS049175-01-A1 and R01 NS051431.

References

1. Haeseleer, F., Palczewski, K. (2002) Calmodulin and Ca^{2+}-binding proteins (CaBPs): variations on a theme. *Adv Exp Med Biol.* **514**, 303–317.
2. Bouché, N., Yellin, A., Snedden, W.A., Fromm, H. (2004) Plant-specific calmodulin-binding proteins. *Annu Rev Plant Biol.* **56**, 435–466.
3. Benaim, G., Villalobo, A. (2002) Phosphorylation of calmodulin. Functional implications. *Eur J Biochem.* **269**, 3619–3631.
4. Xia, Z., Storm, D.R. (2005) The role of calmodulin as a signal integrator for synaptic plasticity. *Nat Rev Neurosci.* **6**, 267–276.
5. Xia, Z., Choi, E.J., Wang, F., Blazynski, C., Storm, D.R. (1993) Type I calmodulin-sensitive adenylyl cyclase is neural specific. *J Neurochem.* **60**, 305–311.
6. O'Day, D.H. (2003) CaMBOT: profiling and characterizing calmodulin-binding proteins. *Cell Signal.* **15**, 347–354.
7. Reddy, V.S., Ali, G.S., Reddy, A.S. (2002) Genes encoding calmodulin-binding proteins in the Arabidopsis genome. *J Biol Chem.* **277**, 9840–9852.
8. Zhu, H., Bilgin, M., Bangham, R., Hall, D., Casamayor, A., Bertone, P., Lan, N., et al. (2001) Global analysis of protein activities using proteome chips. *Science* **293**, 2101–2105.
9. Shen, X., Valencia, C.A., Szostak, J.W., Dong, B., Liu, R. (2005) Scanning the human proteome for calmodulin-binding proteins. *Proc Natl Acad Sci USA* **102**, 5969–5974.
10. Zhang, Z., Ottens, A.K., Golden, E.C., Hayes, R.L., Wang, K.K.W. (2006) Using calmodulin-affinity capture to study the rat brain calmodulin binding proteome and its vulnerability to calpain and caspase proteolysis. *Calcium Binding Proteins* **2**, 125–134.
11. Ottens, A.K., Kobeissy, F.H., Wolper, R.A., Haskins, W.E., Hayes, R.L., Denslow, N.D., Wang, K.K.W. (2005) A multidimensional differential proteomic platform using dual-phase ion-exchange chromatography – polyacrylamide gel electrophoresis/reversed-phase liquid chromatography tandem mass spectrometry. *Anal Chem.* 77, 4836–4845.
12. Tabb, D.L., McDonald, W.H., Yates, J.R. (2002) DTASelect and contrast: tools for assembling and comparing protein identifications from shotgun proteomics. *J Proteome Res.* **1**, 21–26.
13. Wang, K.K.W., Villalobo, A., Roufogalis, B.D. (1989) Calmodulin-binding proteins as calpain substrates. *Biochem J.* **262**, 693–706.
14. O'Day, D.H., Myre, M.A. (2004) Calmodulin-binding domains in Alzheimer's disease proteins: extending the calcium hypothesis. *Biochem Biophys Res Commun.* **320**, 1051–1054.

Part III

Neuroproteomic Methodology and Bioinformatics

Chapter 13

Separation of the Neuroproteome by Ion Exchange Chromatography

Brian F. Fuller and Andrew K. Ottens

Summary

Ion exchange chromatography is a fractionation technique applicable to the separation of brain-derived proteins based on charge. Proteome complexity overwhelms analytical approaches, which is mitigated by fractioning samples into simpler solutions. In this chapter we will cover the design, optimization, and execution of an ion exchange experiment for the separation of a brain lysate. Furthermore, helpful tips and pitfall will be revealed to aid with potential problems that may arise. The discussed proteomic methodology is applicable to multidimensional separations ahead of bottom–up or top–down proteomic strategies for characterizing the neuroproteome.

Key words: Ion exchange chromatography, Neuroproteomics, Proteins separation, Brian, Sample fractionation

1. Introduction

The human proteome is vast and dynamic. For example, the basic human proteome contains 20,000–25,000 predicted protein-coding genes *(1)*. Each gene can produce multiple transcripts, with over 250,000 estimated in humans. Further complexity arises with posttranslational modifications *(2)*. Modifications such as phosphorylation, glycosylation, oxidation, and proteolysis are just a few of the many modifications that can occur for a given protein. The proteome is dynamic, ever changing to adapt and regulate various metabolic functions. Neuroproteomics focuses on the largest fraction of an organism's overall proteome, containing a plethora of proteins *(3)*. Separation science is essential for further study of the neuroproteome.

Andrew K. Ottens and Kevin K.W. Wang (eds.), *Neuroproteomics,* Methods in Molecular Biology, vol. 566
DOI 10.1007/978-1-59745-562-6_13,

This chapter will focus on separation of the neuroproteome by ion exchange chromatography, which separates proteins based on charge distribution. Ion exchange chromatography is a popular prefractionation technique used to reduce sample complexity ahead of further analysis *(4, 5)*. Recent advances in ion exchange chromatography have worked to combine both anion and cation phases in a single separation run similar to two-dimensional gel electrophoresis *(6)*. This novel use of ion exchange chromatography has been previously utilized in a study of traumatic brain injury *(7)*.

The following section outlines procedures for the preparation and separation of brain proteomes by ion exchange chromatography. Emphasis has been placed on design, optimization, and troubleshooting of an ion exchange separation experiment.

2. Materials

1. Cell lysis buffer: 20 mM MOPS, pH 7.0, 3 mM EDTA, 2 mM EGTA, 1 mM DTT; add one fresh protease inhibitor tablet (Roche Life Science, Indianapolis, IN) per 10 mL. Store at 4°C (*see* **Note 1**).
2. Mobile Phase A: 20 mM MOPS, pH 7.0, 1 mM EDTA, 0.1% AcN.
3. Mobile Phase B: 20 mM MOPS, pH 7.0, 1 mM EDTA, 0.1% AcN, 0.5 M NaCl (*see* **Note 2**).
4. Columns: Dionex ProPac® SCX-10 and SAX-10 (2× 250 mm).
5. Standards: BioRad (Hercules, CA) Protein Standards for Anion/Cation Exchange Chromatography.

3. Methods

3.1. Brain Sample Preparation

1. Add snap-frozen brain tissue into a mortar precooled with dry ice or liquid nitrogen.
2. Add liquid nitrogen to the mortar and grind the hard tissue into a fine powder.
3. Add 1 mL of cell lysis buffer to 100 mg of powder.
4. Allow tissue to lyse on a rocker at 4°C for 90 min, vortexing every 20 min.

5. Centrifuge lysate at 17,000 × *g* for 4 min at 4°C.
6. Transfer supernantant to a new tube.
7. Perform protein assay per manufacturer's instructions.
8. Sample can be used immediately or stored at −80°C.

3.2. Column Selection

Ion exchange columns for protein separations are commercially available, although the selection is somewhat limited. Column selection will determine the resultant separation and is discussed later.

1. Casing material and dimensions are important considerations when selecting a column. Most columns will be packed in either stainless steel or polyetheretherketone (PEEK™) tubes. For most ion exchange experiments, PEEK tubing is recommended for compatibilities with concentrated salts. The diameter of a column will directly affect the column's capacity. For protein loads of ≤100 μg, an inner diameter (ID) of 2 mm is sufficient. For larger protein loads ≥1 mg a larger 7–10 mm ID would be required. Column length affects the efficiency by which proteins are separated. Longer columns allow for a greater period of time for the interaction of the lysates and functional groups to increase separation, but require more experimentation time. Select a longer column if greater sample fractionation is desired.
2. Particle size and porosity are important criteria to be examined when selecting a column. Ion exchange columns contain an insoluble matrix coated with a specific functional group, usually in the form of packed particles. Porous particles will allow for greater surface area for protein/functional group interaction and will increase the capacity of the column. Pore size should be 500 Å or larger to allow the inclusion of proteins into the particles. Particle diameter will affect the back pressure on the column and separation efficiency. Generally smaller particles provide better separation, but this is limited by the pressure and flow-rate capabilities of the LC pump.
3. The functional group employed largely defines the separation nature of the column. Ion exchange columns are divided into cation or anion exchangers, selected to either retain positively (cation) or negatively (anion) charged protein motifs, respectively, though mixed mode columns are possible *(6, 8)*. Functional groups are further broken down into strong and weak exchangers, which refer to the exchangers' ability to accept variations in pH and ionization, not the strength of the interaction between the proteins and functional groups. Strong ion exchanges are preferable over weak exchangers since they are more resistant to variations in pH.

4. The last aspect related to column selection is to pick the buffer system (mobile phase). The choice of a suitable buffer will depend primarily on whether an anion or cation exchange column is being used. Buffers suitable for anion exchange chromatography will have a neutral to basic pH, and the counter ion is usually a halide such as the chloride anion. Cation exchange buffers will have an acidic to neutral pH, and the counter ion will be an alkylide anion such as Na^+ or K^+. Buffer concentration should be between 20 and 50 mM. A list of common buffer choices can be found later (**Table 1**). Buffers near physiological pH are recommended to maintain native solubility of sample proteins.

3.3. LC Method Design

The LC method is a program that dictates the operation of the LC system. It controls the mobile phase composition and flow rate as well as triggering of events in the experiment. Program an LC with the following five steps. The flow rate is ordinarily recommended by the column manufacturer.

1. Isocratic step: used to equilibrate the column with mobile phase at a low concentration of salt <5 mM (2–3 min).

Table 1
Common ion exchange buffers

Substance	pK_a (25°C)	Conc. (mM)	Counter ion
Anion exchange buffers			
MOPS*	7.20	20	Cl^-
Tris*	8.06	20	Cl^-
Bis-Tris	6.46	20	Cl^-
Phosphate	12.33	20	Cl^-
Triethanolamine	7.76	20	Cl^-
Cation exchange buffers			
MOPS*	7.20	20	Na^+
HEPES	7.55	50	Na^+
Acetic acid	4.76	50	Na^+/Li^+
Formic acid	3.75	50	Na^+/Li^+
Phosphate	7.20	50	Na^+

*Recommended buffers for initial experimentation

2. Injection step: places sample online with the flow (*see* **Note 3**).
3. Gradient step: used to elute protein from the column, a linear gradient from 5 to 500 mM salt (10–45 min). Refinement is added in **Subheading 3.6** (*see* **Note 4**).
4. High salt step: raises the salt concentration to 900 mM to clean out contaminants that did not elute during the run (1–3 min).
5. Reequilibration step: returns the salt concentration to starting conditions (2–5 min).

3.4. Column Characterization and Quality Control

1. Reconstitute the appropriate ion exchange standard with 2 mL of mobile phase A.
2. Dilute 200 μL of the standard 1:5 with 800 μL of mobile phase A.
3. Run a blank sample (no protein) of mobile phase A.
4. Subsequently, load 25 μL of the standard onto the LC system.
5. Run the LC method (**Subheading 3.3**) with the UV detector set to 280 nm (*see* **Note 5**).
6. Calculate the peak width at half-maximum (PWHM) for each peak (*see* **Note 6**).
7. Save a copy of the chromatogram for future reference.

Using the standards will allow for examination of column separation efficiency. For quality control purposes, the run should be repeated either daily or weekly depending upon usage. PWHM and elution time values can be compared between runs to determine any change in the columns' performance. These values relate to column efficiency, which can be evaluated by the number of theoretical plates (N) for a column. N is determined by taking the elution time of the protein peak divided by PWHM, squaring the result, and multiplying by 5.54. Further, column length (L) divided by N will give the plate height (H), which can be plotted against mobile phase velocity to find the optimal flow rate for the column (**Fig. 1**).

3.5. Fraction Collection

1. Calculate the average PWHM from the column characterization run (*see* **Subheading 3.4**).
2. Multiply the average PWHM in minutes by the flow rate to obtain the fraction volume, and program this value into your LC method.
3. Calculate the delay due to the void volume of your system as the time postinjection to the start of the myoglobin elution. Use this value to offset the start and stop of the fraction collection relative to the gradient step.

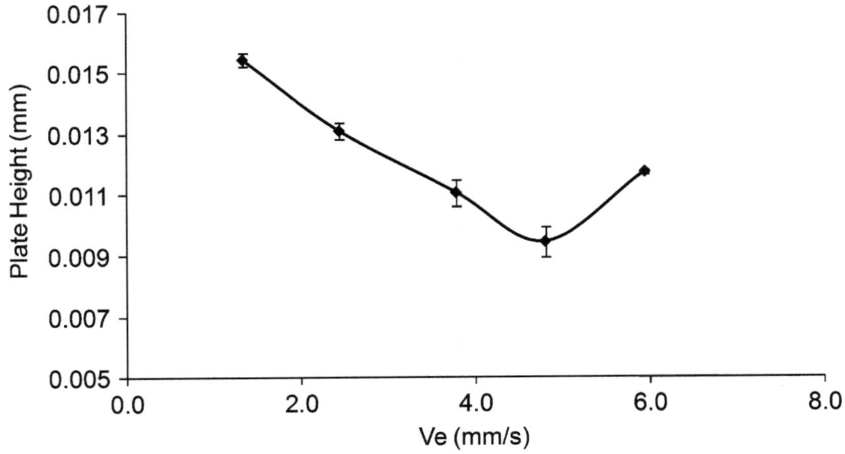

Fig. 1. Flow rate optimization. Shown is a Van Deemter plot calculated from the ribonuclease peak of the cation exchange standards. We optimize flow rate for an analytical column that was 2 mm by 250 mm with a 10-μm particle size. Velocity of the mobile phase (Ve) was calculated by dividing the column length (*L*) by elution time of the first, unretained protein peak (myoglobin) eluted at a given flow rate. The lowest point in the plot reveals the most effiecent flow rate.

3.6. Optimization

1. Run a test brain sample using the method from **Subheading 3.4**.
2. Collect the fractions and concentrate them with a centrifugal filter device with a molecular weight cut-off of 10 kDa.
3. Visualize the protein content of each fraction via SDS-PAGE, stained with Coomassie blue.
4. Using the void volume delay time from **Subheading 3.5**, align each fraction with a gradient elution time and associated salt concentrations.
5. Identify fractions of high and low protein density for optimization.
6. Design a multistep gradient that will reduce the slope of the gradient over regions of high protein density and increase the slope of the gradient for areas of low protein density, based on the correlated time and salt concentration for the identified fractions (**Fig. 2**).
7. Repeat the experiment with the same sample using the optimized method.
8. Make any necessary adjustments to the method.
9. Run samples of interest using the optimized method.

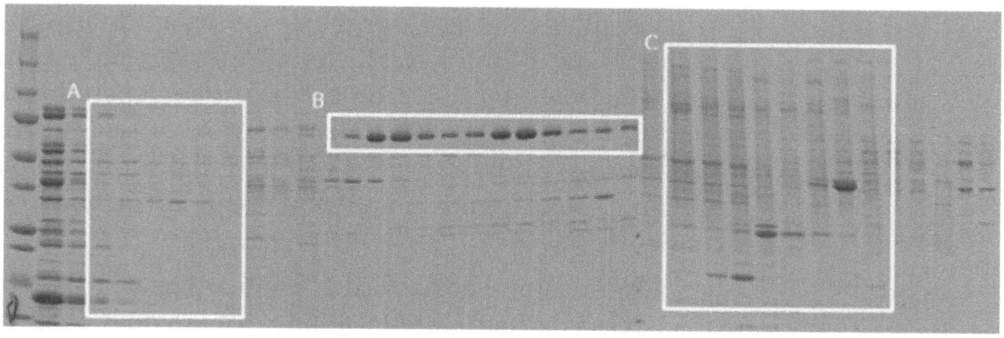

Fig. 2. Deer mouse brain lysate seperated by mixed bed ion exchange chromatography. In this run there are three main areas in need of optimization. The fractions boxed as A have low protein density. The gradient must be steeper to allow the proteins to elute faster over less fractions. The fractions in box B are of low overall density with a single dominant protein. These fractions could be pooled for further analysis. The fractions in box C are of high protein density and must be further resolved by reducing the gradient slope.

3.7. Conclusions

The presented methodology allows for the design and execution of a proteome separation experiment via ion exchange chromatography. Combined with other separations, for example, gel electrophoresis, ion exchange is a powerful tool to separate a proteome. The methodology is useful as a first-dimension fractionation step ahead of 1D and 2D gel electrophoresis, shotgun and LC-MSMS analysis, and for top–down intact protein experiments.

4. Notes

1. A suitable lysis buffer for ion exchange chromatography must contain a very minimal amount of salt, for example, NaCl <5 mM, or ionic detergents, for example, SDS <0.001%. Low concentrations <1% of nonionic detergents can be used but must be tested first for their effects on column performance.
2. The buffers listed in **Subheading 2** were utilized in the experiments for **Figs. 1** and **2**. Selecting a buffer is discussed in **Subheading 3.2** and summarized in **Table 1**.
3. The amount of time needed for the injection step is determined by the volume of the loop and the flow rate of the system. Most injection methods fill the entire loop with sample. Partial loop injection is an option that allows you to vary the amount of sample loaded without having to change the loop. While this can be convenient when you are performing multiple, simultaneous runs it is less precise.

4. A linear gradient has a constant slop value. Starting at 5% mobile phase B is suggested for consistent gradient formation. The final concentration can be varied depending on the sample. A suggested final concentration of salt is found in **Subheading 3.3**; however, if you find that some proteins do not elute, increase the % mobile phase B at the height of the gradient.
5. In solution, amino acids with aromatic rings will absorb ultra violet (UV) light at 280 nm. While this provides an imprecise measure of protein content, it is a quick way to determine relative protein abundance.
6. The measured PWHM is easily converted to a value of two standard deviations from the center of the Gaussian chromatographic peak. Experimental peaks can be affected by tailing at their base. By multiplying PWHM by 5.55 you adjust peak width to standard deviation (σ). 4σ encompasses 95% of the peak area. This is adequate for determining fraction volume and plate height.

References

1. International Human Genome Sequencing Consortium (2004) Finishing the euchromatic sequence of the human genome. *Nature* **431**, 931–945.
2. Garavelli, J. S. (2004) The RESIS database of protein modifications as a resource and annotation tool. *Proteomics* **4**, 1527–1533.
3. Becker, M., Schindler, J., and Nothwang, H. G. (2006) Neuroproteomics – the tasks lying ahead. *Electrophoresis* **27**, 2819–2829.
4. Shin, J., Krapfenbauer, K., and Lubec, G. (2006) Large-scale identification of cytosolic mouse brain proteins by chromatographic prefractionation. *Electrophoresis* **27**, 2799–2813.
5. Sakamoto, T., Uezu, A., Kawauchi, S., Kuramoto, T., Makino, K., Umeda, K., Araki, N., Baba, H., and Nakanishi, H. (2008) Mass spectrometric analysis of microtubule co-sedimented proteins from rat brain. *Genes Cells* **13**, 295–312.
6. Ottens A. K., Kobeissy, F. H., Wolper, R. A., Haskins, W. E., Hayes, R. L., Denslow, N. D., and Wang, K. K. (2005) A multidimensional differential proteomic platform using dual-phase ion-exchange chromatography-polyacrylamide gel electrophoresis/reverse-phase liquid chromatography tandem mass spectrometry. *Anal Chem* **77**, 4836–4845.
7. Kobeissy, F. H., Ottens, A. K., Zhang, Z., Liu M. C., Denslow, N. D., Dave, J. R., Tortella, F. C., Hayes, R. L., and Wang, K. K. (2006) Novel differential neuroproteomics analysis of traumatic brain injury in rats. *Mol Cell Proteomics* **10**, 1887–1898.
8. El Rassi, Z. and Horvath. (1986) Tandem columns and mixed-bed columns in high-performance liquid chromatography of proteins. *J Chromotogr* **359**, 255–264.

Chapter 14

iTRAQ-Based Shotgun Neuroproteomics

Tong Liu, Jun Hu, and Hong Li

Summary

Shotgun proteomics involves the analysis of peptides obtained by enzymatic digestions of proteins and subsequent identification via tandem mass spectrometry. This approach is an effective method for studying global protein expression in neuronal systems. The method described here is a quantitative shotgun neuroproteomics method using amine-specific isobaric tags for a relative and absolute quantitation (iTRAQ)-based workflow. We will provide the technical details for sample preparation, two-dimensional liquid chromatography, tandem mass spectrometry, database search, and statistical analysis to identify differentially expressed proteins. We will use a recent study on a rat model of multiple sclerosis, experimental autoimmune encephalomyelitis to illustrate the successful application of this method.

Key words: Shotgun proteomics, Liquid chromatography, Tandem mass spectrometry, iTRAQ, EAE, Multiple sclerosis, Posttranslational modification

1. Introduction

Shotgun proteomics refers to the direct analysis of complex peptide mixtures to rapidly generate a global expression profile of the proteins within complex systems *(1)*. It is an effective tool for the identification and quantification of a large number of proteins by a workflow sequentially involving 2-dimensional (2D) liquid chromatography (LC), tandem mass spectrometry (MS/MS), and protein database-searching algorithms. In recent years, it has received increasing attentions for neuroproteomics analysis, and has uncovered key proteins involved in the neuronal dysfunction and an inflammatory response associated with neurodegenerative disorders *(2, 3)*. The shotgun approaches, based on 2D-LC-MS/MS, have advantages over 2D gel-based

Andrew K. Ottens and Kevin K.W. Wang (eds.), *Neuroproteomics,* Methods in Molecular Biology, vol. 566
DOI 10.1007/978-1-59745-562-6_14, © Humana Press, a part of Springer Science+Business Media, LLC 2009

proteomics technologies in terms of both detection sensitivity and proteome coverage *(4)*. Quantification of differentially expressed proteins using the shotgun methods can be divided into either labeled or label-free approaches. Several chemical labeling technologies have emerged recently, including ^{18}O-labeling *(5)*, stable isotope-labeled amino acids in culture (SILAC) *(6)*, iTRAQ *(7, 8)* and isotope-coded affinity tag (ICAT) *(9)* methods. Label-free quantification approaches include LC/MS quantification of MS/MS precursors and spectral counting *(10)*. Both approaches have their strengths and limitations. Chemical labeling allows better quantification precision, but suffers from the potential for incomplete peptide labeling or unintended side reactions. Label-free methods are simpler, but quantification outcomes are less precise than labeled methods.

iTRAQ technology provides a means to quantify up to eight different samples simultaneously. In our experimental design, we used four labeling reagents consists of an *N*-methylpiperazine quantification group, a carbonyl stable-isotope mass balance group, and a hydroxyl succinimide ester group that reacts with the primary amines on peptide N-termini and lysine side chains *(11)*. Peptide quantification is based on the relative abundance of the four quantification reporter ions (*m/z* 114, 115, 116, and 117) generated through MS/MS fragmentation of iTRAQ-labeled peptide mixtures. Since iTRAQ reagents are efficient at labeling nearly all peptides *(11)*, this method is more effective at providing higher protein identification sequence coverage than the ICAT method. In addition, the peptide b and y-series ions derived from all four samples are combined in MS/MS spectra, resulting in better ion statistics for more accurate peptide identifications *(12)*.

We will present here an iTRAQ-based shotgun neuroproteomics method that we have used to study EAE, an animal model of multiple sclerosis, with inflammatory and demyelinating symptoms of the central nervous system. Although the pathological manifestations of EAE are well defined, the molecular mechanisms underlying neural deficits remain elusive. To discover novel therapeutic targets for multiple sclerosis, we conducted a proteomics study of EAE animals (*see* **Fig. 1**). Proteins extracted from four rat lumbar spinal cords were first digested with trypsin. Four iTRAQ reagents were utilized to label two control and two EAE samples. Equal amounts of the labeled peptides were combined and separated using strong cation exchange LC (SCXLC) and subsequent reversed-phase LC (RPLC). The resulting peptide fractions were spotted onto matrix-assisted laser desorption/ionization (MALDI) plates and analyzed on a tandem mass spectrometer. Bioinformatic analysis of the MS/MS spectra enabled us to identify 41 differentially expressed proteins in EAE rat spinal cords *(3)*.

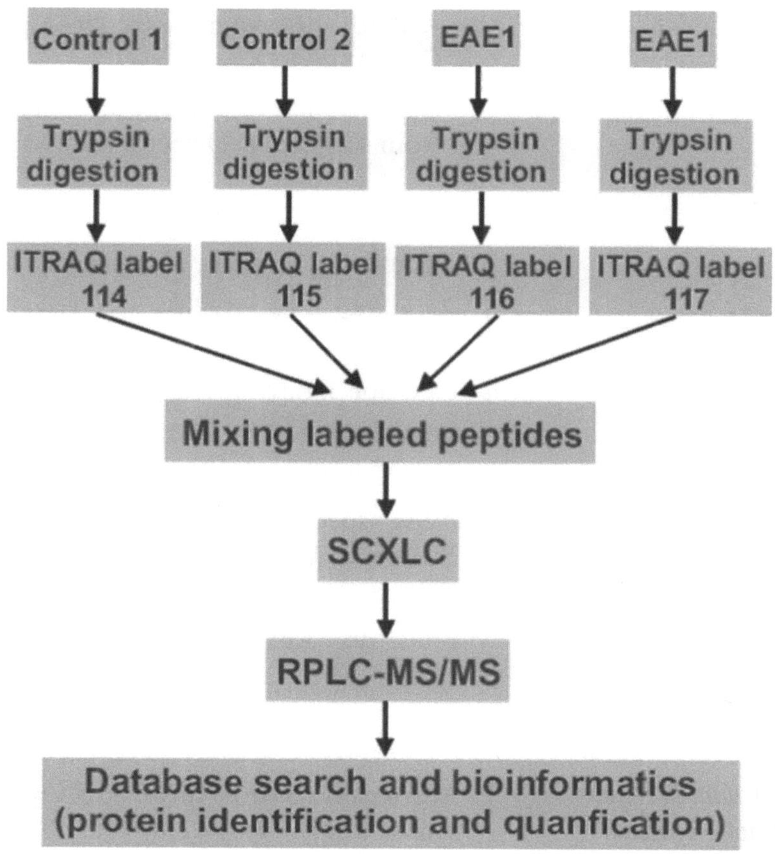

Fig. 1. iTRAQ-based shotgun neuroproteomics workflow. Proteins extracted from the rat lumbar spinal cords are first digested into peptides by trypsin. Two control samples are labeled with the iTRAQ reagents 114 and 115, and two EAE samples are labeled with the iTRAQ reagents 116 and 117. The labeled peptides are combined and separated using SCXLC and RPLC and analyzed on a tandem mass spectrometer. Database search and bioinformatics procedures are used for protein identification and quantification.

2. Materials

2.1. Protein Extraction

1. Dissect lumbar spinal cords from two adjuvant-treated controls and two EAE rats. Rinse well to remove blood completely.
2. Lysis buffer: 25 mM triethylammonium bicarbonate (TEAB), 20 mM Na_2CO_3 and 0.1% (v/v) of protease inhibitor cocktail (Sigma, St Louis, MO), pH 10.0 (*see* **Note 1**).
3. An ultrasonic homogenizer with a microtip that fits a typical 1.5 mL eppendorf tube (e.g., Omni Ruptor 250 ultrasonic homogenizer with a 5/32″ solid titanium microtip, Omni international, Marittta, GA).

4. A Bradford protein assay kit (e.g., from Bio-Rad, Hercules, CA).
5. A microplate spectrophotometer (e.g., Spectra MAX 190 from Molecular Devices, Sunnyvale, CA).

2.2. iTRAQ Labeling

1. Dissolution buffer: 0.5 M TEAB, pH 8.5.
2. Denaturing reagent: 2% SDS (w/v).
3. Reducing reagent: 50 mM Tris-(2-carboxyethyl) phosphine (TCEP).
4. Cysteine blocking reagent: 200 mM methyl methanethiosulfonate (MMTS).
5. HPLC grade ethanol and water.
6. Mass spectrometry grade modified trypsin (e.g., from Promega, Madison, WI).
7. iTRAQ™ Reagents: 114, 115, 116, and 117 (Applied Biosystems Inc. (ABI), Forster City, CA).
8. Vacuum concentrator (e.g., the Eppendorf 5301, Westbury, NY).

2.3. Liquid Chromatography Systems

2.3.1. Strong Cation Exchange Liquid Chromatograph (SCXLC)

1. Mobile phase A: 10 mM KH_2PO_4 and 20% acetonitrile (ACN), adjust to pH 3.0 with 85% H_3PO_4.
2. Mobile phase B: 600 mM KCl, 10 mM KH_2PO_4, and 20% ACN, adjust to pH 3.0 with 85% H_3PO_4.
3. BioCAD Sprint™ perfusion chromatography system (ABI).
4. Column: Polysulfoethyl-A column (4.6 × 200 mm^2, 5 μm, 300 Å, Poly LC Inc., Columbia, MD).
5. pH test paper (Whatman International Ltd. Maidston, England).

2.3.2. Peptide Desalting

1. PepClean C_{18} spin columns (Pierce, Rockford, IL).
2. Loading solution: 5% ACN containing 0.5% trifluoroacetic acid (TFA).
3. Activation solution: 50% ACN containing 0.5% TFA.
4. Elution solvent: 70% ACN.

2.3.3. Reversed-Phase Liquid Chromatography (RPLC)

1. Solvent A: 5% ACN containing 0.1% TFA.
2. Solvent B: 95% ACN containing 0.1% TFA.
3. MALDI matrix solution: 4 mg/mL α-cyano-4-hydroxycinnamic acid (Sigma) in 60% ACN, 5 mM ammonium monobasic phosphate and internal calibrants (50 fmol/mL each of (Glu *(1)*)-fibrinopeptide B (GFP, *m/z* 1570.677, Sigma) and adrenocorticotropic hormone 18–39 (ACTH 18–39, *m/z* 2465.199, Sigma)).

4. LC-Packings Ultimate Chromatography System equipped with a Probot MALDI spotting device (Dionex, Sunnyvale, CA).
5. C_{18} PepMap 0.3 mm i.d. × 5 mm lenth, 5 μm, 100 Å, trapping column (Dionex).
6. C_{18} PepMap 0.1 mm i.d. × 150 mm lenth, 3 μm, 100 Å, capillary column (Dionex).

2.4. Mass Spectrometry

1. 4700 Proteomics Analyzer (ABI).
2. MALDI plates (ABI).
3. Mass Standards Kit containing six peptides mixture (ABI, cat no. 4333604).

2.5. Data Analysis Software

1. 4000 Series Explorer (ABI).
2. GPS Data Explorer v3.5 (ABI).
3. Mascot Search Engine v1.9 (Matrix Science Ltd. London, UK).

3. Methods

3.1. Protein Extraction

To obtain reproducible quantification results, it is important to prepare all four neuronal samples under identical conditions (*see* **Note 2**). In addition, the protein amounts among the four samples prior to trypsin digestion must be made equal.

1. For each sample, place ~15 mg of spinal cord tissue into a 1.5 mL Eppendorf tube on ice. Add 300 μL of the lysis buffer into each tube.
2. Disrupt the tissues on ice via ultrasonication (20 s sonication followed by a 30 s pulse). Repeat five times (*see* **Note 3**).
3. Pellet the insoluble materials by centrifugation for 15 min at 20,000 × *g* at 4°C in a bench-top centrifuge.
4. Transfer the supernatants into a fresh 1.5 mL chilled Eppendorf tube, and kept on ice.
5. Measure the protein concentrations for all four samples using the Bradford assay with bovine serum albumin diluted in the lysis buffer as standards (*see* **Note 4**).
6. Adjust the protein concentration of each sample to the same level with the Lysis Buffer.

3.2. Trypsin Digestion and iTRAQ Labeling

1. Pipette 90 μg of proteins from the four samples into four separate tubes (*see* **Note 5**). Add 1 μL of the denaturing reagent into each tube.

2. To each sample, add 2 µL of reducing reagent and vortex. Bring down the contents with a brief centrifugation. Incubate the sample tubes at 60°C for 1 h. Spin briefly to settle the liquid to the bottom of each tube.
3. To each sample, carefully add 1 µL of the cysteine blocking reagent. Mix by vortexing and centrifuge briefly to collect the solutions to the bottoms of the tube. Incubate at room temperature for 10 min.
4. Reconstitute two vials of trypsin (20 µg/vial) with 25 µL each of HPLC grade water. Vortex briefly.
5. To each sample tube, add 10 µL of the trypsin solution, vortex, and centrifuge briefly to collect the solution to the bottom of the tube. Incubate at 37°C for 12–16 h. Spin briefly to bring the sample solution to the bottom of the tubes (*see* **Note 6**).
6. Bring the iTRAQ reagents to room temperature. Add 70 µL of ethanol into each reagent vial, cap the vial and vortex vigorously, and then centrifuge briefly to settle the iTRAQ reagents to the bottoms of the vials (*see* **Note 7**).
7. Transfer the entire content of one iTRAQ reagent vial into each of the four sample tubes, and vortex to mix thoroughly. Spin briefly to collect the liquid to the bottom of the tubes. Peptides derived from the two control samples are labeled with iTRAQ Reagents 114 and 115 whereas peptides obtained from the two EAE samples are labeled with iTRAQ Reagents 116 and 117. Incubate the reaction vials at room temperature for 1 h.
8. Carefully combine the entire contents of all four iTRAQ-labeled samples into one tube, mix thoroughly by vortexing, then centrifuge briefly (*see* **Note 8**).

3.3. SCXLC

The combined peptide mixture will be first separated by SCXLC on a polysulfoethyl-A column to remove the excess iTRAQ reagents and then fractionate the peptides.

1. To remove both TEAB and the organic solvent from the sample, dry the combined sample completely in a vacuum concentrator (*see* **Note 9**).
2. Reconstitute the peptides by adding 500 µL of SCXLC mobile phase A and confirm the pH value using pH test paper. If the pH is not between 2.5 and 3.0, adjust it by adding additional SCXLC mobile phase A or drops of 1 M phosphoric acid.
3. Before SCXLC separation, centrifuge the sample at 20,000 × *g* for 10 min to pellet any particulates.
4. Equilibrate the column with mobile phase A, then inject the iTRAQ-labeled peptides onto the SCXLC column through a 500 µL sample loading loop.

5. Peptides will be eluted with a 60 min linear gradient from 100% mobile phase A to 100% mobile phase B at a flow rate of 1.0 mL/min. Collect 2-min fractions into Eppendorf tubes.
6. Dry all the SCXLC fractions completely in a vacuum concentrator for subsequent desalting.

3.4. Peptides Desalting Using C_{18} Spin Columns

1. Resuspend the peptides in each fraction in 150 μL of the loading solution (*see* **Note 10**).
2. Add 200 μL of the activation solution into a C_{18} spin column and centrifuge at 1,500 × *g* for 1 min to clean the spin column. Repeat this step once.
3. Equilibrate the spin column with 200 μL of the loading solution and then centrifuge the column at 1,500 × *g* for 1 min. Repeat this step three times.
4. Separately for each SCXLC fraction, load 150 μL of the peptides dissolved in the loading solution onto a spin column and centrifuge at 1,000 × *g* for 1 min to remove the unbound materials. Reload the flow through materials onto the spin column and repeat this step twice (*see* **Note 10**).
5. Wash the bound peptides with 200 μL of the loading solution and centrifuge at 1,500 × *g* for 1 min to remove salts. Repeat this step three times.
6. Elute the peptides using 100 μL of the elution solvent via centrifugation at 1,500 × *g* for 1 min and repeat twice. Collect all three eluents into the same Eppendorf tube.
7. Dry the peptides in a vacuum concentrator before RPLC (*see* **Note 11**).

3.5. Reversed-Phase Liquid Chromatography

1. Reconstitute the samples in 8 μL of solvent A, vortex vigorously, then centrifuge at 10,000 × *g* for 2 min. Transfer the samples into the bottom of auto-sampler vials and place them in cooled autosampler wells.
2. Equilibrate the RPLC column with 5% solvent B for at least 15 min prior to sample injection.
3. Each fraction (6.4 μL) will be loaded onto a C_{18} trapping column using a "microliter pickup" method at a flow rate of 20 μL/min. The bound peptides are subsequently resolved on a C_{18} capillary PepMap column at a flow rate of 200 nL/min with the following gradient:
4. The eluted peptides are mixed with the MALDI matrix in a 1:1 ratio through a 30 nL mixing tee, and spotted onto MALDI plates in an 18 × 18 spot array format (total 324 spots) using the Probot, which collects a spot every 12 s. Repeat the RPLC steps for each SCXLC fraction.

Time (min)	Solvent A	Solvent B
0	95	5
4	92	8
34	82	18
57	62	38
64	5	95
69	5	95
70	95	5
85	95	5

3.6. Mass Spectrometry

1. Load the sample plates into the plate loader of a MALDI tandem mass spectrometer, in our case an ABI 4700 Proteomics Analyzer.
2. Tune the instrument (e.g., deflectors X1, Y1, X2, Y2) using a mass standard mixture (e.g., kit from ABI) for optimal MS/MS sensitivity. For our instrument, we also optimize both the metastable ion suppressor and the timed-ion-selector (TIS) for the most precise precursor ion selection at the maximal resolution of 200, corresponding to ±5 Da at a *m/z* of 1,000. Instrument MS/MS sensitivity should be evaluated daily to ensure optimal iTRAQ quantification outcome.
3. Update the MS calibration file using the masses of the peptides in the mass standard mixture. Update the MS/MS calibration file using GFP MS/MS ion spectra (internal calibrant).
4. Create a new spot set (an Oracle database interface that contains both sample and MS method information in a spread sheet unique to the ABI 4000 Series Explorer software), load and align the iTRAQ sample plates.
5. iTRAQ-labeled peptides from each MALDI spot are analyzed in a data-dependent fashion. All MS/MS data will be acquired using a method optimized with a 1 kV collision energy.
6. Set up the acquisition, processing and job-wide interpretation methods for both MS and MS/MS analysis. In the MS acquisition method, the mass range is set to *m/z* 850–3,000; the focus mass is set at *m/z* 1,900; MS spectra are acquired with 2,500 laser shots at a laser intensity of 3,300. In the MS processing method, GFP (*m/z* 1570.677) and ACTH 18–39 (*m/z* 2465.199) masses are used for internal calibration. In the interpretation method, MS ions that meet the precursor selection criteria (200 ppm spot-to-spot precursor exclusion, S/N ratio ≥ 100 and a maximum of ten most abundant pre-

cursors per spot) are selected for subsequent MS/MS analysis, starting from the weakest to the strongest ion. In the MS/MS acquisition method, spectra are acquired with 3,000 laser shots at a laser intensity of 3,850. The spectra are smoothed with a Savitsky–Golay algorithm (FWHM = 9, polynomial order = 4) (*see* **Note 12**). An MS/MS spectrum example is shown in **Fig. 2**.

3.7. Database Search

1. Peptide identification is performed by searching the MS/MS spectra against rodent sequences in the SwissProt protein database (*see* **Note 13**) using a local Mascot search engine (v1.9) on a GPS server (v3.5, ABI).
2. The following parameters are used for the search: trypsin digestion with maximum of one missed cleavage; precursor mass tolerance set at 50 ppm; MS/MS mass tolerance set at 0.3 Da; iTRAQ-labeled N-termini and lysines as well as MMTS-labeled cysteines are set as fixed modifications; oxidized methionines and iTRAQ-labeled tyrosines are set as variable modifications (*see* **Note 14**).
3. In the GPS Result Browser, choose MS/MS Summary Tab and copy both the isotope carry-over corrected iTRAQ reporter ion peak areas (RPAs) of *m/z* 114–117 and the corresponding Mascot peptide identification matching results to a Microsoft Excel datasheet (*see* **Note 15**).
4. Only peptides identified with a confidence interval (C.I.) value ≥95% are used for protein identification and quantification. Mascot will assign a peptide to a corresponding protein entry with the highest score (usually the protein with the most number of peptides matched) if more than one entry share identical peptide sequence. To reduce the probability of a false identification, we will only quantify proteins containing at least two matched peptides.

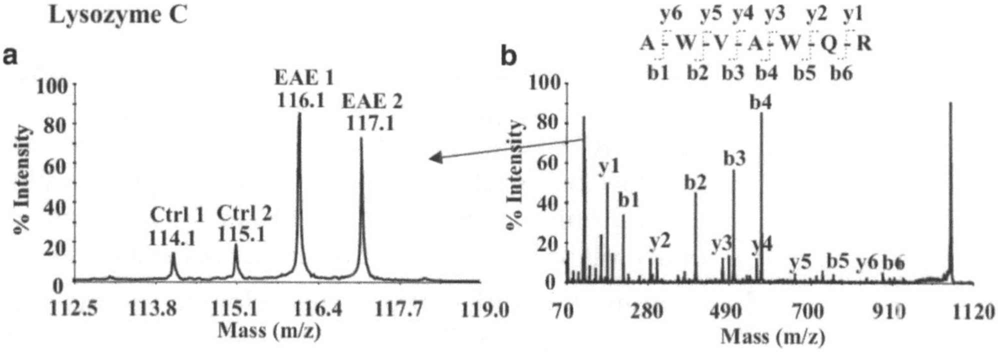

Fig. 2. MS/MS spectrum of a lysozyme C peptide. Lysozyme C is found elevated in EAE samples. (**a**) The peak areas of iTRAQ quantification ions, *m/z* 114 through 117 are used to determine the relative abundance of the peptide. (**b**) Peptide sequence is deduced from the MS/MS spectrum, based on the continuous series of N-terminal (b series) and C-terminal (y series) ions.

5. To estimate the false positive rate (FPR) for protein identification, all spectra are searched against a decoy Swiss-Prot database with all protein sequences reversed using the same database searching criteria outlined above. The FPR is computed as *(13)*:

   $FPR = 2 \times (N_{decoy})/(N_{decoy} + N_{forward})$,
   where N_{decoy} is the number of proteins identified using the decoy database, $N_{forward}$ is the number of proteins identified using the regular database.

6. If FPR is higher than the generally accepted 5%, increase your peptide identification C.I. value cutoff from 95% to 97% or higher for subsequent expression analysis until the calculated FPR value is less than 5%.

3.8. Bioinformatics Analysis to Determine Differentially Expressed Proteins

1. Minor adjustments to the raw RPA values need to be performed prior to protein quantification analysis. For the peptides with RPA values of 0, assign a nominal signal count of 100 to facilitate later mathematical analysis (*see* **Note 16**).
2. For each sample, the peptide population median RPA (M_{114}, M_{115}, M_{116}, M_{117} for control 1, control 2, EAE1 and EAE2 samples, respectively) is calculated in Microsoft Excel.
3. Assuming a comparable overall protein concentration in each sample, individual peptide RPA values are normalized by dividing the raw RPA value with corresponding normalization factors (F_{norm}), calculated using the following formulas (*see* **Note 17**).

$F_{norm114} = M_{114}/M_{114}$ (Control 1)
$F_{norm115} = M_{115}/M_{114}$ (Control 2)
$F_{norm116} = M_{116}/M_{114}$ (EAE 1)
$F_{norm117} = M_{117}/M_{114}$ (EAE 2)

4. A protein expression change is derived from all of its corresponding peptide changes. For each peptide, the mean of the four iTRAQ RPA values ($\overline{RPA}$) is computed in Microsoft Excel. Since low ion signals tend to produce larger quantification inaccuracies, only peptides with $\overline{RPA}$ values larger than 5,000 counts will be included for protein expression analysis *(11)*.
5. To ensure that different peptides belonging to the same protein contribute equally toward the statistical determination of differentially expressed proteins, regardless of their peptide ionization efficiency, relative RPA values are computed using Microsoft Excel. For each peptide, all of its relative RPA values are calculated as the ratios of normalized individual RPA

values divided by the RPA value from the 114-labeled sample. Such ratios are then transformed into $\log_2$ values (R_{114}, R_{115}, R_{116} and R_{117}, *see* **Note 18**). In addition, the relative RPAs between EAE1 (*m/z* 116) and EAE2 (*m/z* 117) are also computed and transformed into $\log_2$ values for the later evaluation of within-group protein variation between the two EAE samples (R_{76}). In cases of multiple MS/MS spectra matched to the same peptide sequence, the peptide ratio is calculated and weighted based on the relative proportion of the $\overline{RPA}$ values of each spectrum.

6. Protein $\log_2$ ratios (P_{114}, P_{115}, P_{116}, P_{117}, and P_{76} (EAE2/EAE1)) are computed as the mean of R values (R_{114}, R_{115}, R_{116}, R_{117}, and R_{76}, respectively) of all its corresponding peptides.
7. The relative protein expression between EAE and the control samples (calculated as the pooled protein $\log_2$ ratios) are computed based on the following equation:

$$P_i = \frac{(P_{116i} + P_{117i})}{2} - \frac{(P_{114i} + P_{115i})}{2},$$

where Pi is the pooled protein $\log_2$ ratio of the ith protein ($i = 1, 2, 3, \ldots N$, where N is the total number of identified proteins).

8. To identify differentially expressed proteins, both p-values in Student's t tests and the pooled standard deviation of within-group protein variation (Sp, *see* **Note 19**) need to be considered (*see* **Note 20**). The p-values are generated by comparing each protein $\log_2$ ratio in the control group (P_{114} and P_{115}) to those in the EAE group (P_{116} and P_{117}) using Microsoft Excel. S_p is computed using the following formula:

$$S_P = \sqrt{\frac{S_C^2 + S_E^2}{2}},$$

where S_P is the pooled standard deviation of within-group protein $\log_2$ ratios; S_C is the standard deviation of all protein $\log_2$ ratios between the two control samples, which is calculated as the standard deviation of P_{115} with Microsoft Excel; S_E is the standard deviation of all protein $\log_2$ ratios between the two EAEs, which is calculated as the standard deviation of P_{76} with Microsoft Excel (*see* **step 6**). Differentially expressed proteins must meet two criteria: (1) p values ≤ 0.05 and (2) Pi values are larger than $1.65S_P$ or smaller than $-1.65S_P$ (corresponding to the top 5% increased and the bottom 5% decrease proteins) (*see* **Note 21** and **22**).

9. Anti-$\log_2$ of P_i values are calculated to produce the exact protein fold change values.

4. Notes

1. The lysis buffer described here is only for extracting soluble proteins. To extract more hydrophobic proteins, an alternative lysis buffer containing SDS, NP40, Triton-100, Tween-20, and urea may be used. For example, 10 mM HEPES (pH 8.0), 150 mM TEAB, 0.02% sodiumazide, 0.1% SDS, 1% NP40, 0.5% deoxycholic acid, 0.2 mM PMSF, 2 μg/mL aprotinin, 2 μg/mL leupeptin, 2 μg/mL pepstatin, and 50 mM NaF may be used. Since high concentrations of detergents will affect trypsin digestion, the lysate should be diluted with H_2O to reduce the detergents concentrations prior to trypsin digestion. The final concentrations of SDS should be ≤0.05%, NP-40 ≤ 0.1%, Triton X-100 ≤ 0.1%, Tween 20 ≤ 0.1%, CHAPS ≤ 0.1%, Urea ≤ 1 M.
2. In the EAE model, 8-week old female Lewis rats were immunized with myelin basic protein emulsified in Complete Freund's Adjuvant (CFA) or CFA/vehicle. The animals exhibiting hind limb paralysis were euthanized by exposure to CO_2. The spinal cords were dissected out, meninges were carefully removed, and the tissues were thoroughly rinsed with saline to remove blood. The lumbar spinal cord was immediately frozen on dry ice and stored at –80°C until use.
3. In most cases, ultrasonication for five times should be enough to break the cells and extract the proteins. However, if the cells do not break completely, ultrasonication can be repeated more times until the proteins can be thoroughly extracted.
4. To obtain accurate iTRAQ quantitative results, it is very important to adjust the protein concentration to be equal among all four samples. If the lysis buffer contains detergents, the Branford assay will not be suitable for accurate protein concentration estimation. A BCA protein assay may be used instead.
5. Based on iTRAQ instructions from ABI, each protein sample should be between 5 and 100 μg for each iTRAQ labeling reaction. To ensure maximum labeling efficiency, sample volumes should be less than 50 μL each. If the sample volume is larger than 50 μL, a speedvac can be used to reduce the sample volume before iTRAQ labeling.
6. It is important to check the protein digestion efficiency before iTRAQ labeling. Take 1 μL of each digested sample and desalt it using a C_{18} ZipTip (Millipore, Billerica, MA). Mix the eluted peptides with the MALDI matrix solution in a 1:1 ratio and spot them onto a MALDI plate. Acquire MS spectra to check if the peptide ion signals are comparable.

7. To maximize labeling efficiency, the concentration of organic reagents (ethanol and iTRAQ reagents) in iTRAQ labeling reactions should be larger than 60% (v/v).
8. In cases of the lysis buffer containing high concentrations of detergents or organic solvents, protein concentration measurements and sample pippetting accuracy maybe affected by the lysis buffer. Additional adjustments in peptide levels may be needed to ensure even peptide mixing in order to obtain reliable protein quantification results. For example, prior to combining all four iTRAQ-labeled samples, a small aliquot (i.e., 1/20) of each iTRAQ-labeled sample can be combined into one tube. The excess iTRAQ reagents are removed by an SCXLC cartridge column (ABI) followed by a C_{18} spin column for desalting (Pierce, cat no. 89870). Dry the eluted peptides in a vacuum concentrator and reconstitute the peptides in RPLC solvent A. Follow steps given in **Subheadings 3.5–3.8**, the peptides can be separated by RPLC and spotted onto a MALDI plate. MS/MS spectra of the peptides are acquired and data are analyzed using the GPS server (ABI). Calculate the population reporter ion peak area (RPA) median of each quantification reporter ion (114, 115, 116, and 117). Assuming most of proteins do not change among the four samples, the RPA medians of the four quantification reporter ions should be close to 1:1:1:1. If one or more samples varies from the others by greater than 10%, the volume(s) of that sample(s) needs to be adjusted accordingly to ensure an equal peptide mixture.
9. To remove all of the TEAB, reconstitute the combined iTRAQ-labeled samples in 100 μL of HPLC grade water and dry the sample in a vacuum concentrator. Repeat this step twice to ensure all the TEAB is evaporated.
10. For high salt fractions, 150 μL of the loading solution may not be enough to resuspend the salts and peptides. More loading solution should be added until all the contents are solubilized. The entire content of the tube should be loaded onto the spin column for desalting.
11. Approximately 30 SCXLC fractions are collected. To reduce the number of subsequent RPLC samples, some fractions can be combined based on the peptide complexity in each fraction. Take 1 μL from each fraction after desalting, mix it with the MALDI matrix in a 1:1 ratio and spot them onto the MALDI plate. Acquire MS spectra, and evaluate MS ion complexity in the spectra. SCXLC factions whose MS spectra contain less than 100 peaks ($S/N \geq 5$) between *m/z* 1,000 and 2,000 may be combined prior to RPLC separations.

12. Different MS instruments have different optimal MS and MS/MS methods for the analysis of iTRAQ-labeled peptides. In the 4700 Proteomic Analyzer, we update both MS and MS/MS calibration files daily and tune the deflector parameters weekly to obtain the maximum sensitivity and mass accuracy.

13. It is important to use the latest version of the protein database to ensure comprehensive peptide identification. Swissprot, IPI, NCBI protein database, or EST (6 frame translation into protein sequences) database can be used, with increasing number of entries and database size. Generally speaking, using bigger databases will likely increase one's chance to match a spectrum to a peptide sequence. However, it will also increase the odds for random matching. We chose the Swissprot database for our study, because of its high protein sequence accuracy and low redundancy.

14. Using too many variable modifications can significantly lower the spectra matching confidence interval (C.I.) values due to increased random matches. Only include modifications relevant to the experiment. iTRAQ labeling is usually very efficient *(11)*; however, it may be prudent to evaluate iTRAQ labeling efficiency for each experiment. To that end, change the fixed iTRAQ modifications into variable modifications and repeat the GPS search. For all the peptide matches with C. I. ≥ 95%, the number of iTRAQ-labeled N-termini (N_{ti}) and lysines (N_{ki}) are compared with the total number of peptide N-termini (N_{tt}) and lysines (N_{kt}). The iTRAQ labeling efficiency is estimated as: $100\% \times (N_{ti} + N_{ki})/(N_{tt} + N_{kt})$.

15. Database search and data display can take hours depending on your computer hardware and the scale of your iTRAQ experiment.

16. The denominators for peptide expression ratio computation cannot be zero. Therefore, use a nominal signal count 100 to represent the background noise level.

17. The assumption of using the median values to normalize RPAs is that four iTRAQ-labeled peptide populations should produce a similar RPA signal distribution profile. In case the RPA signal distribution profiles are dissimilar, LOWESS or other advanced normalization routines can also be used *(14)*.

18. A log transform is used to produce a symmetrical peptide log ratio distribution and provide a means for an accurate estimation of the standard deviations for peptide and protein ratios *(15)*. More importantly, this procedure will ensure that both an increase and decrease in peptide expression levels

contribute equally toward the determination of differentially expressed proteins. We choose 2-based log transformation, because it enables concise interpretations of protein fold changes. For example, a difference of one in this scale corresponds to a twofold change.

19. S_p is a pooled standard deviation to estimate the degree of the protein expression changes caused by random protein-to-protein variations within the control or EAE groups.

20. Student's *t* test is utilized here to account for the relative protein expression value between groups for each protein. Limited by the availability of only 4 versions of iTRAQ reagents at the time of our experiments, the *t* test can be underpowered. However, recent availability of 8 versions of the iTRAQ reagents should substantially improve the *t* test statistical power.

21. Using the methods described here, we acquired 13,834 MS/MS spectra. 2,488 unique peptides from 510 proteins were identified with an estimated false identification rate 3.1%. Forty-one proteins were found differentially expressed in EAE *(3)*. To further identify changes in peptides with post-translational modifications (PTMs), additional PTMs can be set as variable modifications for repeated database PTM searches. The peptides identified have to meet the following criteria: (1) a C.I. value ≥ 80; (2) the PTM peptide matching score is higher than the score when the same spectrum is searched without considering this PTM; and (3) must pass manual spectral inspections. *(16)*

22. It is very important to verify the iTRAQ quantification results using Western blotting, immunohistochemistry, or other biological methods for validation purposes.

References

1. Wu, C. C., and MacCoss, M. J. (2002) Shotgun proteomics: Tools for the analysis of complex biological systems. *Curr. Opin. Mol. Ther.* **4**, 242–250.
2. Liu, T., D'Mello, V., Deng, L., Hu, J., Ricardo, M., Pan, S., Lu, X., Wadsworth, S., Siekierka, J., Birge, R., and Li, H. (2006) A multiplexed proteomics approach to differentiate neurite outgrowth patterns. *J. Neurosci. Methods* **158**, 22–29.
3. Liu, T., Donahue, K. C., Hu, J., Kurnellas, M. P., Grant, J. E., Li, H., and Elkabes, S. (2007) Identification of differentially expressed proteins in experimental autoimmune encephalomyelitis (EAE) by proteomic analysis of the spinal cord. *J. Proteome Res.* **6**, 2565–2575.
4. McDonald, W. H., and Yates, J. R., 3rd (2002) Shotgun proteomics and biomarker discovery. *Dis. Markers* **18**, 99–105.
5. Wu, W. W., Wang, G., Yu, M. J., Knepper, M. A., and Shen, R. F. (2007) Identification and quantification of basic and acidic proteins using solution-based two-dimensional protein fractionation and label-free or (18) o-labeling mass spectrometry. *J. Proteome Res.* **6**, 2447–2459.
6. Mann, M. (2006) Functional and quantitative proteomics using SILAC. *Nat. Rev. Mol. Cell. Biol.* **7**, 952–958.

7. Wiese, S., Reidegeld, K. A., Meyer, H. E., and Warscheid, B. (2007) Protein labeling by iTRAQ: A new tool for quantitative mass spectrometry in proteome research. *Proteomics* 7, 1004.
8. Lund, T. C., Anderson, L. B., McCullar, V., Higgins, L., Yun, G. H., Grzywacz, B., Verneris, M. R., and Miller, J. S. (2007) iTRAQ is a useful method to screen for membrane-bound proteins differentially expressed in human natural killer cell types. *J. Proteome Res.* **6**, 644–653.
9. Shiio, Y., and Aebersold, R. (2006) Quantitative proteome analysis using isotope-coded affinity tags and mass spectrometry. *Nat. Protoc.* **1**, 139–145.
10. Zhang, B., VerBerkmoes, N. C., Langston, M. A., Uberbacher, E., Hettich, R. L., and Samatova, N. F. (2006) Detecting differential and correlated protein expression in label-free shotgun proteomics. *J. Proteome Res.* **5**, 2909–2918.
11. Hu, J., Qian, J., Borisov, O., Pan, S., Li, Y., Liu, T., Deng, L., Wannemacher, K., Kurnellas, M., Patterson, C., Elkabes, S., and Li, H. (2006) Optimized proteomic analysis of a mouse model of cerebellar dysfunction using amine-specific isobaric tags. *Proteomics* **6**, 4321–4334.
12. Shadforth, I. P., Dunkley, T. P., Lilley, K. S., and Bessant, C. (2005) i-Tracker: For quantitative proteomics using iTRAQ. *BMC Genomics* **6**, 145.
13. Peng, J., Elias, J. E., Thoreen, C. C., Licklider, L. J., and Gygi, S. P. (2003) Evaluation of multidimensional chromatography coupled with tandem mass spectrometry (LC/LC-MS/MS) for large-scale protein analysis: The yeast proteome. *J. Proteome Res.* **2**, 43–50.
14. Xia, Q., Wang, T., Park, Y., Lamont, R. J., and Hackett, M. (2007) Differential quantitative proteomics of *Porphyromonas gingivalis* by linear ion trap mass spectrometry: Non-label methods comparison, q-values and LOWESS curve fitting. *Int. J. Mass Spectrom.* **259**, 105–116.
15. Coombes, K. R., Tsavachidis, S., Morris, J. S., Baggerly, K. A., Hung, M. C., and Kuerer, H. M. (2005) Improved peak detection and quantification of mass spectrometry data acquired from surface-enhanced laser desorption and ionization by denoising spectra with the undecimated discrete wavelet transform. *Proteomics* **5**, 4107–4117.
16. Grant, J. E., Hu, J., Liu, T., Jain, M. R., Elkabes, S., and Li, H. (2007) Post-translational modifications in the rat lumbar spinal cord in experimental autoimmune encephalomyelitis. *J. Proteome Res.* **6**, 2786–2791.

Chapter 15

Methods in Drug Abuse Neuroproteomics: Methamphetamine Psychoproteome

Firas H. Kobeissy, Zhiqun Zhang, Shankar Sadasivan, Mark S. Gold, and Kevin K.W. Wang

Summary

Methamphetamine (METH) is recognized as one of the most abused psychostimulants in the USA. METH is an illicit drug that is known to exert neurotoxic effects on both dopaminergic and serotonergic neural systems. Our laboratory has been studying the biochemical mechanisms underlying MDMA and METH-induced neurotoxic effects both in vivo and in vitro. Our substance abuse research focuses on the global alteration of cortical protein expression in rats treated with acute METH. Altered protein expression was identified using a multistep protein separation/proteomic platform. Differential changes of the selected proteins were further confirmed by quantitative immunoblotting. Our study identified 82 differentially expressed proteins, 40 of which were downregulated and 42 of which were upregulated post acute METH treatment. Proteins that were shown to be downregulated included collapsin response mediator protein-2 (CRMP-2), superoxide dismutase 1 (SOD 1), and phosphatidylethanolamine-binding protein-1 (PEBP-1). Proteins that were shown to be upregulated included authophagy-linked microtubule-associated protein light chain 3 (LC3), synapsin-1, and Parkinsonism-linked ubiquitin carboxy-terminal hydroxylase-L1 (UCH-L1). This differential protein expression highlights on the neurotoxic mechanism involved in METH exposure as well as to discover potential markers for METH-induced neurotoxicity. In this chapter, we describe the current protocols for the in vivo rat model of acute METH treatment (40 mg/kg) coupled with the description of the multistep separation platform applied. These methods and protocols are discussed in the paradigm of acute model of methamphetamine drug abuse and can be applied to other models of substance abuse such as to MDMA or cocaine.

Key words: Neurotoxicity, Methamphetamine, Proteomics, Drug of abuse, Proteolysis, Autophagy, Posttranslational modification

Andrew K. Ottens and Kevin K.W. Wang (eds.), *Neuroproteomics,* Methods in Molecular Biology, vol. 566
DOI 10.1007/978-1-59745-562-6_15, © Humana Press, a part of Springer Science+Business Media, LLC 2009

1. Introduction

Methamphetamine (METH) abuse is a growing epidemic worldwide. The use of METH has increased dramatically in the past two decades as evidenced by emergency room visits arising from METH intoxication *(1–3)*. METH is an illicit, addictive psychostimulant drug that is known to be neurotoxic to the dopaminergic and serotonergic systems in the brain. METH neurotoxicity has been demonstrated in the striatum and has also been investigated in other brain regions including the frontal cortex, hippocampus, and cerebellum *(4–8)*. METH abuse-induced neurotoxic events are highly associated with the degeneration of the serotonergic and dopaminergic neural systems; METH abuse leads to widespread, selective axonal terminal damage that is coupled with neuronal degeneration *(9, 10)*.

Recent in vivo and in vitro studies from our laboratory and others have implicated apoptotic and necrotic pathways being activated in METH neurotoxicity *(5, 10–12)*. Neuronal injury involved in METH neurotoxicity involves an increase in intracellular calcium concentrations that result from cross-talk between endoplasmic reticulum stress and the mitochondria *(11)*. The increase in calcium leads to activation of caspases and calcium-dependant protein calpains, culminating in apoptotic and necrotic cell death.

Our laboratory has recently reported that in vivo acute METH treatment induced neurotoxicity in both the cortex and hippocampus brain regions *(13, 14)*. METH treatment leads to the activation of two major protease systems (the pronecrotic calpain and the proapoptotic caspase system), which in turn lead to cytoskeletal damage of structural proteins such as αII-spectrin and microtubule-associated protein tau *(14, 15)*. Recently, Larsen et al. *(16)* and Kanthasamy et al. *(17)* have demonstrated that METH promotes the formation of autophagic bodies within the cell bodies of dopaminergic neurons. This mechanism may represent an additional mechanism involved in METH-induced cell injury, in addition to necrotic and apoptotic mechanisms. Thus, results from these studies suggest that METH-induced neurotoxicity may involve apoptotic, necrotic, and autophagic cell death mechanisms.

In the current study, changes in cortical protein expression in rats treated with acute METH were identified using a multistep protein separation/proteomic platform [cation–anion exchange chromatography/SDS-PAGE (CAX-PAGE)] and confirmed using immunoblotting technique. Proteins were fractionated using a multidimensional protein separation platform comprising nine sequential steps including tandem column chromatography analysis coupled to 1D-gel electrophoresis-LCMSMS proteomic

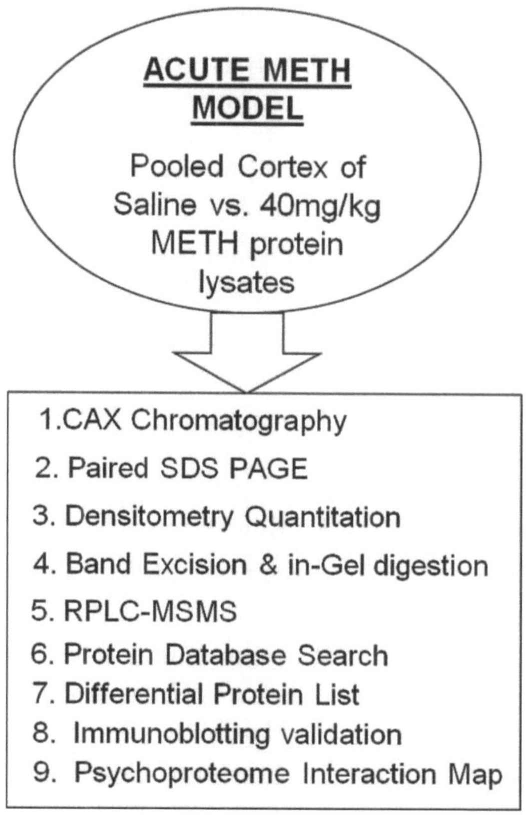

Fig. 1. Schematic of the differential cation–anion exchange chromatography (CAX-PAGE)-tandem mass spectrometry psychoproteomic platform. The schematic illustrates the nine sequential steps following acute administration of METH followed by CAX chromatography and 1D-PAGE separation, as the first and second dimension. After CAX-PAGE separation, selected differential protein bands are excised and in-gel digested followed by RPLC/MSMS generating a differential protein list. Selected protein subsets are then subjected for validation via immunoblotting. Subsequently, using Pathway Architect™ software, a functional interactive map is constructed based on the psychoproteomic data *(18, 20)*.

identification *(18, 19)*. A schematic of the sequential steps used in our multidimensional proteomic platform is shown in **Fig. 1**.

2. Materials

2.1. Animal Model and Methamphetamine

1. Adult male Sprague–Dawley rats (Harlan, Indianapolis, IN, USA) that were of age 60 days and weighed between 250 and 275 g.
2. Pharmacologic agent (+/−) METH hydrochloride (Sigma-Aldrich, St. Louis, MO).
3. 0.9% Physiological saline.
4. Water (HPLC grade, Burdick & Jackson, Muskegon, MI).

2.2. Cortical Tissue Collection and Protein Extraction

1. Dry ice, cold mortar, and pestle (Fisher Scientific).
2. SDS lysis buffer: 0.1% SDS, 150 mM sodium chloride, 1% ethoxylated octylphenol, 1 mM sodium vanadate, 3 mM ethylenediaminetetraacetic acid, 2 mM ethylene glycol bis (2-aminoethyl ether)-tetraacetic acid (EGTA), 1 mM dithiothreitol (DTT); store at 4°C (*see* **Note 1**).
3. Complete Mini protease inhibitor cocktail tablet (Roche Biochemicals, Indianapolis, IN) (*see* **Note 2**).
4. BioRad DC Protein Assay (BIO-RAD Laboratories, Hercules, CA, USA).
5. Albumin standard: 2 mg/ml in ampule (Pierce Cat. #23210).

2.3. 1D-SDS-Polyacrylamide Gel Electrophoresis (1D-SDS-PAGE)

1. BIO-RAD molecular weight markers: Precision Plus Protein All Blue Standards.
2. Precast 10–20% gradient Tris–HCl polyacrylamide gels, 1.0 mm, ten wells (BIO-RAD).
3. Running buffer 10× Tris/tricine/SDS: 100 mM Tris, pH 8.3, 100 mM tricine, 0.1% SDS, kept at room temperature (BIO-RAD).
4. 2× Laemmli sample buffer: BIO-RAD with 5% β mercaptoethanol (*see* **Note 3**).
5. X-Cell Sure Lock Mini Cell Apparatus: Invitrogen Life Technologies, CA.
6. BIO-RAD Power PAC 3000.

2.4. Coomassie Blue Gel Staining

1. Destaining solution: 40% methanol, 50% deionized water, and 10% acetic acid.
2. CBB staining solution: 0.025% (w/v) Coomassie Brilliant Blue R-250 (BIO-RAD) in destaining solution.
3. Methanol, HPLC grade (Fisher Scientific).
4. Glacial acetic acid, HPLC grade (Fisher Scientific).
5. Water (*see* **Note 4**).

2.5. Gel Band Visualization and Image Quantification

1. NIH ImageJ densitometry software (version 1.6, NIH, Bethesda, MD).
2. Epson Expression 8836XL high-resolution flatbed scanner (Epson, Long Beach, CA).
3. SigmaStat software (Version 2.03, Systat Software Inc.).

2.6. Immunoblotting

1. 1× Semidry method in a transfer buffer (39 mM glycine, 48 mM Tris, and 5% methanol).
2. Wash buffer: 1× PBS, 0.1% Tween-20. Store at room temperature.
3. Thick pads (Invitrogen Life Technologies, CA).

4. Thin pads (Invitrogen Life Technologies, CA).
5. Polyvinylidene fluoride (PVDF) membrane (Invitrogen Life Technologies, CA).
6. TBST: 20 mM Tris–HCl, 150 mM NaCl, and 0.003% Tween-20, pH 7.5.
7. Blocking buffer: 5% (w/v) nonfat dry milk in TBST.
8. Primary antibodies: anti-αII-spectrin (Affiniti Research Products, UK), anti-β actin (Sigma Chemical Co., St. Louis, MO), antisynapsin-1 (BD Biosciences, NJ, USA), UCH-L1 (made in-house at Banyan Biomarkers, Alachua, FL), anti-light chain 3 (LC3) (anti-LC3 antibody was raised in rabbits against a synthetic peptide corresponding to the N-terminal of LC3), anti-Map kinase kinase-1 (MKK-1) (Cell Signaling Technology, Beverly, MA), superoxide dismutase1 (SOD1) (gift from Dr. David Borchelt laboratory at the McKnight Brain Institute of the University of Florida, Gainesville, FL), antiphosphatidylethanolamine-binding protein-1 (PEBP-1) (Abcam, Cambridge, UK), and anti-CRMP-2 (IBL, Japan).
9. Secondary biotinylated antibodies (Amersham Biosciences, UK).
10. Streptavidin conjugated alkaline phosphatase (Amersham Biosciences, UK).
11. BIO-RAD Transblot SD Semi-Dry Transfer Cell.

2.7. Psychoproteomic Interaction Map

1. Stratagene PathwayArchitect™ software package version 2.0.1 (Stratagene, La Jolla, CA) (www.stratagene.com).

3. Methods

For a successful profiling of brain psychoproteome after METH administration, special care should be taken upon harvesting the brain tissue to avoid any possible brain tissue proteolysis.

3.1. Animal Habituation and Drug Injection

1. All procedures involving animal handling and processing were done in compliance with the Animal Welfare Act and the University of Florida Institutional Animal Care and Use Committee (IACUC) and the National Institutes of Health guidelines detailed in the Guide for the Care and Use of Laboratory Animals. Adult male (280–300 g) Sprague–Dawley rats were habituated for at least 10 days prior to treatment. Animals were housed in pairs in polyethylene cages containing hardwood bedding in a temperature-controlled (approximately 22°C) room with a 12-h light–dark cycle. Animals were given access to rat chow and tap water ad libitum (*see* **Note 5**).

2. Following habituation, experimental groups were divided into two groups ($n = 7$); each group was injected intraperitoneally (i.p.) with either racemic METH–HCl or an equivalent volume of 0.9% saline. Rats were injected with 10 mg/kg doses of METH four times every hour to achieve the desired dosage of 40 mg/kg in a bolus of 0.3 cc. The saline group (vehicle group) received similar injection schedules of physiological saline.

3.2. Cortical Tissue Collection and Protein Extraction

1. At 24 h postintraperitoneal injection, treated animals were briefly anesthetized with 3–4% isoflurane and were sacrificed by decapitation. METH and saline cortex samples were rapidly dissected, washed with saline solution, snap-frozen in liquid nitrogen, and stored at –80°C for further processing.
2. For cortical protein collection, mortar, pestle, and spatula are kept on dry ice for 20 min. Individual cortical samples are homogenized using the pestle and mortar into a fine powder. After use, make sure to wash (*see* **Note 6**).
3. The pulverized brain tissue powder (homogenates) was then lysed for 90 min at 4°C in 0.1% SDS lysis buffer with complete protease inhibitor cocktail.
4. Vortex the brain cortical lysate every 15 min and then centrifuge at 15,000 × *g* for 10 min at 4°C.
5. Transfer the protein supernatant into a new microcentrifuge tube (*see* **Note 7**).
6. Perform a DC protein assay (BIO-RAD, Hercules, CA) to determine protein concentration (*see* **Note 8**).

3.3. SDS-PAGE

1. Cortical samples are adjusted to a concentration of 2 μg/μl by mixing with distilled water and then adding an equivalent volume of Novex 2× Laemmli sample buffer tricine/glycine-SDS sample buffer to achieve a concentration of 1 μg/μl.
2. Samples are heated for 2–3 min at 90°C and then vortexed prior to loading.
3. Fill the BIO-RAD chamber with 1× Tris/SDS running buffer.
4. Insert the precast gels and wash the ten wells prior to sample loading (*see* **Note 9**).
5. Run the power supply unit at constant voltage (120 V) for 2 h until the tracking dye has just migrated out from the gel (*see* **Note 10**).

3.4. Coomassie Blue Staining

1. Once the electrophoresis step is done, the gels are washed three times with 100-ml distilled water (5 min each wash) (*see* **Note 11**).

2. Fixing step follows washing, and this is achieved by incubating the gel for 1 h in 10% acetic acid on a shaker.
3. Staining step is achieved by incubating the gel in a staining solution for 1 h on a shaker.
4. Destaining step is achieved by the periodical rinsing of the gel in a destaining solution until desired background versus gel band contrast is obtained (*see* **Note 12**).
5. The gel is kept at 4°C in 10% acetic acid for gel band visualization (*see* **Note 13**). Example of a Coomassie blue-stained gel is shown in **Fig. 2** *(18)*.

3.5. Gel Band Visualization and Statistical Analysis of Immunoblotting Quantification

1. For the SDS-1D PAGE visualization, protein fractions were run side-by-side on 10–20% gradient Tris–HCl gels (BIO-RAD) and were visualized with Coomassie Blue staining for differential gel bands selection.
2. NIH ImageJ densitometry software was used for lane and band detection, providing differential comparison between the saline and METH band densitometric analysis.
3. Fold increase or decrease between METH and saline samples was computed by dividing the greater value by the lesser value with a negative sign to indicate a decrease after METH treatment.
4. Densitometric quantification of the immunoblot bands was performed using an Epson Expression 8836XL high-resolution flatbed scanner and NIH ImageJ software.
5. Data were acquired using integrated densitometric values and transformed to percentages of the densitometric levels.

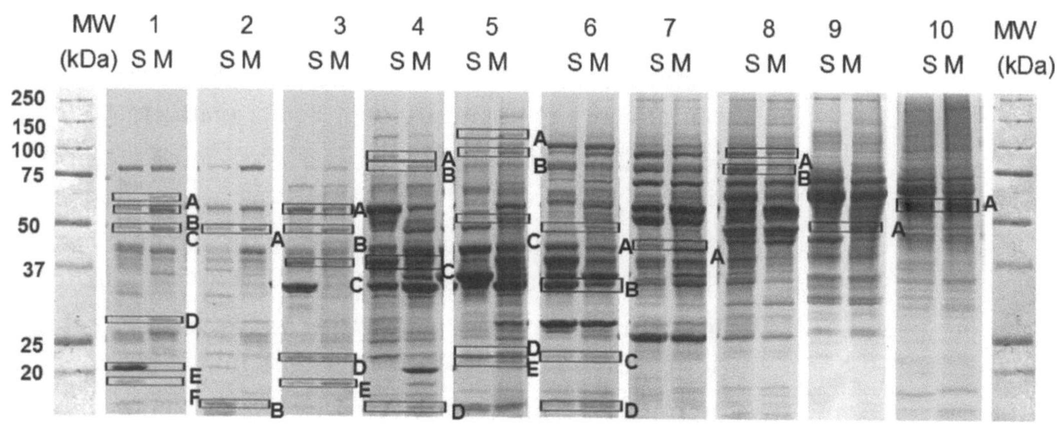

Fig. 2. Comparison of rat METH and saline psychoproteomes by sequential CAX-1D-SDS-PAGE separation. Shown is the side-by-side (saline on the left and METH on the right) pairing of ten fractions of the 23 collected fractions on 1D-SDS-PAGE. This gel is stained with Coomassie blue as described in the **Subheading 3.4**. Reprinted in part with permission from *(18)*, copyright 2008 American Chemical Society.

6. Densitometry values of the four different individuals of saline and METH samples were evaluated for statistical significance using SigmaStat software using a student's *t* test.
7. *P* value of <0.05 was considered to be significant for data acquired in arbitrary density units.

3.6. Immunoblotting Technique

1. For immunoblotting procedure, precast gels are opened from the two plastic covers using a scalpel; the gel is washed with distilled water.
2. PVDF membranes are soaked with 75% transfer buffer and 25% methanol for 30 s.
3. Thick and thin pads are soaked in 1× transfer buffer.
4. Prepare the semidry transfer unit to achieve a sandwich format as follows.
5. Place the sandwich format as follows: a thin membrane followed by a thick pad, PVDF membrane, precast gel, thick pad, and finally add a thin pad.
6. Use a pencil or a pipette to push any air bubbles within the pads, facilitating a uniform protein transfer onto the membrane.
7. Run the transfer unit for 2 h at 20 V at room temperature.
8. Following the transfer, the membrane is blocked in 5% nonfat dry milk in TBST (20 mM Tris–HCl, 150 mM NaCl, and 0.003% Tween-20, pH 7.5) for an hour at room temperature on a shaker platform (*see* **Note 14**).
9. After blocking, incubate the membrane overnight with the primary antibody at 4°C diluted at the proper concentrations on a shaker platform (*see* **Note 15**).
10. On the following day, the membranes were washed with excess TBST three times, each 5 min.
11. Membranes are probed with the secondary antibody for an hour at room temperature on a shaker platform.
12. Primary antibodies were used at a dilution of 1:1,000 in 5% milk.
13. Secondary biotinylated antibodies (Amersham Biosciences, UK) were used at a dilution of 1:3,000 in 5% milk.
14. Discard the secondary antibodies and wash the membranes with excess TBST three times, each 5 min.
15. Immunoreactivity was detected by using streptavidin conjugated alkaline phosphatase.
16. Streptavidin conjugated alkaline phosphatase was used at a dilution of 1:3,000 in 5% milk. Example of one immunoblot of identified upregulated proteins is shown in **Fig. 3** *(18)*.

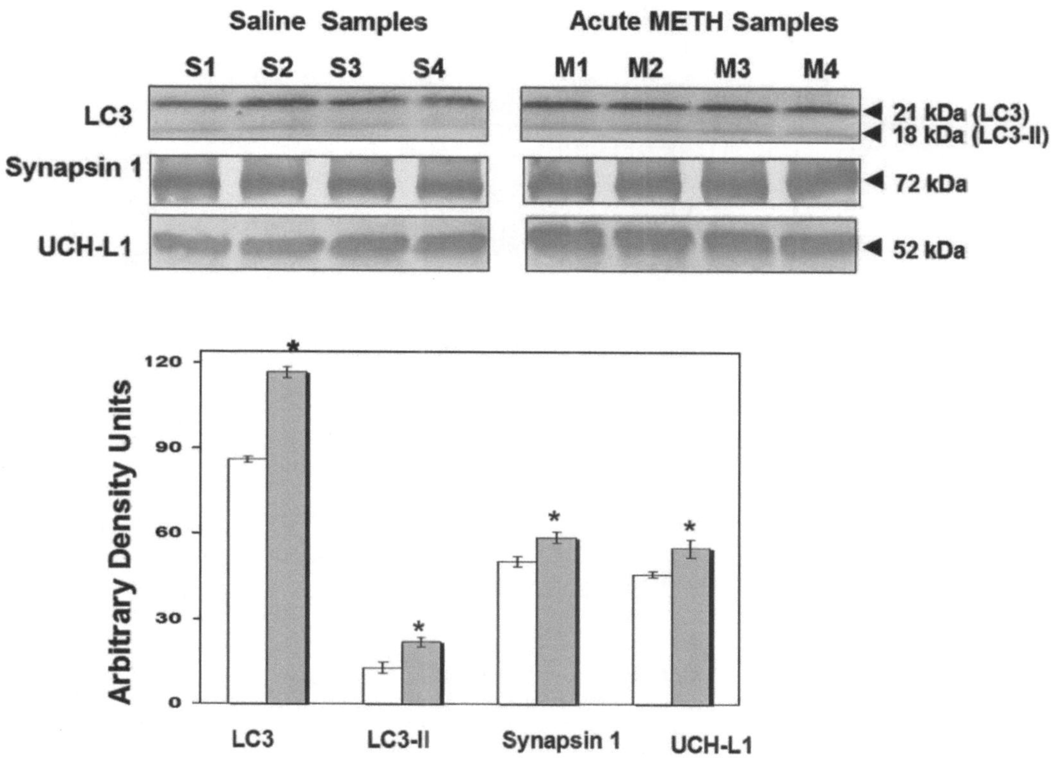

Fig. 3. Immunoblotting validation of acute METH proteins in individual saline and METH cortex. Immunoblot analysis of intact UCHL-1 (52 kDa), synapsin-1 (72 kDa), and intact LC3 (21 kDa) and LC3-II (18 kDa) proteins from four cortical METH-treated samples and saline control samples ($n = 4$). These immunoblots show higher protein abundance in acute METH samples than in the saline samples. Graphical densitometric analysis shows elevated proteins (UCH-L1, synapsin-1, LC3, and LC3-II) in METH-treated samples than in the saline controls. Saline (S) samples are represented by *white bars* and acute METH (M) samples are represented by *gray bars*. Reprinted in part with permission from *(18)*, copyright 2008 American Chemical Society.

3.7. Psychoproteomic Interaction Map of the METH Differential Proteins

Protein interaction map of acute METH proteins was searched using the Stratagene database (www.stratagene.com) and was built by using Stratagene PathwayArchitect™ software package version 2.0.1 (Stratagene, La Jolla, CA). PathwayArchitect™ is a software for analyzing functional, posttranslational modifications and metabolism interactions. It allows for the identification and visualization of pathways, constructing regulation networks and proteomic interaction maps. An example of the generated psychoproteomic map for acute METH treatment is shown in **Fig. 4** *(18, 20)*.

4. Notes

1. Prepare fresh DTT for the lysis buffer.
2. Use a fresh protease inhibitor cocktail tablet every time you prepare the lysis buffer.

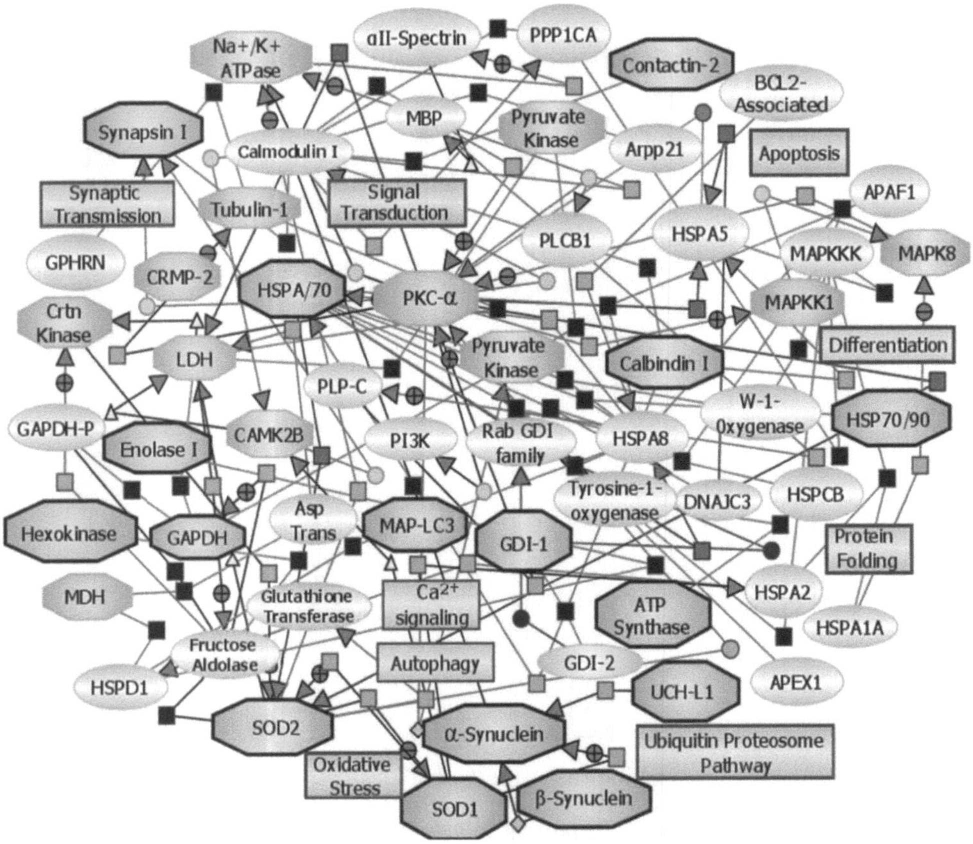

Fig. 4. A schematic of the interaction psychoproteome map of differential proteins following acute METH treatment. Shown is a schematic of the interaction functional psychoproteome map post acute METH treatment. Proposed functional processes of altered protein expression have been highlighted. The constructed psychoproteomic interaction map highlights the potential role of oxidative stress, apoptosis, neural differentiation, and the autophagic pathways as major events involved following METH treatment. Reprinted in part with permission from *(18)*, copyright 2008 American Chemical Society.

3. Always add β-mercaptoethanol freshly and make sure that you are working under a hood.
4. Double-distilled water with resistivity up to 18.2 MΩ cm and total organic content less than 1 ppb can be used to prepare some solutions.
5. The bedding was removed during the challenge phase of METH treatment because preliminary experiments had shown that high doses of METH cause rats to ingest excessive amounts of the bedding. This contributed to adverse health effects or mortality.
6. Washing of the mortar, pestle, and spatula should be performed when you are using different animal groups such as saline versus METH treated. No need for washing when homogenizing tissues within animals of the same group.
7. The protein supernatant is always kept at 4°C or in ice until collection so as to prevent proteolysis.

8. Brain sample lysates used in the DC protein assay were diluted up to 1/5 and 1/10 of original volume (with HPLC water) so that they can lie within the linear range of DC protein assay (0.125–2 μg/μl).
9. Check precast gel expiration date due to inconsistent results in expired gel.
10. Preferably, run the gel at 4°C or on ice to have a better running of the proteins.
11. The gels are washed with distilled water to remove any residual SDS left from the running buffer.
12. Destaining of the solution can be done by varying the methanol concentration. It can be started with 50% (v/v) methanol; 7% (v/v) acetic acid for 30 min, followed with 30% (v/v) methanol; 7% (v/v) acetic acid for 30 min, followed with 10% (v/v) methanol; 7% (v/v) acetic acid.
13. 10% Acetic acid is used as a fixative that helps retain proteins in the gel band.
14. Blocking solution using 5% nonfat dry milk in TBST is used so as to prevent nonspecific antibody binding and avoid nonspecific bands. It can be achieved at room temperature for 1 or 24 h at 4°C.
15. Primary antibody solutions can be stored at 4°C, to be used another time; however, 0.05% of sodium azide should be added to prevent bacterial contamination.

Acknowledgments

The author would like to thank Dr. Matthew Warren for the technical assistance. This work was supported in part by the Donald and Irene Dizney Eminent Scholar Chair, held by Mark S. Gold, M.D., Distinguished Professor, McKnight Brain Institute and also by the Department of Defense (DOD) grant #DAMD17-03-1-0066, NIH grant #*R01 NS049175-01 A1*.

References

1. Perez, J. A., Jr., Arsura, E. L., and Strategos, S. (1999) Methamphetamine-related stroke: four cases, *J. Emerg. Med.* **17**, 469–471.
2. Lan, K. C., Lin, Y. F., Yu, F. C., Lin, C. S., and Chu, P. (1998) Clinical manifestations and prognostic features of acute methamphetamine intoxication, *J. Formos. Med. Assoc.* **97**, 528–533.
3. NSDUH Report (2006) Methamphetamine use, abuse, and dependence: 2002–2004, *National Survey on Drug Use and Health*, http://www.oas.samhsa.gov/2 k7/meth/meth.cfm (accessed on November 11, 2008).
4. Sokolov, B. P., and Cadet, J. L. (2006) Methamphetamine causes alterations in the MAP kinase-related pathways in the brains of mice

that display increased aggressiveness, *Neuropsychopharmacology* **31**, 956–966.

5. Warren, M. W., Zheng, W., Kobeissy, F. H., Cheng Liu, M., Hayes, R. L., Gold, M. S., Larner, S. F., and Wang, K. K. (2006) Calpain- and caspase-mediated alphaII-spectrin and tau proteolysis in rat cerebrocortical neuronal cultures after ecstasy or methamphetamine exposure, *Int. J. Neuropsychopharmacol.* **9**, 1–11.
6. Jimenez, A., Jorda, E. G., Verdaguer, E., Pubill, D., Sureda, F. X., Canudas, A. M., Escubedo, E., Camarasa, J., Camins, A., and Pallas, M. (2004) Neurotoxicity of amphetamine derivatives is mediated by caspase pathway activation in rat cerebellar granule cells, *Toxicol. Appl. Pharmacol.* **196**, 223–234.
7. Pu, C., Broening, H. W., and Vorhees, C. V. (1996) Effect of methamphetamine on glutamate-positive neurons in the adult and developing rat somatosensory cortex, *Synapse* **23**, 328–334.
8. Bowyer, J. F., Pogge, A. R., Delongchamp, R. R., O'Callaghan, J. P., Patel, K. M., Vrana, K. E., and Freeman, W. M. (2007) A threshold neurotoxic amphetamine exposure inhibits parietal cortex expression of synaptic plasticity-related genes, *Neuroscience* **144**, 66–76.
9. Cadet, J. L., Ordonez, S. V., and Ordonez, J. V. (1997) Methamphetamine induces apoptosis in immortalized neural cells: protection by the proto-oncogene, *bcl-2*, *Synapse* **25**, 176–184.
10. Cadet, J. L., Jayanthi, S., and Deng, X. (2003) Speed kills: cellular and molecular bases of methamphetamine-induced nerve terminal degeneration and neuronal apoptosis, *FASEB J.* **17**, 1775–1788.
11. Jayanthi, S., Deng, X., Noailles, P. A., Ladenheim, B., and Cadet, J. L. (2004) Methamphetamine induces neuronal apoptosis via cross-talks between endoplasmic reticulum and mitochondria-dependent death cascades, *FASEB J.* **18**, 238–251.
12. Staszewski, R. D., and Yamamoto, B. K. (2006) Methamphetamine-induced spectrin proteolysis in the rat striatum, *J. Neurochem.* **96**, 1267–1276.
13. Warren, M. W., Kobeissy, F. H., Liu, M. C., Hayes, R. L., Gold, M. S., and Wang, K. K. (2006) Ecstasy toxicity: a comparison to methamphetamine and traumatic brain injury, *J. Addict. Dis.* **25**, 115–123.
14. Warren, M. W., Kobeissy, F. H., Liu, M. C., Hayes, R. L., Gold, M. S., and Wang, K. K. (2005) Concurrent calpain and caspase-3 mediated proteolysis of alpha II-spectrin and tau in rat brain after methamphetamine exposure: a similar profile to traumatic brain injury, *Life Sci.* **78**, 301–309.
15. Wallace, T. L., Vorhees, C. V., Zemlan, F. P., and Gudelsky, G. A. (2003) Methamphetamine enhances the cleavage of the cytoskeletal protein tau in the rat brain, *Neuroscience* **116**, 1063–1068.
16. Larsen, K. E., Fon, E. A., Hastings, T. G., Edwards, R. H., and Sulzer, D. (2002) Methamphetamine-induced degeneration of dopaminergic neurons involves autophagy and upregulation of dopamine synthesis, *J. Neurosci.* **22**, 8951–8960.
17. Kanthasamy, A., Anantharam, V., Ali, S. F., and Kanthasamy, A. G. (2006) Methamphetamine induces autophagy and apoptosis in a mesencephalic dopaminergic neuronal culture model: role of cathepsin-D in methamphetamine-induced apoptotic cell death, *Ann. NY Acad. Sci.* **1074**, 234–244.
18. Kobeissy, F. H., Warren, M. W., Ottens, A. K., Sadasivan, S., Zhang, Z., Gold, M. S., and Wang, K. K. (2008) Psychoproteomic analysis of rat cortex following acute methamphetamine exposure, *J. Proteome Res.* **7**, 1971–1983.
19. Ottens, A. K., Kobeissy, F. H., Wolper, R. A., Haskins, W. E., Hayes, R. L., Denslow, N. D., and Wang, K. K. (2005) A multidimensional differential proteomic platform using dual-phase ion-exchange chromatography-polyacrylamide gel electrophoresis/reversed-phase liquid chromatography tandem mass spectrometry, *Anal. Chem.* **77**, 4836–4845.
20. Kobeissy, F. H., Sadasivan, S., Zheng, Z., Gold, M. S., and Wang, K. K. W. (2008) Psychiatric research: psychoproteomics, degradomics and systems biology. *Expert Rev. Proteomics* **5**, 293–314.

Chapter 16

Shotgun Protein Identification and Quantification by Mass Spectrometry in Neuroproteomics

Bingwen Lu, Tao Xu, Sung Kyu Park, Daniel B. McClatchy, Lujian Liao, and John R. Yates, III

Summary

Shotgun proteomics is based on identification and quantification of peptides from digested proteins using tandem mass spectrometry. In this chapter, we discuss computational methods to analyze tandem mass spectra of peptides, including database searching, de novo peptide sequencing, hybrid approaches, library searching, and unrestricted modification search. A special focus is given to database searching programs, since they are the most widely used. The process of inferring proteins from identified peptides is then discussed. We also provide description of key steps in the quantitative analysis of mass spectrometry proteomics data. These methods are valuable tools for discovery and hypothesis-driven analyses in neuroproteomics.

Key words: Shotgun proteomics, Mass spectrometry, Quantification, Neuroproteomics, Bioinformatics, Database searching

1. Introduction

The mammalian brain possesses stunning complexity with hundreds of anatomical and functional distinct brain regions, each with a comparable number of distinct neuronal cell types. To obtain a comprehensive picture of brain function from these intricate neural networks, large-scale analyses are required. Several large-scale gene analysis projects have begun to provide tens of thousands of gene expression profiles throughout the brain *(1)*. Essential neuronal information, however, is absent from these genomic analyses, including cellular localization and posttranslational modifications (PTMs). In a neuron, a protein may be localized to different com-

Andrew K. Ottens and Kevin K.W. Wang (eds.), *Neuroproteomics,* Methods in Molecular Biology, vol. 566
DOI 10.1007/978-1-59745-562-6_16, © Humana Press, a part of Springer Science+Business Media, LLC 2009

partments, i.e., the cell body or the synapse, and in a diseased brain, proteins can aggregate, altering its localization to inclusion bodies. The disruption of protein localization can lead to brain malfunction. For example, Fragile-X syndrome, the most common inherited mental retardation, is caused by a mutation in the fmr1 gene that encodes the Fragile -X mental retardation protein (FMRP), which transports RNA from the cell body to the synapse for local translation (2). The mutation leads to transcriptional silencing of the gene, and therefore the disruption of local expression of a group of synaptic proteins *(2)*. PTMs, such as phosphorylation, glycoslyation, and ubiquitation, are directly related to a protein's activity, interactions, and localization. Alterations in PTM are reported in many neurological disorders. For instance, the hyperphosphorylation of the tau protein is a pathological hallmark of many human dementias, and mutations in the ubiquitin system have been identified in Parkinson's disease, a neurodegenerative movement disorder *(3, 4)*. Both cellular localization and PTMs can be studied using shotgun proteomics. Thus, the application of mass spectrometry to analyze the brain proteome is an essential counterpart to large scale genomic analyses.

A proteome is the collection of all proteins of an organism. Proteomics is the study of the diverse properties and functions of proteins as a function of time and physiological conditions. Proteomes are dynamic, responding to cellular and environmental challenges. Many different technologies have been developed for proteomic studies. These technologies include two-dimensional polyacrylamide gel electrophoresis (2D PAGE) *(5)*, protein microarrays *(6)*, mass spectrometric (MS) characterization and identification of intact proteins (also known as "top-down" proteomics) *(7, 8)*, MS identification of digested proteins ("bottom-up" proteomics), and methods based on the use of genetics. Among these tools, bottom-up mass spectrometry currently plays a key role in qualitative and quantitative analysis in proteomics.

An overview of shotgun proteomic analysis is schematically illustrated in **Fig. 1** [or *see* **ref.** *9*]. A protein mixture is digested using proteolytic enzymes such as trypsin. The resulting peptides are then separated by one- or multidimensional liquid chromatography (LC). The separated peptides are then ionized and introduced into the instrument for mass spectrometry analysis. Traditional MS/MS data acquisition consists of two stages. The first stage is the full scanning of all peptide ions introduced into the instrument at a given time to obtain an MS spectrum. During the second stage, peptide ions, either selected by intensity from the full scan analysis (data-dependent acquisition), or selected by a predefined mass range (data-independent acquisition) *(10)*, are subjected to fragmentation by collision-induced dissociation (CID) before the MS/MS spectrum is acquired. The MS/MS spectrum acquired at the second stage is a record of all m/z values with associated intensities of the resulting fragment ions

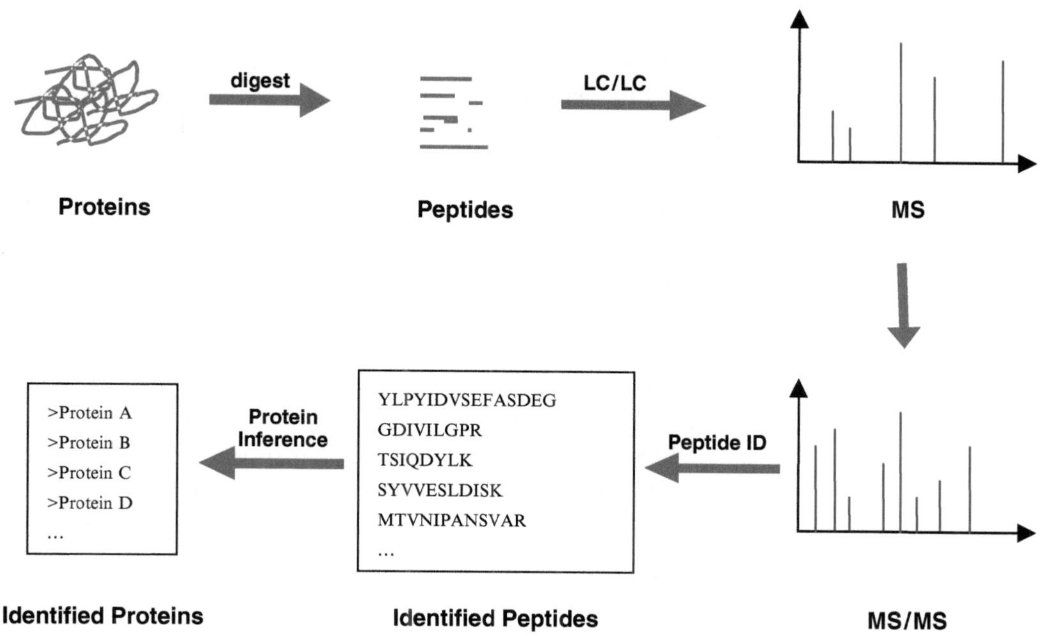

Fig. 1. Experimental procedure for MudPIT and data analysis. A protein mixture is digested with an endoprotease, such as trypsin, to generate digested peptides. The digested peptides are separated by multiple dimensional liquid chromatography before being sprayed into a mass spectrometer. MS and MS/MS spectra are acquired by the mass spectrometer. Computer software tools are employed to infer peptide sequences, and finally a list of identified proteins is produced.

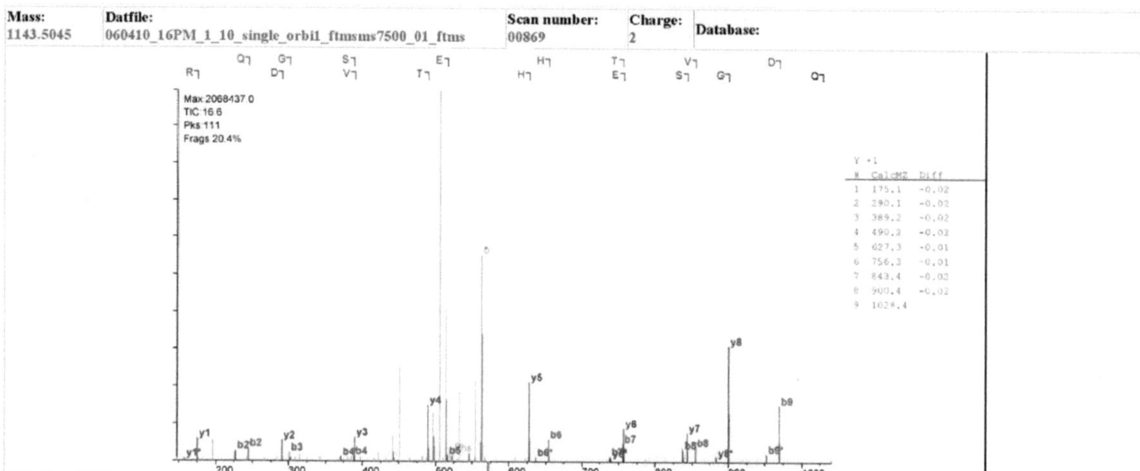

Rank/Sp	(M+H)+	deltCn	XCorr	Sp	Ions	Reference	Peptide
1/1	1143.50305	0.0000	3.2761	7.612	15/18	17PM_#10|body(1:1)|sp|P00921|CAH2_BOVIN	D.DQGSEHTVDR.K
2/41	1143.49182	0.4302	1.8668	3.448	10/18	SPBC19G7.02	L.ENGSIHEDDK.G
3/20	1143.50707	0.4420	1.8282	3.334	10/20	SPAC664.01c	E.FNENGPPSGQP.N
4/78	1143.54346	0.4588	1.7730	3.171	9/16	SPCC757.12	T.DANYYQTIR.T
5/57	1143.53944	0.4656	1.7508	3.105	9/18	SPBC1604.15	T.SDNHTVTVDR.S

Fig. 2. An example of tandem mass spectrum.

under a mass range such as 200–2,000 (**Fig. 2**). After the MS and MS/MS data are acquired, efforts are directed to the identification and quantification of peptides by computational analysis.

The sequences of the peptides that gave rise to the tandem mass spectra are inferred by using various software tools, such as database searching tools (SEQUEST *(11)*, Mascot *(12)*), de novo peptide sequencing tools (PepNovo *(13)*, PEAKS *(14)*), sequence tag database search tools [GutenTag *(15)*, Inspect *(16)*], spectrum library search tools [LIBQUEST *(17)*, BiblioSpec *(18)*], or an unrestricted modification search [MSAlign *(19)*, TwinPeaks *(20)*]. Among these tools, the most commonly used approach for peptide identification nowadays is still database searching. After peptides are identified, protein inference is carried out to obtain protein identifications. Peptide and protein quantification follows by examination of the mass spectral information.

There are currently many approaches to identify peptides from MS/MS data. One needs to bear in mind that each of these peptide identification approaches has its own advantages and limitations. We discuss these approaches in the following sections, beginning with the discussion of database searching methods.

2. Peptide Identification Using MS/MS Database Searching

2.1. Basic Principles

The goal of a tandem mass spectral database search is to identify the best sequence match to the spectrum. All MS/MS database search tools operate in a similar way. They return the best matching peptide found in the database for each input spectrum, unless there is no candidate peptide in the database that satisfy the search parameters specified by the user. SEQUEST *(11)* and Mascot *(12)* are the two most widely used database search engines.

Generally there are three major steps for a database search program to find the peptide in the database that is a best match to a given tandem mass spectrum (a) candidate peptide selection; (b) preliminary scoring; and (c) final scoring and peptide ranking.

The first step is to select candidate peptides that are within some mass tolerance of the peptide represented in the tandem mass spectrum, enzyme specificity, and other constraints that are specified in the search parameter file. Some programs begin by placing an enzyme specificity constraint, while others begin with the peptide mass constraint, or short stretches of sequence interpreted from the fragmentation pattern. In order to speed up the candidate peptide selection process, some programs preindex the sequence database based on peptide mass and the enzyme specificity. Preprocessing a database makes a search go much faster. For example, tryptic peptides can be indexed and mass lists can be premade. The downside is that a separate indexed database needs to be created for each enzyme or each time a database is updated.

The second step is a preliminary scoring step that is used to create a short list of candidate peptides. This preliminary scoring step is important for the speed of the identification process, since the final scoring algorithms are usually slow, making it impractical to score every candidate sequence. Preliminiary scoring is typically performed based on the number of shared peaks between the experimental spectrum and the theoretical spectrum.

The third step is final scoring and ranking sequence matches. A final score is computed for each of the peptide in the candidate list selected by the preliminary scoring routine. The final scoring uses one of the following two methods to measure closeness of fit between spectra and peptide sequences. The first method uses a shared peak count model to generate a quantitative measure of the fit. The second method uses fragment ion frequency to generate the probability that the sequence and the spectrum are the best fit. The final scoring is usually more sophisticated and sensitive than the preliminary scoring method. All the candidate peptides are ranked by the final score, and the sequence with the top final score is considered as the best match to the experimental spectrum.

A list of common MS/MS database search engines is given in **Table 1**. Two additional reviews *(27, 28)* are also helpful if the reader would like to know more about MS/MS database search algorithms.

2.2. Preliminary Scoring

The most commonly used preliminary scoring method is the number of shared peaks between the experimental spectrum and the theoretical spectrum. This overly simplified approach is used by most probability-based database search programs, such as PepProbe *(25)*, OMSSA *(23)*, and X!Tandem *(26)*.

SEQUEST employs a different preliminary scoring method to derive preliminary score (S_p). S_p sums the peak intensity of fragment ions matching the predicted sequence ions and accounts for the continuity of an ion series and the length of a peptide. The original S_p score is

$$S_p = \frac{\left(\sum_k I_k\right) m(1+\beta)(1+\rho)}{L},$$

where the first term in the product is the sum of ion abundances of all matched peaks, m is the number of matches, β is a "reward" for each consecutive match of an ion series (for example, 0.075), ρ is a "reward" for the presence of an ammonium ion (for example 0.15), and L is the number of all theoretical fragment ions for an amino acid sequence.

ProLuCID *(24)*, a database search algorithm recently developed in the Yates laboratory uses an approximated binomial

Table 1
MS/MS database search programs

Program	Category	Web site, reference, or company
SEQUEST	C	http://www.thermo.com/ Eng et al. *(11)*
Mascot	C free online	http://www.matrixscience.com/ Perkins et al. *(12)*
MasLynx	C	http://www.waters.com/ Clauser et al. *(21)*
MS-Tag/MS-Seq	P	http://prospector.ucsf.edu/ Clauser et al. *(21)*
Sonar	C	Field et al. *(22)*
OMSSA	P	http://pubchem.ncbi.nlm.nih.gov/omssa/ Geer et al. *(23)*
ProLuCID	N	http://fields.scripps.edu/prolucid/ Xu et al. *(24)*
PepProbe	P	http://bart.scripps.edu/public/search/pep_probe/search.jsp Sadygov and Yates *(25)*
X!Tandem	P	http://www.thegpm.org/TANDEM/ Craig and Beavis *(26)*
SpectrumMill	C	http://www.chem.agilent.com/

C Commercial, *P* publicly available, *N* need to contact authors for availability

probability score for preliminary scoring. This binomial probability score is computed using the following formulae

$$P(x >= m) = \sum_{k=m}^{n} P(x = k)$$
$$\text{where } P(x = k) = \frac{n!}{k!(n-k)!} p^k (1-p)^{(n-k)}, \tag{1}$$

where n is the number of theoretical peaks of the candidate peptide tested, which is determined by the peptide length together with the minimum and maximum m/z in the tandem mass spectrum;

m is the number of theoretical peaks that match to a peak in the experimental spectrum and is guaranteed to be less than or equal to n; p is the probability that any fragment ion matches a peak in the spectrum, which is determined by the mass tolerance for a fragment ion match and the density and distribution of peaks in the experimental spectrum. The binomial probability score $P(x >= m)$ is the probability of getting m or more matches when n theoretical peaks are tested. By design, the binomial probability score computed by ProLuCID is database independent and is solely dependent on characteristics of the spectrum and the peptide sequence.

2.3. Final Scoring

There are two major types of final scores. One is a similarity score that measures the closeness of fit between the theoretical spectrum and the experimental spectrum. A cross-correlation score (used in SEQUEST) or dot-product score (used in Sonar) *(22)* have been used for this purpose. Another type of final score is a probability or statistical score that measures how likely a given hit is a random hit.

The cross-correlation score is used by SEQUEST to measure the similarity between the theoretical spectrum and the experimental spectrum:

$$\mathrm{Corr}(E,T) = \sum X_i y_i + \tau.$$

The correlation is processed and averaged to remove the periodic noise in the interval (–75 to +75).

The hypergeometric probability score and Poisson probability score are used by PepProbe *(27)* for the final scoring. The hypergeometric probability score can be expressed as

$$P_{K,N}(K_1, N_1) = \frac{C_K^{K_1} X C_{N-K}^{N_1-K_1}}{C_N^{N_1}}.$$

The Poisson probability score is

$$P(x,u) = \frac{u^x}{x!}\mathrm{e}^{-u} \quad \text{where } u = \frac{C_K^{K_1} X C_{N-K}^{N_1-K_1}}{C_N^{N_1}}.$$

Two recently developed open source database search program, X!Tandem *(26)* and OMSSA *(23)*, also compute an E-value as its final score. An E-value for a hit is a score that is the expected number of random hits from a database to a given spectrum such that the random hits have an equal or better score than the hit as follows:

$$E = n \sum_{s \geq s_m} P(s).$$

Alternatively, ProLuCID computes a *Z* score for each final candidate peptide. For each spectrum, there should only be one correct answer and all the other candidate peptides can be considered as random matches. We have found that the distribution of XCorrs for the top 500 peptide hits for each spectrum to be very close to a normal distribution with the true hit being an obvious outlier and statistically significantly different from the other final candidates. There are many ways to detect outliers from normal distributions and the *Z* score of the Grubbs' test *(29)* is used in ProLuCID. The *Z* score is calculated as the difference between the outlier and the mean divided by the standard deviation SD (**Eq. 2**). A large *Z* score means that the XCorr of the top hit is significantly different from the other hits so that the peptide is more likely to be a true hit,

$$Z = \frac{X - \mu}{\mathrm{SD}} \quad \text{where } \mathrm{SD} = \sqrt{\frac{\sum_{i=1}^{n}(x_i - \mu)^2}{n-1}}. \tag{2}$$

X is the XCorr of the top hit, μ is the mean XCorr of all the final candidate peptides, and *n* is the number of final candidate peptides.

2.4. Database Search Parameters

Following are some parameters that are important for database search programs. The users of database search programs are usually allowed to configure these parameters.

1. Protein database: A protein database has to be specified by the user for the database search program to use. The search program will consider each peptide in the protein database that fulfills the user specified search conditions, such as peptide mass tolerance, enzyme specificity, possible modifications, etc. The selection of a protein database is important and will be discussed in more details in **Subheading 2.5**.
2. Types of sequence ions: Precursor ions can be fragmented using different dissociation mechanisms, such as CID, electron capture dissociation (ECD) *(30)*, or electron transfer dissociation (ETD) *(31)*. Different dissociation methods will produce different types of fragment ions. For example, the major fragment ion series in CID spectra are B- and Y-ions, while fragment ions in ETD spectra are mostly C- and Z-ions. The database search programs need to know the spectrum type in order to generate the theoretical spectrum and compare the theoretical spectrum with the experimental spectrum.
3. Peptide and fragment mass type: The user also needs to specify the method of calculating the peptide (precursor) and the fragment ion mass. This can be done using either monoisotopic or average masses.

4. Peptide and fragment ion mass tolerances: In order to select candidate peptides, the search program compares calculated peptide molecular weight with the measured precursor molecular weight. Generally, the narrower you can set the parent mass tolerance, the faster the search will go because fewer peptides will need to be correlated in the next step. Similarly, the fragment ion mass tolerance is used when the search program compares the peaks in the theoretical spectrum with the peaks in the experimental spectrum. One needs to be certain that the accuracy of the mass spectrometer is not exceeded when setting the parent and fragment ion mass tolerances in the search (e.g., tolerance set too narrow).
5. Enzyme specificity: Most biological samples are digested with a protease before mass spectrometry, and typically trypsin is used. Accordingly a database search program can be configured to search for peptides that fulfill the enzyme specificity used to create the peptides, and any peptide in the database that does not fulfill both the molecular weight and enzyme specificity constraints will not be considered.
6. Static and differential modifications:
   (a) A static or fixed modification is a mass shift that occurs on every specified residue in all proteins, e.g., C +57 for carboamidomethylation on Cysteine.
   (b) Differential or variable modification is a mass shift of a PTM that possibly is present on certain amino acid residues. The user usually needs to specify the mass shift, residues where the mass shift may occur, together with the maximum number of differential modifications to consider per candidate peptide. The number of differential modifications to be considered will usually dramatically affect the database search speed. The more modifications considered, the longer the search will take as the process scales 2^n, where n is the number of potential modifications sites in the peptide.

2.5. Selection of Protein Sequence Database

Identification of proteins using mass spectrometry relies on the existence of reliable protein sequence databases. Ideally, an organism-specific database should be used if the protein database is relatively complete. This is probably true for most of the model organisms. The most commonly used public databases are the International Protein Index (IPI, http://www.ebi.ac.uk/IPI/) database collection maintained by European Bioinformatics Institute (EBI), the National Center for Biotechnology Information's (NCBI) Entrez protein database (http://www.ncbi.nlm.nih.gov/sites/entrez?db = Protein), the NCBI Reference Sequence (RefSeq, http://www.ncbi.nlm.nih.gov/RefSeq/) database, and the Universal Protein Resource (UniProt, http://www.pir.uniprot.org/) protein database that combines Swiss-Prot, TrEMBL, and PIR. Currently, the EBI_IPI database provides protein FASTA databases of seven model organisms, including human, mouse, rat, zebrafish, Arabidopsis, chicken, and cow. The flybase (http://flybase.bio.indiana.edu/), wormbase (http://www.wormbase.org/), and Saccharomyces Genome Database (SGD, http://www.yeastgenome.org/) are other commonly used public databases.

If the protein sequence of the target organism is incomplete but the genome is sequenced, six-frame translation or gene-finding programs can be used to generate predicted proteins. For most organisms, however, the vast majority of sequence information for any given organism remains as cDNA sequences in the form of either single-pass sequenced expressed sequence tags (ESTs) or (more rare) complete full-length cDNA sequences. The use of EST databases should be cautioned, since single-pass EST sequence data are usually of poorer quality.

If there is no genomic or transcript data available for the organism, the last resort would be using the protein nonredundant (NR) database from NCBI or the UniProt protein database. Alternatively, one can use de novo peptide sequencing and MS-BLAST to identify the proteins, as discussed below.

2.6. Evaluation of the Search Results

Hundreds of thousands of tandem mass spectra can be routinely generated in a shotgun proteomics experiment. A database search program will find one best matching peptide for each of these spectra even though many of the spectra may be from sequencing attempts on chemical noise, may be nonpeptides, or are of poor quality. It is known that after the database searching is completed, but before any filtering is performed, the majority of the identifications are false positives that result from random hits to the database or matches of MS/MS of background noise to sequences, and only a small proportion of peptide identification results are true hits. Postdatabase search-filtering programs, such as DTASelect *(32)* and PeptideProphet *(33)*, are essential for the optimal separation of true peptide/protein matches from random matches. For a peptide to be successfully identified by a database search algorithm, it has to pass the following three tests (1) it must be ranked high enough in the preliminary scoring to be selected for final scoring, (2) it must be assigned to the top rank during the final scoring, and (3) its score or scores have to be high enough to pass the postsearch filtering criteria. The major challenge to improving the overall performance of a database search algorithm is to increase the sensitivity of searches while maintaining adequate discrimination between correct answers and false positives.

The DTASelect algorithm is a powerful tool for filtering database search results. Traditionally, filtering of SEQUEST results by DTASelect used threshold cutoffs for XCorr and DeltaCN where DeltaCN is the difference between the top hit XCorr and the second best hit XCorr divided by the XCorr of the top hit. In the latest version of DTASelect (DTASelect2) *(34)*, these two measurements are combined using a quadratic discriminant function to compute a confidence score to achieve a user-specified false discovery rate.

2.7. Sources of Failure

Here we discuss some sources of failure to correctly assign peptide sequences to MS/MS spectra that truly represent peptides and not simply chemical noise or other background contaminants. The reader is also referred to the review by Nesvizhskii *(28)* for a similar discussion.

1. *Poor MS/MS quality.* Low abundant peptide precursors usually result in poor quality MS/MS spectra, and some peptides may not fragment well because of their chemical properties.
2. *Search constraints.* In most of the shotgun proteomics experiments, it is true that the majority of peptides are fully cleaved by trypsin digestion. However, half- and nontryptic peptides can be detected in biological samples, in part, because of nonspecific cleavage by the enzyme, the presence of in vivo protease activities, or in-source fragmentation of peptides. Thus, fully tryptic searches will miss half- or nontryptic peptides, especially the peptides close to the N terminus/amino terminus of the protein, since the protein sequence in the database usually may or may not contain the initial Met, and often protein sequences in the database may contain additional amino acids residues like that of a signaling peptide.
3. *Unanticipated modifications.* If the spectrum is generated from a peptide with a modification(s) that is not specified in the search parameters, then the mass of the peptide will not match the precursor mass, thus it will not be considered by the database search algorithm and will not be identified. These modifications include chemical modifications (such as sodium and potassium adducts, in vitro carbamylation, loss of water) and biological PTMs (phosphorylation, in vivo carbamylation, methylation, acetylation, etc.).
4. *Spectrum charge state determination errors.* For high mass accuracy data, such as those produced by LTQ-Orbitrap instruments, the charge state of a precursor ion usually can be correctly assigned. However, for low resolution data, the charge state of multiply charged precursor ions may be ambiguous. A common practice is searching the spectrum twice, once assuming a +2 charge and then a +3 charge. However, this approach will miss the peptides with precursor charge states higher than +3.
5. *Peptide sequence not in the database.* Most protein databases are created by gene translation. It is possible that some of the

proteins may not be present in the database or the stop and start sites were incorrectly interpreted, or alternate splicing was incorrectly predicted. If a protein sequence is not in the database, due to a SNP, a mutation, sequencing or prediction errors, then there is no way that the database search program can identify its peptide sequences.

6. *Deficiencies of the scoring scheme or postdatabase search filtering*. Even for high quality MS/MS peptide spectra, the database search program may still fail to identify its sequence.

3. Other Methods for Peptide Identification from MS/MS

If database searching fails to identify a sequence for a tandem mass spectrum, there are currently several other approaches to identify the corresponding peptide. De novo sequencing, combining sequence tags and database searching, library searching, and an unrestricted blind search can be performed.

3.1. De Novo Peptide Sequencing

De novo peptide sequencing is used to infer peptide sequences from tandem mass spectra data without using a sequence database to aid in the interpretation. Usually, searching against a sequence database is the first choice for peptide identification. However, de novo peptide sequencing comes into play in various situations. First, the protein of interest might not be present in the sequence database. For example, the sequence database might be incomplete, which is still the situation for many animals and plants, or the protein of interest might be a novel protein whose sequence information is not available from the sequence database. Second, there could be prediction errors in gene-finding programs. Thus, it might not be possible to find the true protein from the predicted protein database. Third, genes might undergo alternative splicing, which would result in novel proteins. The occurrence of single nucleotide polymorphisms (SNPs) in coding regions may also lead to different protein variants. In a fourth case, de novo sequencing can be helpful for studying amino acid mutations and protein modifications. Finally, when a database search generates ambiguous results, de novo sequencing can be used as a potential validation tool.

De novo peptide sequencing has its limitations, one being the necessity for a reasonably complete backbone cleavage between each pair of adjacent amino acids. In other words, one needs a complete ion series (b, y, c, or z) to infer the full peptide sequence accurately, and CID does not usually produce a complete ion series. Furthermore, internal peptide fragmentation is common.

The recent fragmentation technology, ECD *(30)* and ETD *(31)*, tends to generate more complete ion series, and could potentially be more suitable for de novo peptide sequencing. Researchers have also tried to combine complementary fragmentation techniques, CID and ECD, for de novo sequencing *(35)*. Mass accuracy plays an important role in de novo peptide sequencing *(36, 37)*, and since high accuracy mass spectrometers are usually more expensive, this is another limitation on de novo peptide sequencing.

Follow up verification of peptide sequences from de novo analysis takes additional effort, because high confidence de novo peptide sequencing from tandem mass spectra is a difficult problem. For this reason, de novo peptide sequencing is rarely applied to complex mixtures. Only recently has de novo peptide sequencing been shown to conceptually work on a complex mixture by employing both high mass accuracy CID and high mass accuracy ECD techniques *(35)*.

Various de novo sequencing programs have been developed over the years (**Table 2**). Some of these programs are publicly

Table 2
De novo peptide sequencing programs

Program	Category	Reference, Web site, or company
Lutefisk	P	http://www.hairyfatguy.com/lutefisk/ Taylor and Johnson *(38)*
MSNovo	P	http://msms.cmb.usc.edu/supplementary/msnovo/ Mo et al. *(39)*
NovoHMM	P	http://people.inf.ethz.ch/befische/proteomics Fischer et al. *(40)*
PepNovo	P	http://peptide.ucsd.edu/pepnovo.html Frank et al. *(13)*
PEAKS	C	http://www.bioinfor.com/products/peaks/ Ma et al. *(14)*
MassSeq	C	Micromass/Waters
DeNovoX	C	http://www.thermo.com
EigenMS	N	Bern et al. *(41)*
PILOT	N	DiMaggio and Floudas *(42)*

C Commercial, *P* publicly available, *N* need to contact authors for availability

available, such as PepNovo and Lutefisk. Some are only available as commercial software, such as PEAKS. Most of these programs are geared toward tandem mass spectra generated by CID fragmentation technology. Some of the earlier algorithms have previously been reviewed *(43)*.

3.2. Hybrid Approach

The hybrid approach combines de novo peptide sequencing and database searching for peptide identification. This approach starts by inferring short peptide sequences from MS/MS data by de novo sequencing, followed by a database search constrained by the sequence tag information. The short peptide sequences are usually referred to as sequence tags. De novo interpretation of short stretches of sequence is often more accurate and reliable than trying to sequence full length peptides from a tandem mass spectrum. There are two methods that use sequence tags for database searching.

For the first method, the sequence tags are used to eliminate candidate peptides for MS/MS database searching *(15, 16, 44)*. The database searching process can be potentially sped up since the sequence tags obtained from de novo inference can eliminate unwanted candidate peptides thus saving time scoring the experimental spectrum against these candidate peptides. Another advantage of the sequence tag approach is that it allows error-tolerant database searches as explored by Mann and Wilm *(44)*. Tanner and colleagues showed that the filtering using sequence tags obtained from de novo sequencing can be very efficient so that it allows for fast identification of posttranslationally modified peptides *(16)*. Hybrid approaches using sequence tags and MS/MS database searching were implemented in the software tools such as GutenTag *(15)*, Inspect *(16)*, SPIDER *(45)*, and OpenSea *(46)*.

For the second method, the sequence tags obtained are not used for MS/MS database searching. Instead, the sequence tags are used for sequence-based database searching, such as BLAST *(47)* or FASTA *(48)*. In practice, even if the sequence of the studied protein is not in a protein database, chances are that the homologs of this protein are in the database. In such a situation, a sequence-based database searching program that handles homologous mutations will be able to use the sequence tags to identify the protein homologs. Two common homology search programs, FASTA and BLAST, are usually employed for sequence-based database searching using sequence tags inferred from tandem mass spectra. For example, FASTA has been employed in CIDentify *(38)* and FASTS *(49)*, while BLAST has been implemented in MS-BLAST *(50)*. Similar to de novo sequencing, the follow up verification of identified peptides from sequence-based homology searches will usually take additional effort.

A list of software tools for MS/MS data analysis using the hybrid approach is given in **Table 3**.

3.3. Library Search

Library searching matches experimental spectra to a library of peptide spectra collected from previous peptide identification studies. The library contains peptide MS/MS data that have been matched to peptides and filtered using stringent filtering criteria. Spectral library searching is a natural way to take advantage of previous peptide identifications. Actually, library searching is a commonly practiced method for identifying mass spectra of small organic molecules, by comparisons of the newly collected experimental spectra with spectra of known molecules *(52)*. About 10 years ago Yates et al. *(17)* proposed and explored the possibility of carrying out peptide spectra library searching using tandem mass spectra of peptides. Recently, with more and more MS/MS data collected and identified, this approach has drawn attention from other research groups *(18, 53–55)*.

For a traditional database search, experimental spectra are matched to theoretical mass spectra generated from a sequence database. The theoretical mass spectra are generated anew each time an experimental MS/MS spectrum is searched, thus wasting a large amount of computational time. Furthermore, even

Table 3
Programs employing hybrid approaches

Program	Category	Reference, Web site, or company
GutenTag	P	http://fields.scripps.edu/GutenTag/ Tabb et al. *(15)*
Inspect	P	http://peptide.ucsd.edu/inspect.html Tanner et al. *(16)*
OpenSea	N	Searle et al. *(46)*
SPIDER	P	http://bif.csd.uwo.ca/spider/ Han et al. *(45)*
ByOnic	P	http://bio.parc.xerox.com/home Bern et al. *(51)*
CIDentify	P	http://www.hairyfatguy.com/lutefisk/ Taylor and Johnson *(38)*
MS-BLAST	P	http://genetics.bwh.harvard.edu/msblast/Shevchenko et al. *(50)*

C Commercial, *P* publicly available, *N* need to contact authors for availability

though efforts have been made to model peptide fragmentation patterns in a mass spectrometer *(56)*, theoretical mass spectra are usually generated with the peptide fragmentation patterns not being fully modeled. Library searching provides a solution to address the above issues. By searching against the peptide spectra library, which is usually smaller, the computational time is saved. Furthermore, there is no need to model the fragmentation patterns of respective identified peptides, since they are intrinsic to the experimental spectra. Another advantage of library searching is the inclusion of spectra corresponding to modified peptides in the library. Identification of modified peptides by database search is known to be very time consuming. Furthermore, modified peptides usually go through a more sophisticated fragmentation pattern, for example, precursor neutral losses and product ion neutral losses are frequently observed from CID phosphorylated peptides *(57)*. However, once the posttranslationally modified peptides are identified, the spectra can be included in the library. Searching against the spectra of modified peptides and spectra of unmodified peptides should take the same amount of time, and the fragmentation pattern of the modified peptides is again modeled in the experimental spectra.

Library searching has its limitations. Since the search is against a library of identified peptide spectra, this approach is not suitable for discovery of previously unobserved peptides. Furthermore, constructing a spectral library for library searching is not a trivial process. Extreme precautions should be taken to not include incorrectly identified peptide spectra. Additional efforts should also be spent to reannotate the peptide spectra in the library once errors are found.

Nevertheless, once the peptide spectra libraries are established, library searching could be very useful for some studies such as targeted proteomics, where the primary goal is to repeatedly investigate a predefined set of proteins and there is no need to discover novel peptides. A list of library search tools is given in **Table 4**.

3.4. Unrestricted Modification Search

Unrestricted modification search is the identification of posttranslationally modified peptides without knowing the modification types. Protein PTMs greatly increase the complexity of the proteome. PTMs often regulate activity and function of proteins. The study of protein PTMs is very important and is one of the current focuses of proteomic studies. There are two common ways of identifying modified peptides from MS/MS data: restricted modification search and unrestricted modification search.

The first approach to identify PTMs is by restricted modification search based on a selected set of modifications by multiple considerations of potentially modified database peptides *(58)*. This approach has been widely applied in the identification

Table 4
Library searching programs

Program	Category	Reference, Web site, or company
BiblioSpec	P	http://proteome.gs.washington.edu/bibliospec/documentation/ Frewen et al. *(18)*
LIBQUEST	N	Yates et al. *(17)*
SpectraST	P	http://www.peptideatlas.org/spectrast/ Lam et al. *(54)*
X! Hunter	P	http://h201.thegpm.org/tandem/thegpm_hunter.html Craig et al. *(53)*

C Commercial, *P* publicly available, *N* need to contact authors for availability

of peptides with known modifications, and are implemented in common database search engines such as SEQUEST, Mascot, X!Tandem, and OMSSA. One limitation of restricted modification search is that only known modifications (mass shifts) can be considered.

The second approach, unrestricted modification search, has the capability of identifying peptides of unknown modifications. The first unrestricted modification search was demonstrated by Pevzner and colleagues, using a blind search based on a Smith–Waterman-like spectral alignment algorithm called MSAlign *(19)*. Havilio and Wool also showed that a variant of the SEQUEST algorithm, TwinPeaks, has the capability to carry out unrestricted modification search *(20)*. An issue arising in unrestricted modification searches is justifying the mass shifts observed. The follow-up to such findings is not usually clear.

It should be noted that identification of PTMs is computationally expensive, for either restricted or unrestricted modification searches. The unrestricted modification search will usually take even more computational time.

4. Protein Summary and Protein Inference

In most proteomic studies, the interest of researchers lies in the identification of proteins, although the interpretation starts with peptide inference from tandem mass spectra. Thus it is necessary to group peptides together to produce the identified proteins list, a process usually referred to as protein inference.

There are several issues associated with protein inference. First of all, a small false positive rate at the spectrum level usually will translate into a higher false positive rate at the peptide level, which can become even higher again at the protein level. The reason for this phenomenon is that the incorrectly assigned peptide-spectral matches can be described as random matches. Due to the random nature of this process, almost every new incorrectly assigned spectrum adds one new incorrect peptide identification. However, the correctly assigned peptide-spectrum matches are not random matches. Thus correctly identified peptides tend to have multiple copies of associated MS/MS spectra, while peptides incorrectly identified tend to have one or a few associated MS/MS spectra. The propagation of error rate at the peptide level to the protein level can be explained similarly. This issue can be addressed to some extent by requiring two or more identified peptides for each identified protein. If a user chooses to use reverse databases for the estimation of false positive rates, one can see the protein identification false positive rate drop dramatically when one increases the number of identified peptides for each identified protein **(Table 5)**.

The second issue in protein inference is the shared peptides problem. Due to the presence of homologous proteins, alternatively spliced isoforms, or redundant protein entries in the database, an identified peptide can correspond to multiple entries

Table 5
False positive rates drop when the required number of identified peptides increases to infer a protein

Required peptides #		1	2	3	4
Spectrum	FP	5.00%	1.01%	0.12%	0.00%
	Forward	14,736	13,872	13,298	12,743
	Reverse	737	140	16	0
Peptide	FP	11.43%	2.32%	0.32%	0.00%
	Forward	5,861	5,140	4,727	4,424
	Reverse	670	119	15	0
Protein	FP	40.21%	7.21%	0.86%	0.00%
	Forward	1,512	791	580	475
	Reverse	608	57	5	0

in the protein database. One approach is to use the principle of Occam's razor to address this issue. The principle of Occam's razor states that the explanation of any phenomenon should make as few assumptions as possible, eliminating those that make no difference in the observable predictions of the explanatory hypothesis or theory. Both DTASelect *(32, 34)* and ProteinProphet *(59)* implement this principle. Under this principle, if two different proteins have identical sequence coverage, the proteins should be grouped together. For example, if both protein A and protein B have three peptides identified as 1, 2, and 3, these two proteins should be treated the same. Under this principle, subset proteins should also be removed. For example, if protein A has three peptides identified as 1, 2, and 3, while protein B only has two peptides identified as 1 and 2, then B is treated as a subset protein of A. However, to allow users to be able to identify a processed protein as well as its precursor, DTASelect provides an option to show subset proteins on the inferred proteins report. **Figure 3** shows an example of DTASelect output where several proteins are grouped together since the identified peptides are common to all the proteins. See **Table 6** for a list of protein summary tools.

Locus Key:

Validation Status	Locus	Sequence Count	Spectrum Count	Sequence Coverage	Length	MolWt	pI	Descriptive Name

Similarity Key:

Locus	# of identical peptides	# of differing peptides

U	YDR098C-B	12	19	7.6%	1755	198594	8.4	YDR098C-B SGDID:S000007391, Chr IV from 651121-649817,649815-645853, reverse complement, transposable_element_gene, "TyB Gag-Pol protein; proteolytically processed to make the Gag, RT, PR, and IN proteins that are required for retrotransposition"
U	YER160C	12	19	7.6%	1755	198568	8.5	YER160C SGDID:S000000962, Chr V from 498119-496818,496816-492851, reverse complement, transposable_element_gene, "TyB Gag-Pol protein; proteolytically processed to make the Gag, RT, PR, and IN proteins that are required for retrotransposition"
U	YER138C	12	19	7.6%	1755	198561	8.1	YER138C SGDID:S000000940, Chr V from 449020-447719,447717-443752, reverse complement, transposable_element_gene, "TyB Gag-Pol protein; proteolytically processed to make the Gag, RT, PR, and IN proteins that are required for retrotransposition"
U	YDR365W-B	12	19	7.6%	1755	198658	8.0	YDR365W-B SGDID:S000007401, Chr IV from 1206987-1208291,1208293-1212255, transposable_element_gene, "TyB Gag-Pol protein; proteolytically processed to make the Gag, RT, PR, and IN proteins that are required for retrotransposition"
U	YDR316W-B	12	19	7.6%	1755	198526	8.2	YDR316W-B SGDID:S000007399, Chr IV from 1096059-1097363,1097365-1101327, transposable_element_gene, "TyB Gag-Pol protein; proteolytically processed to make the Gag, RT, PR, and IN proteins that are required for retrotransposition"
U	YDR261C-D	12	19	8.4%	1604	181660	7.5	YDR261C-D SGDID:S000007395, Chr IV from 992343-991039,991037-987528, reverse complement, transposable_element_gene, "TyB Gag-Pol protein; proteolytically processed to make the Gag, RT, PR, and IN proteins that are required for retrotransposition"

Filename	XCorr	DeltCN	Conf%	ObsM+H+	CalcM+H+	PPM	SpR	SpScore	Ion%	#	Sequence
0331602 SCX 1 05 itms.04042.04042.2	3.4392	0.4852	100.0%	1639.7972	1639.7927	2.8	2	368.9	53.6%	1	K.EVHTNQDPLDVSASK.I
0331602 SCX 1 13 itms.06207.06207.2	2.4487	0.1061	98.6%	2125.186	2124.1929	-4.8	5	309.5	36.8%	1	K.FLQNSNLGGIIPTVNGKI
0331602 SCX 1 08 itms.06936.06936.2	3.6085	0.2443	100.0%	1412.7021	1412.6984	2.7	1	1303.9	81.8%	5	K.EILSVDYTDIMK.I
0331602 SCX 1 08 itms.04182.04182.2	3.3061	0.3657	100.0%	990.5224	990.5229	-0.5	1	1058.0	92.9%	3	K.VACQLIMR.G
0331602 SCX 1 10 itms.01627.01627.2	1.7524	0.1134	98.0%	1039.543	1039.5424	0.6	12	406.2	62.5%	1	R.SYTNTTKPK.V
0331602 SCX 1 04 itms.03302.03302.2	5.2663	0.3421	100.0%	1866.8362	1866.8317	2.4	1	1188.5	67.6%	1	H.NVSTSNNSPSTDNDSISI
0331602 SCX 1 04 itms.03264.03264.2	2.9523	0.2433	100.0%	1752.7915	1752.7888	1.5	1	332.5	46.9%	1	N.VSTSNNSPSTDNDSISK.
0331602 SCX 1 05 itms.03624.03624.2	2.2295	0.1547	99.9%	802.41296	802.41327	-0.4	2	924.9	91.7%	1	C.PDCLIGK.S
0331602 SCX 1 04 itms.03463.03463.1	1.9144	0.2281	100.0%	841.39496	841.39435	0.7	6	342.6	66.7%	1	F.TDETTKF.R
0331602 SCX 1 06 itms.04601.04601.2	2.9741	0.3023	100.0%	1451.7769	1451.7745	1.6	1	693.0	66.7%	1	K.TVDTTNYVILQGK.E
0331602 SCX 1 10 itms.06885.06885.2	4.6191	0.5087	100.0%	2088.963	2088.9514	5.5	1	1007.8	59.4%	2	R.LDQFNYDALTFDEDLNR.
0331602 SCX 1 09 itms.03978.03978.2	2.323	0.2628	100.0%	894.5405	894.54126	-0.8	12	498.7	78.6%	1	K.LNVPLNPK.G

Fig. 3. DTASelect example output where several proteins are grouped together with the identified peptides common to all the proteins.

Table 6
Protein summary tools

Program	Category	Reference, Web site, or company
DTASelect	P	http://fields.scripps.edu/DTASelect/ Tabb et al. *(15)*
ProteinProphet	P	http://tools.proteomecenter.org/software.php Nesvizhskii et al. *(59)*
Scaffold	C	http://www.proteomesoftware.com/

C Commercial, *P* publicly available, *N* need to contact authors for availability

5. Protein Quantification

5.1. Basic Principles

Protein quantification is used to measure the amount of a protein in a sample. Mass spectrometry-based protein quantification is achieved either through stable isotope labeling or through label-free strategies. Stable isotope labeling strategies typically compare naturally abundant stable isotope peptides to physicochemically identical peptides with atoms enriched with a heavy stable isotope *(60)*. Several quantitative labeling technologies have been developed by in vivo labeling via metabolic incorporation or in vitro labeling via chemical reactions. The most popular in vivo labeling technologies are ^{15}N labeling and stable isotope labeling with amino acids in cell culture (SILAC) *(61)*. ^{15}N labeling requires cells be grown in either ^{14}N (light) or ^{15}N (heavy) media to incorporate each isotope into the proteins in the two different samples, respectively. Mixtures of two cell cultures are analyzed by mass spectrometry and the mass shift of heavy isotope labeled peptides are observed from light peptides. Both elution profiles from light and heavy peptides are compared to determine relative abundance. SILAC is gaining popularity utilizing a single isotope or a combination of ^{2}H, ^{13}C, and ^{15}N isotopes.

A limitation of metabolic labeling is that tissues or body fluids that cannot be grown in culture usually are not applicable to this technology. In vitro labeling via chemical reaction is an alternative approach to incorporate stable isotope tags onto specific sites such as N-terminal, C-terminal, cysteine, lysine, tyrosine, etc. Chemical reaction labeling strategies include ICAT *(62)*, iTRAQ *(63)*, and ^{18}O labeling *(64)*.

Figure 4 illustrates a general scheme for stable isotope labeling combined with mass spectrometry. Labeling with stable isotopes is expensive, and sometimes achieving highly enriched samples takes a large amount of time and effort. As an alternative approach, label-free strategies have been presented by a number of groups, and with its simplicity and cost effectiveness, it is gaining momentum. There are several different label-free quantitative strategies including spectral counting *(65, 66)*, LC-MS peak area based strategies *(67–70)*, LC-MS strategies *(71, 72)*, and differential mass spectrometry *(73)*.

In the following, we discuss the computational analysis of quantitative proteomics data using stable isotope labeling. The analysis usually contains the following steps (1) extraction of peptide ion chromatograms; (2) smoothing of chromatograms and noise reduction; (3) ion current ratio calculation to infer peptide abundance ratio; and (4) expression of peptide ratios as relative

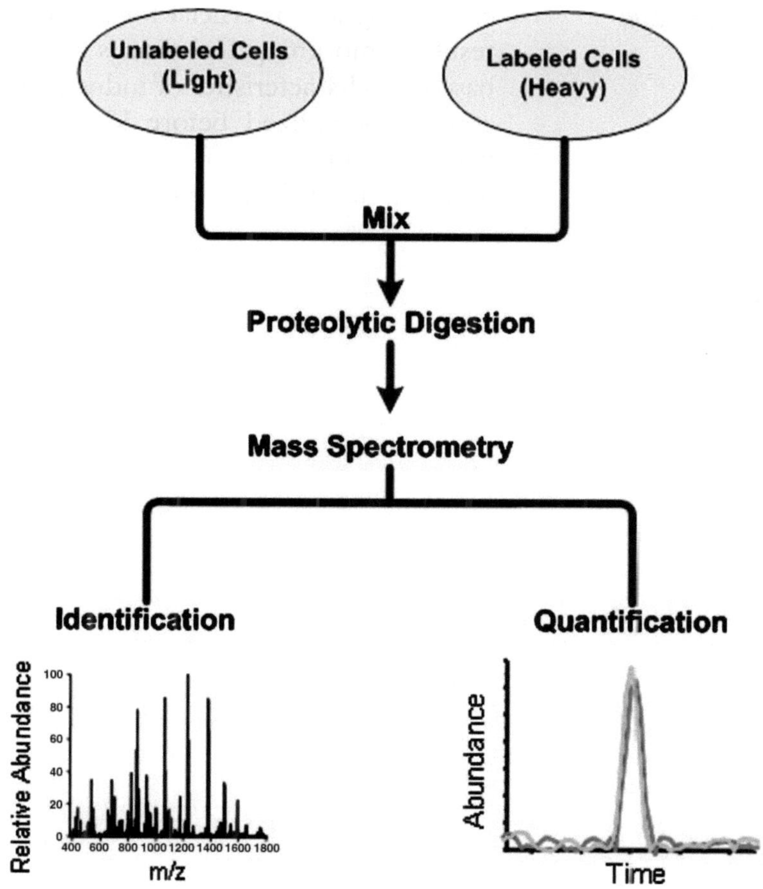

Fig. 4. General schematic of quantitative proteomics using stable isotope labeling.

protein abundance. We then discuss two special quantification techniques using tandem mass spectra (instead of using full scan MS) and spectral counting.

5.2. Extraction of Peptide Ion Chromatograms

Ion chromatogram extraction process is a computationally intensive step in quantitative analysis. Before the quantitative analysis, protein database search results are generally required, because the identified peptide information from database search results including scan number, retention time, and sequence are often used for quantitative applications to locate specific peptide elution time in chromatograms to rebuild a peak profile. In addition, the amino acid elemental composition data is also needed to calculate both light and heavy isotope masses. **Figure 5** shows an example of reconstructed chromatogram generated by Census. In contrast, label-free strategies often extract chromatograms from spectral files of multiple samples without a preidentification database search, and perform chromatogram alignments to accommodate peak shifts using several different types of algorithms including dynamic time warping, correlation optimized warping, parametric time warping, and peak alignment by a genetic algorithm.

5.3. Smoothing and Noise Reduction

Finding correct start and end points of peptide elution peaks in chromatograms is crucial for accurate and precise quantification results. Smoothing techniques and baseline removal algorithms based on characteristics of individual peptide elution profiles are commonly utilized before identifying peptide peaks. In cases where the intensity of a peptide is low while its corresponding light or heavy isotopomer is high, algorithms using a single

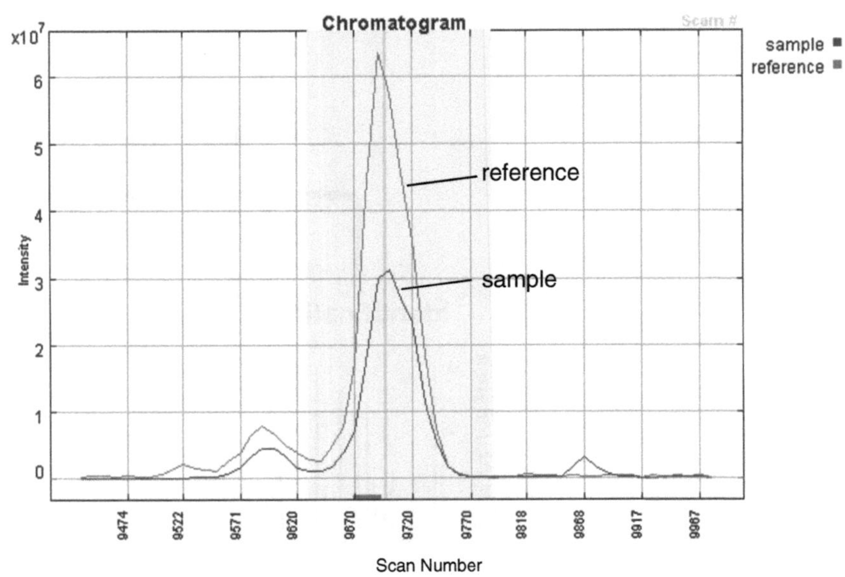

Fig. 5. Reconstructed chromatograms of both labeled and unlabeled peptides.

isotopomer may not correctly find the peak range to accommodate both light and heavy peaks. An improvement in accuracy for peak finding can be achieved with normalization of both light and heavy elution profiles *(74, 75)*.

5.4. Ion Current Ratio Calculation

After defining start and end points of peptide elution peaks, the next step is to calculate relative peptide abundances for the light and heavy peptides. There are several different algorithms to calculate relative peptide abundances – peak area *(76–78)*, linear regression *(75, 79)*, and principal component analysis *(74)*. The peak area approach simply calculates the area of both light and heavy peaks and compares their ratios. The linear regression approach converts peak profiles of the two isotopologues into a scatter plot based on their ion intensities. The slope of linear regression of the data points indicates the relative abundance of light and heavy peaks and the correlation coefficient shows the quality of the elution profiles. The advantage of linear regression is that it is tolerant to poor signal-to-noise ratio (S/N) data. Principal component analysis generates a similar scatter plot of ion intensities from both light and heavy isotopologues, and calculates two principal components and their eigenvalues. The slope of the first principal component indicates the peptide abundance ratio.

5.5. Expression of Peptide Ratios as Relative Protein Abundances

After the peptide ratios are obtained, there are two different approaches for calculating protein ratios. One approach is to simply calculate the mean of all peptide measurements. The second method is a weighted average where peptides with given weight based on scores such as the quality or the standard deviation are used to derive a protein abundance ratio. By using known mixtures of ^{15}N-labeled yeast samples, Park et al. *(75)* showed that weighted average using quality scores of peptides provides more accurate protein abundance measurements. **Figure 6** shows an example of protein quantification output using the software Census *(75)*.

A list of protein quantification software tools is provided in **Table 7**.

5.6. Quantification from Tandem Mass Spectra

While full scan mass spectra have been used predominantly for quantitative analyses, the use of tandem mass spectra has some benefits over single-stage MS. The use of tandem mass spectra can improve sensitivity, specificity, enhance dynamic range, and increase the signal-to-noise ratio. There are several different types of strategies that take advantage of tandem mass spectrometry for protein quantification including selective reaction monitoring (SRM), and data-independent acquisition *(80)*. SRM analyses are achieved by monitoring user-defined precursors and collecting a fragment ion in the tandem mass spectra. Another approach is a

```
H   CenSus v. 0.9 census-out.txt file
H   Created by
H   John Venable jvenable@scripps.edu
H   "Robin, Sung Kyu Park rpark@scripps.edu"
H   Michael J.MacCoss
H   "The Scripps Research Institute, La Jolla, CA"
H   created date    Wed Mar 08 14:06:17 PST 2006
H   PLINE   LOCUS   AVERAGE_RATIO   STANDARD_DEVIATION   PEPTIDE_NUM   DESCRIPTION
H   SLINE   UNIQUE   SEQUENCE   RATIO   REGRESSION_FACTOR   DETERMINANT_FACTOR   FILE_NAME
H   Total peptides   528
H   Quantified peptides   384
H   Quantification efficiency   72.73%
P   YGR192C   0.792   0.05   14   "TDH3 SGDID:S0003424, Chr VII from 883815-882817, reverse complement, Verified ORF "
                                   FALSE   FALSE
S   U   R.IALSRPNVEVVALNDPFITNDYAAYMFK.Y   0.822   0.998   0.995   120205_yeast_1tol_LTQ_FTMSComp-06_itms.07951.07951.2
                                   TRUE   TRUE
S   U   R.IALSRPNVEVVALNDPFITNDYAAYMFK.Y   0.878   0.999   0.999   120205_yeast_1tol_LTQ_FTMSComp-06_itms.07948.07948.3
                                   TRUE   TRUE
S   U   R.PNVEVVALNDPFITNDYAAYMFK.Y   0.738   0.895   0.8   120205_yeast_1tol_LTQ_FTMSComp-02_itms.11067.11067.2
                                   TRUE   TRUE
S   U   K.KIATYQERDPANLPWGSSNVDIAIDSTGVFK.E   0.77   0.852   0.725   120205_yeast_1tol_LTQ_FTMSComp-06_itms.17830.17830.3
                                   TRUE   TRUE
S   U   K.IATYQERDPANLPWGSSNVDIAIDSTGVFK.E   0.737   0.851   0.724   120205_yeast_1tol_LTQ_FTMSComp-03_itms.17972.17972.3
                                   TRUE   TRUE
S       K.VVITAPSSTAPMFVMGVNEEK.Y   0.876   0.986   0.972   120205_yeast_1tol_LTQ_FTMSComp-02_itms.09330.09330.2
                                   TRUE   TRUE
S       K.IVSNASCTTNCLAPLAK.V   0.804   0.968   0.938   120205_yeast_1tol_LTQ_FTMSComp - 02_itms.07738.07738.2
                            TRUE   TRUE
S   U   K.VINDAFGIEEGLMTTVHSLTATQK.T   0.81   0.979   0.959   120205_yeast_1tol_LTQ_FTMSComp-03_itms.08284.08284.3
                                   TRUE   TRUE
S   U   K.VINDAFGIEEGLMTTVHSLTATQK.T   0.818   0.977   0.955   120205_yeast_1tol_LTQ_FTMSComp-03_itms.08234.08234.2
                                   TRUE   TRUE
S       R.TASGNIIPSSTGAAK.A   0.785   1   0.999   120205_yeast_1tol_LTQ_FTMSComp - 02_itms.21642.21642.2
                            TRUE   TRUE
S       R.TASGNIIPSSTGAAK.A   0.758   0.977   0.955   120205_yeast_1tol_LTQ_FTMSComp - 02_itms.07405.07405.1
                            TRUE   TRUE
S       R.VPTVDVSVVDLTVK.L   0.783   0.994   0.988   120205_yeast_1tol_LTQ_FTMSComp - 04_itms.08023.08023.1
                            TRUE   TRUE
S   U   K.GVLGYTEDAVVSSDFLGDSHSSIFDASAGIQLSPK.F   0.704   0.865   0.748   120205_yeast_1tol_LTQ_FTMSComp-06_itms.11258.11258.3
                                   TRUE   TRUE
S       R.VVDLVEHVAK.A   0.806   0.975   0.951   120205_yeast_1tol_LTQ_FTMSComp - 02_itms.07633.07633.2
                            TRUE   TRUE
P   YBR072W   1.249   0.458   5   "HSP26 SGDID:S0000276, ChrII from 382027-382671, Verified ORF "
                                   FALSE   FALSE
S   U   R.PNNYAGALYDPRDETLDDWFDNDLSLFPSGFGFPR.S   1.524   0.901   0.812   120205_yeast_1tol_LTQ_FTMSComp-02_itms.12068.12068.3
                                   TRUE   TRUE
S   U   R.DETLDDWFDNDLSLFPSGFGFPR.S   1.449   0.988   0.976   120205_yeast_1tol_LTQ_FTMSComp-02_itms.12622.12622.3
                                   TRUE   TRUE
S   U   R.DETLDDWFDNDLSLFPSGFGFPR.S   1.467   0.99   0.981   120205_yeast_1tol_LTQ_FTMSComp-02_itms.12819.12819.2
                                   TRUE   TRUE
S   U   R.SVAVPVDILDHDNNYELK.V1.369   0.851   0.725   120205_yeast_1tol_LTQ_FTMSComp-02_itms.21313.21313.2
                            TRUE   TRUE
S   U   K.NQILVSGEIPSTLNEESKDK.V0.436   0.943   0.889   120205_yeast_1tol_LTQ_FTMSComp-02_itms.08459.08459.2
                            TRUE   FALSE
```

Fig. 6. Census example output.

Table 7
List of protein quantification tools

Program	Category	Reference, Web site, or company
Census	P	http://fields.scripps.edu/download.php Park et al. *(75)*
ProRata	P	http://www.msprorata.org/ Pan et al. *(74)*
MSQuant	P	http://msquant.sourceforge.net/
XPRESS	P	http://tools.proteomecenter.org/XPRESS.php Li et al. *(72)*
ASAPRatio	P	http://tools.proteomecenter.org/ASAPRatio.php Li et al. *(78)*

C Commercial, *P* publicly available

data-independent strategy that uses successive isolation windows (typically in the range of 10~20 m/z) throughout the mass range, and the ions in each isolation window are fragmented to create a tandem mass spectrum.

5.7. Spectral Counting

The total number of spectra matching to each protein obtained from data-dependent data acquisition is called "spectral count." Washburn et al. noted in 2001 that more spectra matched to abundant proteins than to less-abundant proteins *(9)*. Pang et al. *(81)* and Gao et al. *(82)* suggested that spectral count is related to protein abundance and therefore could be used as a measure of abundance. A number of groups now use spectral count as a semiquantitative approach, and even in complex mixtures it correlates well with protein abundances *(64)*. Because small proteins tend to have fewer spectral counts than large proteins, spectral count normalization is often used to minimize quantitative errors. Zybailov et al. *(83)* proposed a normalization approach using normalized spectral abundance factors (NSAFs)

$$(\mathrm{NSAF})_k = \frac{(\mathrm{SpC})_k \,/\, (\mathrm{Length})_k}{\sum_{i=1}^{N} (\mathrm{SpC})_i \,/\, (\mathrm{Length})_i},$$

where $(SpC)i$ is the number of spectral counts for a protein i, $(Length)i$ is the length of protein i, and N is the total number of proteins in the experiment. Briefly, NSAF for a protein k is a normalized spectral count taking into consideration the spectral counts and protein lengths of all proteins identified in an experiment. The relative abundance proteins within an experiment can be compared by using the NSAF of each protein.

Spectral counting offers a way of protein quantification without carrying out stable isotope labeling. However, spectral counting is usually less sensitive and less accurate, compared to protein quantification using stable isotope labeling.

6. Concluding Remarks

With significant progress in protein chemistry, protein separation methods, and mass spectrometry technologies, it is not uncommon for a mass spectrometer to acquire mass spectral data that contains information about thousands of proteins in a single day. Data analysis becomes a critical step in mass spectrometry-based shotgun proteomics. With the development of accurate and automated algorithms to identify and quantify proteins in a large scale, shotgun proteomics has become a powerful analytical approach to tackle significant problems in neuroscience.

Analysis of shotgun proteomics data usually begins with peptide identifications from tandem mass spectrometry data. A variety of approaches, such as database searching, de novo peptide sequencing, hybrid approach combining de novo peptide sequencing and database searching, spectral library searching, and unrestricted PTM searching, can be employed to identify peptides from tandem mass spectra. Each approach has its own strengths and weaknesses. Generally speaking, database searching is still the most widely used approach for identifying peptides from tandem mass spectra. Sometimes, a combination of several peptide identification approaches can help mine the data more thoroughly. Proteins are then inferred from the peptide information derived from the MS/MS data.

With steady advances in MS technologies, quantitative proteomics also has progressed dramatically. Current approaches for quantitative proteomics depend primarily on stable isotope labeling, including in vivo metabolic labeling and in vitro chemical labeling. In some situations, label-free strategies are also employed in quantitative proteomics study. These strategies could potentially generate valuable information for differential expression analysis of proteins.

Although a comprehensive understanding of the brain proteome does not yet exist, many of these strategies have already been applied to the analysis of brain tissue. Several groups have extensively analyzed the postsynaptic density (PSD), which is a dynamic compartment in neurons essential for communication between neurons. Valuable information has been obtained regarding the identities of PSD protein components, as well as relative and absolute quantities of the major components *(84–86)*. Pathological entities that form key features of neurological diseases, such as amyloid plaques *(87)* and neurofibrillary tangles *(88)* in Alzheimer's disease, Lewy bodies *(89)* in Parkinson's disease, have recently been enriched or purified, and their protein contents have been defined. The tremendous complexity of the brain proteome, however, does provide a challenge to these strategies. The biggest obstacle is that the brain has the highest alternative splicing activity than any other tissue with thousands of splice isoforms of a single protein. Overall, using shotgun proteomics in discovery-oriented and hypothesis-driven analyses will ultimately lead to a more thorough understanding of outstanding questions in neuroscience and will help generate new hypotheses to move forward our understanding of our own brain.

References

1. Sunkin, S. M., and Hohmann, J. G. (2007) Insights from spatially mapped gene expression in the mouse brain. *Hum. Mol. Genet.* **16**, R209–R219.
2. Vanderklish, P. W., and Edelman, G. M. (2005) Differential translation and fragile X syndrome. *Genes Brain Behav.* **4**, 360–384.
3. Betarbet, R., Sherer, T. B., and Greenamyre, J. T. (2005) Ubiquitin-proteasome system and Parkinson's diseases. *Exp. Neurol.* **191**, S17–S27.
4. Iqbal, K., and Grundke-Iqbal, I. (2007) Developing pharmacological therapies for Alzheimer disease. Cell Mol. Life Sci. **64**, 2234–2244.
5. Herbert, B. R., Sanchez, J. C., and Bini, L. (1997) Two-dimensional electrophoresis: The state of art and future directions. In *Proteome Research: New Frontiers in Functional Genomics*, Springer, New York, NY, pp. 13–33.
6. Zhu, H., and Snyder, M. (2003) Protein chip technology. *Curr. Opin. Chem. Biol.* **7**, 55–63.
7. Little, D. P., Speir, J. P., Senko, M. W., O'Connor, P. B., and McLafferty, F. W. (1994) Infrared multiphoton dissociation of large multiply charged ions for biomolecule sequencing. *Anal. Chem.* **66**, 2809–2815.
8. Senko, M. W., Speir, J. P., and McLafferty, F. W. (1994) Collisional activation of large multiply charged ions using Fourier transform mass spectrometry. *Anal. Chem.* **66**, 2801–2808.
9. Washburn, M. P., Wolters, D., and Yates, J. R. 3rd (2001) Large-scale analysis of the yeast proteome by multidimensional protein identification technology. *Nat. Biotechnol.* **19**, 242–247.
10. Venable, J. D., Dong, M. Q., Wohlschlegel, J., Dillin, A., and Yates, J. R. 3rd. (2004) Automated approach for quantitative analysis of complex peptide mixtures from tandem mass spectra. Nat. Methods **1**, 39–45.
11. Eng, J. K., McCormack, A. L., and Yates, J. R. 3rd (1994) An approach to correlate tandem mass spectral data of peptides with amino acid sequences in a protein database. *J. Am. Soc. Mass Spectrom.* **5**, 976–989.
12. Perkins, D. N., Pappin, D. J. C., Creasy, D. M., and Cottrell, J. S. (1999) Probability-based protein identification by searching sequence databases using mass spectrometry data. *Electrophoresis* **20**, 3551–3567.

13. Frank, A., and Pevzner, P. (2005) PepNovo: De novo peptide sequencing via probabilistic network modeling. *Anal. Chem.* **77**, 964–973.
14. Ma, B., Zhang, K., Hendrie, C., Liang, C., Li, M., Doherty-Kirby, A., and Lajoie, G. (2003) PEAKS: Powerful software for peptide de novo sequencing by MS/MS. *Rapid Commun. Mass Spectrom.* **17**, 2337–2342.
15. Tabb, D. L., Saraf, A., and Yates, J. R. 3rd (2003) GutenTag: High-throughput sequence tagging via an empirically derived fragmentation model. *Anal. Chem.* **75**, 6415–6421.
16. Tanner, S., Shu, H., Frank, A., Wang, L. C., Zandi, E., Mumby, M., Pevzner, P. A., and Bafna, V. (2005) InsPecT: Identification of posttranslationally modified peptides from tandem mass spectra. *Anal. Chem.* **77**, 4626–4639.
17. Yates, J. R. 3rd, Morgan, S. F., Gatlin, C. L., Griffin, P. R., and Eng, J. K. (1998) Method to compare collision-induced dissociation spectra of peptides: Potential for library searching and subtractive analysis. *Anal. Chem.* **70**, 3557–3565.
18. Frewen, B. E., Merrihew, G. E., Wu, C. C., Noble, W. S., and MacCoss, M. J. (2006) Analysis of peptide MS/MS spectra from large-scale proteomics experiments using spectrum libraries. *Anal. Chem.* **78**, 5678–5684.
19. Tsur, D., Tanner, S., Zandi, E., Bafna, V., and Pevzner, P. A. (2005) Identification of post-translational modifications by blind search of mass spectra. *Nat. Biotechnol.* **23**, 1562–1567.
20. Havilio, M., and Wool A. (2007) Large-scale unrestricted identification of post-translation modifications using tandem mass spectrometry. *Anal. Chem.* **79**, 1362–1368.
21. Clauser, K. R., Baker, P., and Burlingame, A. L. (1999) Role of accurate mass measurement (+/- 10 ppm) in protein identification strategies employing MS or MS MS and database searching. *Anal. Chem.* **71**, 2871–2882.
22. Field, H. I., Fenyo, D., and Beavis, R. C. (2002) RADARS, a bioinformatics solution that automates proteome mass spectral analysis, optimises protein identification, and archives data in a relational database. *Proteomics* **2**, 36–47.
23. Geer, L. Y., Markey, S. P., Kowalak, J. A., Wagner, L., Maynard, D. M., Yang, X. Y., Shi, W. Y., and Bryant, S. H. (2004) Open mass spectrometry search algorithm. *J. Proteome Res.* **3**, 958–964.
24. Xu, T., Venable, J. D., Park, S. K., Cociorva, D., Lu, B., Liao, L., Wohlschlegel, J., Hewel, J., and Yates, J. R. 3rd (2006) ProLuCID, a fast and sensitive tandem mass spectra-based protein identification program. *Mol. Cell. Proteomics* **5**, S174.
25. Sadygov, R. G., and Yates, J. R. 3rd (2003) A hypergeometric probability model for protein identification and validation using tandem mass spectral data and protein sequence databases. *Anal. Chem.* **75**, 3792–3798.
26. Craig, R., and Beavis, R. C. (2004) TANDEM: Matching proteins with tandem mass spectra. *Bioinformatics* **20**, 1466–1467.
27. Sadygov, R. G., Cociorva, D., and Yates, J. R. 3rd (2004) Large-scale database searching using tandem mass spectra: Looking up the answer in the back of the book. *Nat. Methods* **1**, 195–202.
28. Nesvizhskii, A. I. (2007) Protein identification by tandem mass spectrometry and sequence database searching. *Methods Mol. Biol.* **367**, 87–119.
29. Grubbs, F. E. (1969) Procedures for detecting outlying observations in samples. *Technometrics* 11, 1–21.
30. Zubarev, R. A., Horn, D. M., Fridriksson, E. K., Kelleher, N. L., Kruger, N. A., Lewis, M. A., Carpenter, B. K., and McLafferty, F. W. (2000) Electron capture dissociation for structural characterization of multiply charged protein cations. *Anal. Chem.* **72**, 563–573.
31. Syka, J. E., Coon, J. J., Schroeder, M. J., Shabanowitz, J., and Hunt, D. F. (2004) Peptide and protein sequence analysis by electron transfer dissociation mass spectrometry. *Proc. Natl Acad. Sci. USA* **101**, 9528–9533.
32. Tabb, D. L., McDonald, W. H., and Yates, J. R. 3rd (2002) DTASelect and contrast: Tools for assembling and comparing protein identifications from shotgun proteomics. *J. Proteome Res.* **1**, 21–26.
33. Keller, A., Nesvizhskii, A. I., Kolker, E., and Aebersold, R. (2002) Empirical statistical model to estimate the accuracy of peptide identifications made by MS/MS and database search. *Anal. Chem.* **74**, 5383–5392.
34. Cociorva, D., Tabb, D., and Yates, J. R. 3rd (2006) Validation of tandem mass spectrometry database search results using DTASelect. In *Current Protocols in Bioinformatics, Supplement 16*, Wiley, Hoboken, NJ, pp. 13.4.1–13.4.14.
35. Savitski, M. M., Nielsen, M. L., Kjeldsen, F., and Zubarev, R. A. (2005) Proteomics-grade de novo sequencing approach. *J. Proteome Res.* **4**, 2348–2354.
36. Horn, D. M., Zubarev, R. A., and McLafferty, F. W. (2000) Automated reduction and

interpretation of high resolution electrospray mass spectra of large molecules. *J. Am. Soc. Mass Spectrom.* **11**, 320–332.

37. Frank, A. M., Savitski, M. M., Nielsen, M. L., Zubarev, R. A., and Pevzner, P. A. (2007) De novo peptide sequencing and identification with precision mass spectrometry. *J. Proteome Res.* **6**, 114–123.
38. Taylor, J. A., and Johnson, R. S. (1997) Sequence database searches via de novo peptide sequencing by tandem mass spectrometry. *Rapid Commun. Mass Spectrom.* **11**, 1067–1075.
39. Mo, L., Dutta, D., Wan, Y., and Chen, T. (2007) MSNovo: A dynamic programming algorithm for de novo peptide sequencing via tandem mass spectrometry. *Anal. Chem.* **79**, 4870–4878.
40. Fischer, B., Roth, V., Roos, F., Grossmann, J., Baginsky, S., Widmayer, P., Gruissem, W., and Buhmann, J. M. (2005) Novo-HMM: A hidden Markov model for de novo peptide sequencing. *Anal. Chem.* **77**, 7265–7273.
41. Bern, M., and Goldberg, D. (2006) De novo analysis of peptide tandem mass spectra by spectral graph partitioning. *J. Comput. Biol.* **13**, 364–378.
42. DiMaggio, P. A. Jr, and Floudas, C. A. (2007) De novo peptide identification via tandem mass spectrometry and integer linear optimization. *Anal. Chem.* **79**, 1433–1446.
43. Lu, B., and Chen, T. (2004) Algorithms for de novo peptide sequencing via tandem mass spectrometry. *Drug Discov. Today: Biosilico* **2**, 85–90.
44. Mann, M., and Wilm, M. (1994) Error-tolerant identification of peptides in sequence databases by peptide sequence tags. *Anal. Chem.* **66**, 4390–4399.
45. Han, Y., Ma, B., and K. Zhang. (2005) SPIDER: Software for protein identification from sequence tags containing de novo sequencing error. *J. Bioinform. Comput. Biol.* **3**, 697–716.
46. Searle, B. C., Dasari, S., Wilmarth, P. A., Turner, M., Reddy, A. P., David, L. L., and Nagalla, S. R. (2005) Identification of protein modifications using MS/MS de novo sequencing and the OpenSea alignment algorithm. *J. Proteome Res.* **4**, 546–554.
47. Altschul, S. F., Gish, W., Miller, W., Myers, E. W., and Lipman, D. J. (1990) Basic local alignment search tool. *J. Mol. Biol.* **215**, 403–410.
48. Pearson, W. R., and Lipman, D. J. (1988) Improved tools for biological sequence comparison. *Proc. Natl Acad. Sci. USA* **85**, 2444–2448.
49. Mackey, A. J., Haystead, T. A., and Pearson, W. R. (2002) Getting more from less: Algorithms for rapid protein identification with multiple short peptide sequences. *Mol. Cell. Proteomics.* **1**, 139–147.
50. Shevchenko, A., Sunyaev, S., Loboda, A., Shevchenko, A., Bork, P., Ens, W., and Standing, K. G. (2001) Charting the proteomes of organisms with unsequenced genomes by MALDI-quadrupole time-of-flight mass spectrometry and BLAST homology searching. *Anal. Chem.* **73**, 1917–1926.
51. Bern, M., Cai, Y., and Goldberg, D. (2007) Lookup peaks: A hybrid of de novo sequencing and database search for protein identification by tandem mass spectrometry. *Anal. Chem.* **79**, 1393–1400.
52. Heller, S. (1999) The history of the NIST/EPA/NIH mass spectral database. *Today's Chemist at Work* **8**, 45–50.
53. Craig, R., Cortens, J. C., Fenyo, D., and Beavis, R. C. (2006) Using annotated peptide mass spectrum libraries for protein identification. *J. Proteome Res.* **5**, 1843–1849.
54. Lam, H., Deutsch, E. W., Eddes, J. S., Eng, J. K., King, N., Stein, S. E., and Aebersold, R. (2007) Development and validation of a spectral library searching method for peptide identification from MS/MS. *Proteomics* **7**, 655–667.
55. Liu, J., Bell, A. W., Bergeron, J. J., Yanofsky, C. M., Carrillo, B., Beaudrie, C. E., and Kearney, R. E. (2007) Methods for peptide identification by spectral comparison. *Proteome Sci.* **5**, 3.
56. Zhang, Z. (2004) Prediction of low-energy collision-induced dissociation spectra of peptides. *Anal. Chem.* **76**, 3908–3922.
57. DeGnore, J. P., and Qin, J. (1998) Fragmentation of phosphopeptides in an ion trap mass spectrometer. *J. Am. Soc. Mass Spectrom.* **9**, 1175–1188.
58. Yates, J. R. 3rd, Eng, J. K., McCormack, A. L., and Schieltz, D. (1995) Method to correlate tandem mass spectra of modified peptides to amino acid sequences in the protein database. *Anal. Chem.* **67**, 1426–1436.
59. Nesvizhskii, A. I, Keller, A., Kolker, E., and Aebersold, R. (2003) A statistical model for identifying proteins by tandem mass spectrometry. *Anal. Chem.* **75**, 4646–4658.
60. Julka, S., and Regnier, F. (2004) Quantification in proteomics through stable isotope coding: A review. *J. Proteome Res.* **3**, 350–363.

61. Ong, S. E., Blagoev, B., Kratchmarova, I., Kristensen, D. B., Steen, H., Pandey, A., and Mann, M. (2002) Stable isotope labeling by amino acids in cell culture, SILAC, as a simple and accurate approach to expression proteomics. *Mol. Cell. Proteomics* **1**, 376–386.
62. Gygi, S. P., Rist, B., Gerber, S. A., Turecek, F., Gelb, M. H., and Aebersold, R. (1999) Quantitative analysis of complex protein mixtures using isotope-coded affinity tags. *Nat. Biotechnol.* **17**, 994–999.
63. Ross, P. L., Huang, Y. N., Marchese, J. N., Williamson, B., Parker, K., Hattan, S., Khainovski, N., Pillai, S., Dey, S., Daniels, S., Purkayastha, S., Juhasz, P., Martin, S., Bartlet-Jones, M., He, F., Jacobson, A., and Pappin, D. J. (2004) Multiplexed protein quantitation in *Saccharomyces cerevisiae* using amine-reactive isobaric tagging reagents. *Mol. Cell. Proteomics* **3**, 1154–1169.
64. Mirgorodskaya, O. A., Kozmin, Y. P., Titov, M. I., Korner, R., Sonksen, C. P., and Roepstorff, P. (2000) Quantitation of peptides and proteins by matrix-assisted laser desorption/ionization mass spectrometry using (18) O-labeled internal standards. *Rapid Commun. Mass Spectrom.* **14**, 1226–1232.
65. Liu, H., Sadygov, R. G., and Yates, J. R. 3rd (2004) A model for random sampling and estimation of relative protein abundance in shotgun proteomics. *Anal. Chem.* **76**, 4193–4201.
66. Blondeau, F., Ritter, B., Allaire, P. D., Wasiak, S., Girard, M., Hussain, N. K., Angers, A., Legendre-Guillemin, V., Roy, L., Boismenu, D., Kearney, R. E., Bell, A. W., Bergeron, J. J., and McPherson, P. S. (2004) Tandem MS analysis of brain clathrin-coated vesicles reveals their critical involvement in synaptic vesicle recycling. *Proc. Natl Acad. Sci. USA* **101**, 3833–3838.
67. Bondarenko, P. V., Chelius, D., and Shaler, T. A. (2002) Identification and relative quantitation of protein mixtures by enzymatic digestion followed by capillary reversed-phase liquid chromatography-tandem mass spectrometry. *Anal. Chem* **74**, 4741–4749.
68. Chelius, D., and Bondarenko, P. V. (2002) Quantitative profiling of proteins in complex mixtures using liquid chromatography and mass spectrometry. *J. Proteome Res.* **1**, 317–323.
69. Chelius, D., Zhang, T., Wang, G., and Shen, R. F. (2003) Global protein identification and quantification technology using two-dimensional liquid chromatography nanospray mass spectrometry. *Anal. Chem.* **75**, 6658–6665.
70. Higgs, R. E., Knierman, M. D., Gelfanova, V., Butler, J. P., and Hale, J. E. (2005) Comprehensive label-free method for the relative quantification of proteins from biological samples. *J. Proteome Res.* **4**, 1442–1450.
71. Wang, W., Zhou, H., Lin, H., Roy, S., Shaler, T. A., Hill, L. R., Norton, S., Kumar, P., Anderle, M., and Becker, C. H. (2003) Quantification of proteins and metabolites by mass spectrometry without isotopic labeling or spiked standards. *Anal. Chem.* **75**, 4818–4826.
72. Li, X. J., Yi, E. C., Kemp, C. J., Zhang, H., and Aebersold, R. (2005) A software suite for the generation and comparison of peptide arrays from sets of data collected by liquid chromatography-mass spectrometry. *Mol. Cell. Proteomics* **4**, 1328–1340.
73. Wiener, M. `C., Sachs, J. R., Deyanova, E. G., and Yates, N. A. (2004) Differential mass spectrometry: A label-free LC-MS method for finding significant differences in complex peptide and protein mixtures. *Anal. Chem.* **76**, 6085–6096.
74. Pan, C., Kora, G., McDonald, W. H., Tabb, D. L., VerBerkmoes, N. C., Hurst, G. B., Pelletier, D. A., Samatova, N. F., and Hettich, R. L. (2006) ProRata: A quantitative proteomics program for accurate protein abundance ratio estimation with confidence interval evaluation. *Anal. Chem.* **15**, 7121–7131.
75. Park, S. K., Venable, J. D., Xu, T., Yates, J. R. 3rd (2008) A quantitative analysis software tool for mass spectrometry-based proteomics. *Nat. Methods* **5**, 319–322.
76. Schulze, W. X., and Mann, M. (2004) A novel proteomic screen for peptide-protein interactions. *J. Biol. Chem.* **279**, 10756–10764.
77. Han, D. K., Eng, J., Zhou, H., and Aebersold, R. (2001) Quantitative profiling of differentiation-induced microsomal proteins using isotope-coded affinity tags and mass spectrometry. *Nat. Biotechnol.* **19**, 946–951.
78. Li, X. J., Zhang, H., Ranish, J. A., and Aebersold, R. (2003) Automated statistical analysis of protein abundance ratios from data generated by stable-isotope dilution and tandem mass spectrometry. *Anal. Chem.* **75**, 6648–6657.
79. MacCoss, M. J., Wu, C. C., Yates, J. R. 3rd (2003) A correlation algorithm for the automated analysis of quantitative 'shotgun' proteomics data. *Anal. Chem.* **75**, 6912–6921.
80. Venable, J. D., Dong, M. Q., Wohlschlegel, J., Dillin, A., and Yates, J. R. 3rd (2004) Automated approach for quantitative analysis of complex peptide mixtures from tandem mass spectra. *Nat. Methods.* **1**, 39–45.

81. Pang, J. X., Ginanni, N., Dongre, A. R., Hefta, S. A., and Opiteck, G. J. J. (2002) Biomarker discovery in urine by proteomics. *J. Proteome Res.* **1**, 161–169.
82. Gao, J., Opiteck, G. J., Friedrichs, M. S., Dongre, A. R., and Hefta, S. A. J. (2003) Changes in the protein expression of yeast as a function of carbon source. *J. Proteome Res.* **2**, 643–649.
83. Zybailov, B. L., Florens, L., and Washburn, M. P. (2007) Quantitative shotgun proteomics using a protease with broad specificity and normalized spectral abundance factors. *Mol. Biosyst.* **3**, 354–360.
84. Jordan, B. A., Fernholz, B. D., Boussac, M., Xu, C. F., Grigorean, G., Ziff, E. B., and Neubert, T. A. (2004) Identification and verification of novel rodent postsynaptic density proteins. *Mol. Cell. Proteomics* **3**, 857–871.
85. Yoshimura, Y., Yamauchi, Y., Shinkawa, T., Taoka, M., Donai, H., Takahashi, N., Isobe, T., and Yamauchi, T. (2004) Molecular constituents of the postsynaptic density fraction revealed by proteomic analysis using multidimensional liquid chromatography-tandem mass spectrometry. *J. Neurochem.* **88**, 759–768.
86. Cheng, D., Hoogenraad, C. C., Rush, J., Ramm, E., Schlager, M. A., Duong, D. M., Xu, P., Wijayawardana, S. R., Hanfelt, J., Nakagawa, T., Sheng, M., and Peng, J. (2006) Relative and absolute quantification of postsynaptic density proteome isolated from rat forebrain and cerebellum. *Mol. Cell. Proteomics* **5**, 1158–1170.
87. Liao, L., Cheng, D., Wang, J., Duong, D. M., Losik, T. G., Gearing, M., Rees, H. D., Lah, J. J., Levey, A. I., and Peng, J. (2004) Proteomic characterization of postmortem amyloid plaques isolated by laser capture microdissection. *J. Biol. Chem.* **279**, 37061–37068.
88. Wang, Q., Woltjer, R. L., Cimino, P. J., Pan, C., Montine, K. S., Zhang, J., and Montine, T. J. (2005) Proteomic analysis of neurofibrillary tangles in Alzheimer disease identifies GAPDH as a detergent-insoluble paired helical filament tau binding protein. *FASEB J.* **19**, 869–871.
89. Leverenz, J. B., I. Umar, Wang, Q., Montine, T. J., McMillan, P. J., Tsuang, D. W., Jin, J. H., Pan, C., Shin, J., Zhu, D., and Zhang, J. (2007) Proteomic identification of novel proteins in cortical Lewy bodies. *Brain Pathol.* **17**, 139–145.

Part IV

Biofluid Analysis and Clinical Translation

Chapter 17

Identification of Glycoproteins in Human Cerebrospinal Fluid

Hye Jin Hwang, Thomas Quinn, and Jing Zhang

Summary

Human cerebrospinal fluid (CSF), which circulates within the ventricles of the brain and the subarachnoid space of the central nervous system (CNS), is an excellent source for proteomic discovery of biomarkers in neurodegenerative disorders, including Alzheimer's and Parkinson's disease. Protein glycosylation is an abundant and biologically significant posttranslational modification. Glycoproteins, commonly associated with membrane and secreted proteins, are highly enriched in body fluids, including CSF. Focusing on glycoproteins also improves the dynamic range of proteomic profiling of the CSF, where low abundance proteins are difficult to identify because of the CSF's enormous complexity. As an ongoing process to define the human CSF proteome, we have recently employed a complementary proteomic approach, with integrated lectin affinity column and hydrazide chemistry, for CSF glycoprotein identification. This investigation has revealed many proteins of low abundance that are related to the CNS structurally and/or functionally. This review centers on the technical details involved in various steps in sample preparation as well as proteomic analysis of CSF glycoproteins.

Key words: Cerebrospinal fluid, Glycoprotein, Hydrazide chemistry, Lectin affinity column, Mass spectrometry, Proteomics, Alzheimer's disease, Parkinson's disease

1. Introduction

Neurodegenerative disorders, e.g., Alzheimer's disease (AD) and Parkinson's disease (PD), are one of the major causes of death worldwide. The diagnosis of most, if not all, neurodegenerative diseases is based on clinical assessments which are currently unsatisfactory. Therefore, biomarkers that reflect disease states and/or stages are pressingly needed. Proteomic profiling of human cerebrospinal fluid (CSF) for neurodegenerative biomarkers is appealing because unique markers discovered in the CSF,

Andrew K. Ottens and Kevin K.W. Wang (eds.), *Neuroproteomics*, Methods in Molecular Biology, vol. 566
DOI 10.1007/978-1-59745-562-6_17, © Humana Press, a part of Springer Science+Business Media, LLC 2009

which are intimately associated with the tissue of pathology, likely reflect the pathogenesis of the disease of interest, thereby shedding more light on novel mechanisms of each disease *(1, 2)*. However, proteins derived from the central nervous system (CNS) are typically low in abundance and are overshadowed by abundant proteins during the analysis due to the sample complexity and significant dynamic range in protein concentration. To circumvent this difficulty, one of the approaches used is to focus on a "subproteome," e.g., phosphorylated or glycosylated proteins, in a complex sample *(3–6)*.

One subproteome of particular interest is that of glycoproteins. Protein glycosylation is an abundant and biologically significant posttranslational modification (PTM), most commonly associated with membrane and secreted proteins *(7, 8)*. Perturbation of cellular processes is known to be associated with structural changes in glycoproteins. In fact, many clinical biomarkers and therapeutic targets are glycoproteins *(9–11)*. Consequently, it is expected that a systematic and detailed analysis of glycosylated proteins in human CSF will provide biomarkers that can assist with the clinical diagnosis of AD and PD, monitoring disease progression, and evaluating the effects of existing and future therapeutic drugs. Additionally, unique disease biomarkers may reveal novel mechanisms underlying various neurodegenerative diseases, which are currently largely unknown, and may also provide new therapeutic targets.

To date, there are two major methods of isolating glycoproteins: (a) lectin affinity purification and (b) hydrazide chemistry. While lectin affinity enrichment can isolate multiple types of glycoproteins in complex biological samples *(12, 13)*, hydrazide chemistry for glycoprotein extraction is specific for *N*-glycopeptide capture *(14)*. Using these two methods, in combination with integrated two-dimensional (2D) liquid chromatography (LC) separation and tandem mass spectrometry (MS), we have identified a total of 216 glycoproteins/peptides in human CSF, including many low abundance proteins. This investigation not only, for the first time, categorized many glycoproteins in human CSF but also expanded the existing overall CSF protein database *(15)*. This report mainly focuses on the detailed methods utilized to prepare CSF samples before proteomics analysis. A few more recent methodology improvements (over those used in the original publication) *(16)* are also incorporated.

2. Materials

2.1. Collection of Human CSF

1. 24-gauge bullet-tip Sprotte spinal needle (Pajunk GmbH Medecin Technik, West Germany).

2. 1% lidocaine.
3. Protease inhibitor cocktail (Sigma-Aldrich, St. Louis, MO) dissolved in 10 mL of water. Aliquot the mixture and store at −80°C (*see* **Note 1**).

2.2. Hemoglobin Assay

1. Obtain a Human Hemoglobin ELISA Quantitation Kit (Bethyl, Montgomery, TX). Coating antibody (1 mg/mL): sheep anti-human hemoglobin affinity purified at a working dilution of 1/100; calibrator (2.0 mg/mL): human hemoglobin calibrator at a working range of 6.25–400 ng/mL; HRP detection antibody (1 mg/mL): sheep anti-human hemoglobin-HRP conjugate at a working dilution of 1/10,000.
2. Polystyrene 96-well microplate.
3. Coating buffer: 150 mM sodium carbonate, 350 mM sodium bicarbonate, 30 mM sodium azide. Adjust pH to 9.6 with HCl. Pass through a 0.22-μm pore size filter. Store at 4°C.
4. Wash solution: 50 mM Tris–HCl, pH 8.0, 140 mM NaCl, 0.05% Tween 20. Store at room temperature.
5. Blocking (Postcoat) solution: 50 mM Tris–HCl, pH 8.0, 140 mM NaCl, 1% BSA. Store at 4°C.
6. Sample/conjugate diluent: 50 mM Tris–HCl, pH 8.0, 140 mM NaCl, 0.05% Tween 20, 1% BSA. Store at 4°C.
7. Enzyme substrate solution: SureBlue TMB (3,3′,5,5′-tetramethylbenzidene) Microwell peroxidase substrate (1-component) (Kirkegaard and Perry, Gaithersburg, MD; *see* **Note 2**). Store at 4°C.
8. Stopping solution: 2 M HCl (2 M H_2SO_4 can be used). Store at room temperature.

2.3. Protein Determination with Bradford Assay in CSF Without Precipitation (See Note 3)

1. Use Coomassie Plus-The Better Bradford Assay Kit (Pierce, Rockford, IL).
2. Coomassie Plus-The Better Bradford Assay Reagent contains coomassie G-250 dye, methanol, phosphoric acid, and solubilizing agents in water. Store at 4°C.
3. Albumin standard: bovine serum albumin (BSA) at 2.0 mg/mL in 0.9% saline and 0.05% sodium azide. Store at room temperature.

2.4. Protein Determination with Bicinchoninic Acid Assay in CSF After Precipitation (See Note 3)

1. Use Bicinchoninic acid (BCA) protein assay reagent Kit (Pierce, Rockford, IL).
2. 96-well microplate.
3. Prepare a working reagent by mixing 50 parts of BCA Reagent A with 1 part of BCA Reagent B (50:1, Reagent A:B) (*see* **Note 4**). BCA Reagent A contains sodium carbonate, sodium bicarbonate, bicinchoninic acid, and sodium tartrate in 0.1 M NaOH. BCA Reagent B contains 4% cupric sulfate.
4. Albumin standard from **Subheading 2.3**.

2.5. Isolation of Glycoprotein by Lectin Affinity Column

1. Use Qproteome Total Glycoprotein Kit (Qiagen, Valencia, CA) for glycoprotein isolation with lectin column. The binding and elution buffer are provided.
2. Binding buffer: For each lectin spin column you plan to use, supplement a 3.5-mL aliquot of binding buffer with 35 μL of protease inhibitor solution (100×).
3. Elution buffer: Add 6 μL of protease inhibitor solution (100×) to a 600-μL aliquot of elution buffer.
4. Acetone: Chill acetone to –20°C.
5. Dissolving buffer: 50 mM NH_4HCO_3, pH 8.3, 0.05% SDS.
6. Reducing solution: 10 mM tris-(2-carboxyethyl)phosphine (TCEP).
7. Blocking solution: 6.7 mM iodoacetamide (IAA).
8. Sequencing grade modified trypsin (Promega, Madison, WI).

2.6. Isolation of Glycoprotein by Hydrazide Resin

1. Zeba Desalt Spin column (Pierce, Rockford, IL).
2. Hydrazide resin in isopropanol (Bio-Rad, Hercules, CA).
3. Coupling buffer: 100 mM sodium acetate, 150 mM NaCl, pH 5.5.
4. Sodium periodate (10×): 150 mM sodium periodate.
5. Urea buffer: 8 M urea, 0.4 M NH_4HCO_3, pH 8.3.
6. Reducing solution: 10 mM tris-(2-carboxyethyl)phosphine (TCEP).
7. Blocking solution: 6.7 mM iodoacetamide (IAA).
8. Dissolving buffer: 50 mM NH_4HCO_3, pH 8.3, 0.05% SDS.
9. Sequencing grade modified trypsin.
10. 1.5 M NaCl.
11. 80% acetonitrile (ACN).
12. 100% methanol.
13. PNGase F (New England BioLabs, Beverly, MA): 1 μL diluted in 0.3 mL of 0.1 M NH_4HCO_3, pH 8.3.

2.7. Desalting

1. Waters Oasis MCX cartridge 1 cc/30 mg (Waters GmbH, Eschborn, Germany) is a strong cation-exchange (SCX) resin with hydrophobic characteristics.
2. 5 mL of 1% TFA: 50-μL TFA and 4.95-mL HPLC water.
3. Wetting solution: 100% methanol.
4. Equilibration solution: 10 mL of 0.1% TFA (1 mL of 1% TFA, 9 mL of HPLC water).

5. Wash solution: 5 mL of 80% ACN/0.1% TFA (500 μL of 1% TFA, 4 mL of ACN, 500 μL of HPLC water).
6. Elution solution: 2 mL of 10% NH_4OH/90% methanol (1.8-mL methanol, 200-μL NH_4OH).
7. 0.4% formic acid.

2.8. Automated 2D LC-MS/MS Analysis of Glycopeptides

1. SCX column: 100 mm in length × 320 μm in i.d., with 5-μm particles.
2. Reversed-phase (RP) column: 100 mm in length × 180 μm in i.d., with 3-μm C18 particles.
3. Solvent A: 0.1% formic acid in water.
4. Solvent B: 0.1% formic acid in ACN.
5. Solvent C: 5% ACN, 0.1% formic acid in water.
6. Solvent D: 1 M ammonium chloride, 5% ACN, 0.1% formic acid in water.

3. Methods

3.1. Collection of Human CSF

1. Obtain written informed consent.
2. Place individuals in the lateral decubitus position and infiltrate the L4-5 interspace with 1% lidocaine to provide local anesthesia.
3. Perform a lumbar puncture atraumatically with a 24-gauge bullet-tip Sprotte spinal needle and draw CSF with sterile syringe(s).
4. Individuals must remain in bed for 1 h following lumbar puncture.
5. Collect all CSF samples in the morning after overnight fasting, and store at −80°C until further use (*see* **Note 5**).

3.2. Hemoglobin Assay (*See* Note 6)

1. Analyze standards, samples, blanks, and/or controls in triplicate. Insert the required number of microtiter well strips into a holder. Dilute 1 μL capture antibody (A80–135A) to 100 μL with coating buffer for each well to be coated. Incubate coated plate for 60 min.
2. After incubation, aspirate the capture antibody solution from each well.
3. Wash each well with wash solution as follows: Fill each well with wash solution. Remove wash solution by aspiration. Repeat for a total of three washes.

4. Add 200 μL of blocking (Postcoat) solution to each well. Incubate for 30 min. After incubation, remove the blocking (Postcoat) solution and wash each well three times as in **step 3**.
5. Dilute the standards and samples in sample/conjugate diluent. Standard: 6.25, 12.5, 25, 50, 100, and 200 ng/mL; sample: 1–20 diluent.
6. Transfer 100 μL of the standard or sample to assigned wells. Incubate plate for 60 min. After incubation, remove samples and standards, and wash each well five times as in **step 3**.
7. Dilute the HRP conjugate (A80–135P) in conjugate diluent 1:10,000. Transfer 100 μL to each well. Incubate for 60 min. After incubation, remove the HRP conjugate and wash each well five times as in **step 3**.
8. Transfer 100 μL of enzyme substrate solution to each well. Incubate the plate for 20 min.
9. To stop the TMB reaction, apply 100 μL of 2 M HCl to each well. If using another substrate, use the stop solution recommended by the manufacturer.
10. Using a microtiter plate reader, read the plate at the wavelength that is appropriate for the substrate used (450 nm for TMB).

3.3. CSF Protein Assay

3.3.1. Bradford Assay

1. Pipette 10 μL of each standard or unknown sample into the appropriate microplate wells.
2. Add 300 μL of the Coomassie Plus Reagent to each well, and mix with a plate shaker for 30 s.
3. Remove plate from shaker and then incubate plate for 10 min at room temperature.
4. Measure the absorbance at or near 595 nm with a plate reader.
5. Prepare a standard curve by plotting the average Blank-corrected 595 nm measurement for each BSA standard vs. its concentration in μg/mL. Use the standard curve to determine the protein concentration of each unknown sample.

3.3.2. BCA Assay

1. Pipette 25 μL of each standard or unknown sample replicate into a microplate well.
2. Add 200 μL of the working reagent to each well and mix plate thoroughly on a plate shaker for 30 s.
3. Cover plate and incubate at 37°C for 60 min.
4. Cool plate to room temperature. Measure the absorbance at or near 562 nm on a plate reader.

5. Prepare a standard curve by plotting the average Blank-corrected 562-nm measurement for each BSA standard vs. its concentration in μg/mL. Use the standard curve to determine the protein concentration of each unknown sample.

3.4. Isolation of Glycoprotein (*See* Note 7)

3.4.1. Isolation of Glycoprotein by Lectin Affinity Column

1. Use a 2 mL sample of pooled human CSF for glycoprotein enrichment via lectin affinity column in each experiment. Dry the sample to 200 μL with a SpeedVac and then centrifuge at 1,000 × *g* for 5 min. Next, transfer the supernatant to another tube, and dissolve the pellet with 100 μL of detergent supplied by the kit accompanying the lectin column. After that, combine the solutions from these two tubes and add 1,500 μL of binding buffer along with Protease Inhibitor Solution (100×). Next, vortex the solution to ensure even mixing.
2. Prepare lectin spin columns by loosening the screw cap of the column a quarter of a turn, snap off the bottom closure, and place the lectin spin column in a 2-mL collection tube. Centrifuge the lectin spin column for 2 min at 500 × *g* in a microcentrifuge.
3. Discard the flow-through and pipet 500-μL binding buffer supplemented with protease inhibitor solution (100×) onto the lectin spin column. Centrifuge for 2 min at 500 × *g* in a microcentrifuge and discard the flow-through.
4. Load the resultant samples from **step 1** onto the lectin spin column and then incubate for 1 min, with subsequent centrifugation for 2 min at 500 × *g* (*see* **Note 8**).
5. Apply 750 μL of binding buffer supplemented with protease inhibitor solution (100×) to wash the lectin spin column, centrifuge for 2 min at 500 × *g*, and discard the flow-through. Repeat this step. Transfer the lectin spin column to a clean microcentrifuge tube.
6. Apply 100 μL of the elution buffer to the prepared lectin spin column, incubate for 1 min, and centrifuge for 1 min at 500 × *g*. Next, elute the glycoproteins with 100 μL of elution buffer with protease inhibitor solution for a total of six times. Pool the elute fractions and determine the protein concentration (e.g., using the Bradford method or BCA assay).
7. Clean the sample by chilling acetone to –20°C (acetone precipitation) and the tube with the sample to 4°C (*see* **Note 9**).
8. Add six volumes of cold acetone to the cold sample tube. Invert the tube three times. Incubate the tube at –20°C until a precipitate forms (30 min or overnight).

9. Decant the acetone. Dissolve the precipitated pellet in dissolving buffer (*see* **Note 10**).
10. Add 10 mM TCEP to the sample, and incubate at room temperature for 30 min, and then add 6.7 mM IAA to the sample and incubate for another 30 min in the dark at room temperature. The proteins on the resin will be denatured and alkylated. Save ~1 μg of protein to verify with SDS-PAGE (*see* **Note 11**).
11. Add trypsin at a concentration of 1 μg of trypsin/200 μg of CSF protein, and digest at 37°C overnight. Next, dry the peptides by SpeedVac before MS analysis (*see* **Note 12**).

3.4.2. Isolation of Glycoprotein by Hydrazide Resin

1. In each experiment, exchange 2 mL of the pooled CSF sample with coupling buffer using a Zeba desalt spin columns. Equilibrate two Zeba desalt spin column three times with 2.5 mL of coupling buffer, and then remove the buffer by spinning at 1,000 × *g* for 2 min. Next, add 2.5 mL of coupling buffer (it is now ready for desalting or buffer exchange). Spin off the coupling buffer from Zeba desalt spin column into a new collection tube, and then add 2 mL of pooled CSF sample to the center of the compacted resin bed of the Zeba desalt spin column. Collect the CSF sample by spinning at 1,000 × *g* for 2 min.
2. Before adding sodium periodate solution, check the pH of the sample (≤5.5). Once the given pH has been reached, add sodium periodate solution (final concentration: 15 mM), and then incubate the sample at room temperature for 1 h in the dark.
3. Prepare 400 μL of moist hydrazide resin (800 μL of 50% slurry) per sample by washing three times. Remove three resin volumes of deionized water by spinning at 3,000 × *g* for 5 min, followed by washing/equilibration with coupling buffer three times. After equilibration, resuspend the hydrazide resin in 50% slurry using coupling buffer. Separate the resin into two 2-mL tubes and 200 μL of moist resin (400 μL 50% slurry in each one).
4. Remove the added sodium periodate by using the Zeba desalt spin column. Add the oxidized CSF to a prepared column, and collect the CSF using a new collection tube by spinning at 1,000 × *g* for 2 min.
5. Add half of the oxidized sample to one of the 200-μL hydrazide resin tubes equilibrated in coupling buffer, and add the rest of it to the other resin tube. Conjugate the glycoproteins from CSF to hydrazide resin at room temperature by rotating the sample overnight (*see* **Note 13**).

6. After the coupling reaction, collect the resin by centrifugation at 3,000 × *g* for 5 min. Remove nonglycoproteins by washing the resin five times with 1 mL of urea solution. When you wash the resin, combine the two tubes of resin together. After the last wash and removal of the urea buffer, resuspend the hydrazide resin in 400 μL of urea buffer.
7. Add 10 mM TCEP to the hydrazide resin, and incubate at room temperature for 30 min, and then add 6.7 mM IAA to the hydrazide resin and incubate for another 30 min in the dark at room temperature (denatured and alkylated). Now, wash the resin three times with urea solution and 50 mM of NH_4HCO_3. Subsequently, suspend the resin in two bed volumes of dissolving buffer.
8. Add trypsin at a concentration of 1 μg of trypsin/200 μg of CSF protein, and digest at 37°C overnight. Remove the trypsin-released peptides by washing the resin three times with three bed volumes of each of the following: 1.5 M NaCl, 80% ACN, 100% methanol, water, and 50 mM NH_4HCO_3.
9. Release N-linked glycopeptides with PNGase F (1 μg diluted in 0.3 mL of 50 mM NH_4HCO_3) at 37°C, and place the mixture on a shaker or rotate overnight. Collect the supernatant and deposit it to a glass vial, and wash the hydrazide resin twice with 200 μL of 80% ACN. Combine the washes with the supernatant in the glass vial. Dry the released glycopeptides via a SpeedVac before MS analysis.

3.5. Desalting

1. Wet a MCX 1 cc cartridge with 2 mL methanol. Next, equilibrate with 2 mL of 0.1% TFA.
2. Acidify the sample with 1% TFA in less than 100 μL (adjust to pH ≤ 3). Load the sample slowly onto the cartridge (*see* **Note 14**).
3. Wash the sample with 5 mL of 0.1% TFA buffer, followed by 5 mL of 80% ACN/0.1% TFA buffer 2-mL water.
4. Elute with 1 mL of 10% NH_4OH/90% methanol. Dry the sample in a SpeedVac (no acidic sample should be present). Reconstitute with 24 μL or appropriate amount of 0.4% formic acid. The sample is now ready for mass spectrometry (*see* **Note 15**).

3.6. Automated 2D LC-MS/MS Analysis of Glycopeptides

1. The samples prepared by both methods were separated by a 2D microcapillary high-performance LC system, with an integrated SCX column with two alternating RP columns, followed by MS/MS analysis (*see* **Note 16**).
2. Six fractions were eluted from the SCX column using a binary gradient of 2–90% solvent D vs. solvent C. Each fraction was injected onto RP columns automatically with the peptides

being resolved using a 300-min binary gradient of 5–80% solvent B vs. solvent A.

3. A flow rate of 160 μL/min with a split ratio of 1/80 was used.
4. The MS acquisition was operated in a data-dependent MS/MS mode where each survey scan mass spectrum was followed by MS/MS analysis of one of the available precursor ions from the prior survey scan. Ions selected for collision-induced dissociation (CID) were dynamically excluded for 3 min.

3.7. MS/MS Database Search and Protein Identification

1. MS/MS data were searched against the International Protein Index (IPI) human protein database version 3.01 from the European Bioinformatics Institute (EBI) using the SEQUEST algorithm *(17)*.
2. For the MS/MS database search, the search criteria were set to expect the following modifications. (1) For the hydrazide chemistry approach: carboxymethylated cysteines (fixed), oxidized methionines (variable), and an enzyme-catalyzed conversion of asparagine to aspartic acid at the site of carbohydrate attachment (variable) *(5)*. (2) For the lectin affinity approach: carboxymethylated cysteines (fixed) and oxidized methionines (variable) were the expected modifications.
3. The search results were validated using PeptideProphet *(18)* and ProteinProphet *(19)* for peptide and protein identification, respectively. In validation of protein identification, a Protein-Prophet probability score of 0.9 was used as the filtering criteria to ensure a <1% overall error rate for protein identification (*see* **Note 17**).

4. Notes

1. Protease inhibitor is not routinely included in a typical clinical CSF tap; thus, it needs to be added as soon as the CSF is thawed for the first time, if the sample will be reused. However, there are also caveats associated with adding protease inhibitors. For a more detailed discussion, see a recent review article *(20)*.
2. TMB (3,3′,5,5′-tetramethylbenzidene) is highly recommended but OPD (o-phenylenediamine dihydrochloride) or ABTS (2,2′-azino-bis-(3-ethylbenzthiazoline-6-sulfonic acid) can also be used. Wavelength should be 450 nm for TMB, 490 nm for OPD, and 405 nm for ABTS, respectively.

3. Bradford assay is preferred when CSF proteins are not precipitated first. The reason for this is that the concentration of amino acids, also picked up by BCA assay, in the CSF is high and variable. However, after proteins are precipitated, BCA assay is better than the Bradford assay as less protein is needed for the BCA assay.
4. One should prepare sufficient volume of working reagent based on the number of samples to be assayed. The working reagent is stable for several days when stored in a closed container at room temperature.
5. Human CSF is closely regulated via balanced secretion and absorption with an average circulating volume between 125 and 150 mL in an adult; as a result, the amount of CSF that can be obtained is usually limited to less than 25–30 mL. Additionally, there is a significant variation in CSF production during the day as well as a rostro-caudal gradient of protein concentration *(21–23)*. Thus, it is critical to match CSF samples not only for the timing of CSF taps, but also the fractions of CSF obtained.
6. Protein concentration in the CSF is relatively low compared to plasma (CSF/plasma: <1/200), and the protein profiles in CSF are similar to those in plasma *(23)*; as a result, even a minor contamination of CSF with blood could significantly confound the interpretation of quantitative or qualitative proteomic analysis of CSF. Two criteria are commonly used in our lab to control for blood contamination (1) CSF red blood cell (RBC) count, as determined by standard clinical chemistry laboratory, have to be less than 10 RBC/μL and (2) hemoglobin level in CSF have to be less than 280 ng/mL, which is roughly equivalent to 1:500,000 dilution of human plasma (14 g/dL) or 10 RBC/μL (five million/μL).
7. Several methods have been described for the isolation of glycopeptides from digests of purified glycoproteins and from more complex protein mixtures. Affinity chromatography using lectins, naturally occurring proteins that bind to specific glycans, is used for enriching glycopeptides *(24)*. In this study, the total glycoprotein spin column in the Qproteome total glycoprotein kit contains concanavalin A (ConA) and wheat germ agglutinin (WGA) lectins. They are used for the general enrichment of the total glycoprotein from a CSF sample. ConA is a lectin that binds mannosyl and glucosyl residues containing unmodified hydroxyl groups at positions C3, C4, and C6 and can be utilized for the targeted binding of certain oligosaccharide structures of N-glycosylated proteins. The use of WGA allows for the isolation of glycostructures comprising *N*-acetylglucosamine and sialic acids. However, each lectin species isolates only a subset of glycan

species. Hydrazide chemistry is used to selectively isolate, identify, and quantify N-linked glycopeptides in a specific and efficient manner *(14)*. Combining these two methods significantly improves CSF glycoprotein identification (*see* **Fig. 1** for more details).

8. One should collect the flow-through if analysis of other CSF proteins, e.g., nonglycoprotein, is desired.
9. If acetone precipitation is performed after trypsin digestion, the sample can be lost.
10. Do not overdry the pellet, or it may not dissolve properly.
11. This protocol can be adapted to a quantification method (e.g., iTRAQ labeling method). When the iTRAQ method is used, MMTS should be used for blocking.

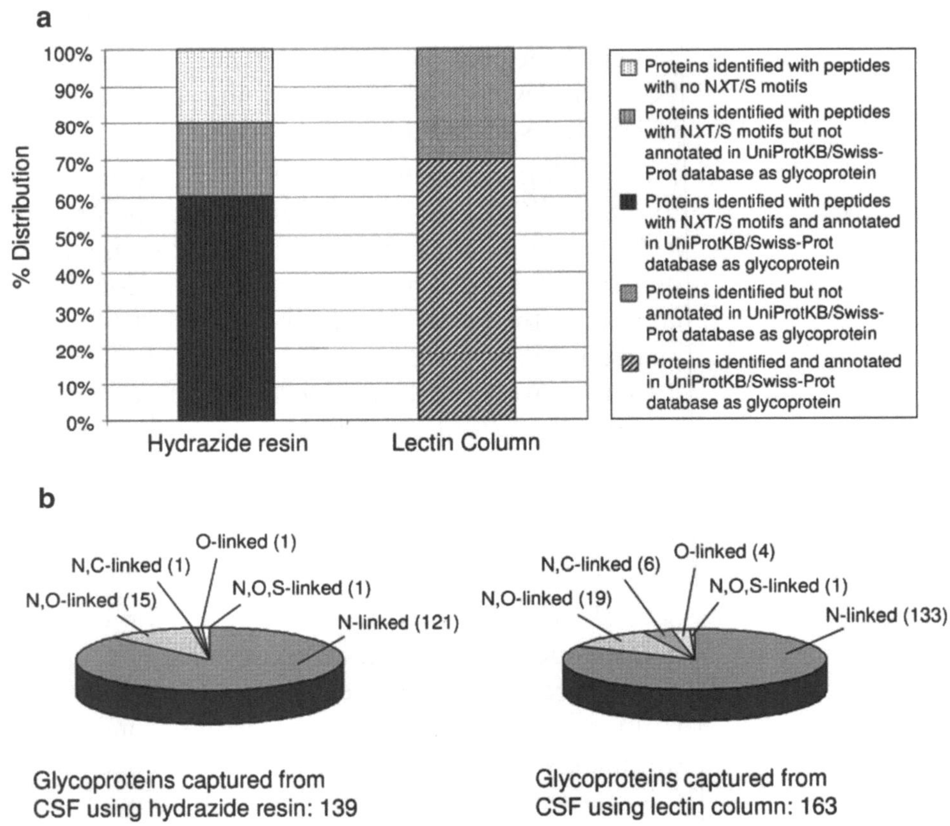

Fig. 1. Comparison of hydrazide chemistry and lectin affinity column for glycoprotein enrichment and identification from human CSF. (**a**) The specificity and efficiency of glycoprotein capture using hydrazide chemistry and lectin affinity column. (**b**) The categorization of the identified proteins isolated from human CSF using hydrazide chemistry and lectin affinity column, respectively. With respect to the lectin affinity approach, 236 proteins were identified, and 163 of those proteins were annotated in UniProtKB/Swiss-Prot database as glycoproteins; that is, the specificity of this approach was approximately 69%. The diversity of the 163 glycoproteins that were identified and annotated was similar to that identified using hydrazide chemistry. For the hydrazide chemistry method, 235 proteins were identified. Further analysis indicated that among the 235 identified proteins, 45 proteins (19%) were identified with peptides that had no NXT/S motif and, therefore, were considered as nonspecific bound proteins captured by hydrazide chemistry (reproduced from **(16)** with permission from ACS publications).

12. Analyze ~1 μg of peptides after digestion, then save 1 μg of protein from **step 10** of **Subheading 3.4.1** and utilize SDS-PAGE to check for the degree of completion of tryptic digestion. Digest proteins completely for high specificity and yield of glycopeptides. If digestion is not complete, another batch of enzyme can be added, and digest the samples for an additional 4 h at 37°C. Digested peptides can be stored frozen at −20°C for several weeks.
13. Coupling can be left for 24 h at room temperature and immobilized glycopeptides on hydrazide resin can be stored at 4°C for up to a month.
14. The sample can be loaded repeatedly to ensure binding. One can use a vacuum for wetting, equilibration, and washing. Supposedly MCX cartridges can be run dry without loss of recovery.
15. The sample elute tubes are 1.5-mL plastic microcentrifuge tubes. All tubes are rinsed in 100% USP-grade ethanol and water, and then air-dried prior to use. It appears that thorough washing removes acetonitrile soluble materials, which forms a layer in aqueous solutions and interferes with evaporation.
16. With advances in technology, we now commonly use off-line SCX separation with higher end mass spectrometers, e.g., MALDI-TOF-TOF (Applied Biosystem, Foster City, CA), Orbitrap or LTQ-FT (ThermoFisher Scientific, San Jose, CA).
17. It is our experience that very few proteins overlap those searched against a decoy database when PeptideProphet and ProteinProphet are used to filter the candidate proteins.

Acknowledgments

The research is supported by following grants from the NIH (AG025327, ES012703, and NS060252) as well as the Michael J. Fox Foundation and a Shaw Endowment to Dr. Jing Zhang.

References

1. Blennow, K. (2005) CSF biomarkers for Alzheimer's disease: Use in early diagnosis and evaluation of drug treatment. *Expert. Rev. Mol. Diagn.* **5**, 661–672.
2. Zhang, J., Goodlett, D. R., and Montine, T. J. (2005) Proteomic biomarker discovery in cerebrospinal fluid for neurodegenerative diseases. *J. Alzheimers Dis.* **8**, 377–386.
3. Gronborg, M., Kristiansen, T. Z., Stensballe, A., Andersen, J. S., Ohara, O., Mann, M., Jensen, O. N., and Pandey, A. (2002) A mass spectrometry-based proteomic approach for identification of serine/threonine-phosphorylated proteins by enrichment with phospho-specific antibodies: Identification of a novel protein, Frigg, as a protein kinase A substrate. *Mol. Cell Proteomics* **1**, 517–527.
4. Yang, Z., and Hancock W. S. (2004) Approach to the comprehensive analysis of glycoproteins isolated from human serum using a multi-lectin

affinity column. *J. Chromatogr. A* **1053**, 79–88.

5. Liu, T., Qian, W. J., Gritsenko, M. A., Camp, D. G., 2nd, Monroe, M. E., Moore, R. J., and Smith, R. D. (2005) Human plasma N-glycoproteome analysis by immunoaffinity subtraction, hydrazide chemistry, and mass spectrometry. *J. Proteome. Res.* **4**, 2070–2080.
6. Yang, Z., Hancock, W. S., Chew, T. R., and Bonilla, L. (2005) A study of glycoproteins in human serum and plasma reference standards (HUPO) using multilectin affinity chromatography coupled with RPLC-MS/MS. *Proteomics* **5**, 3353–3366.
7. Jaeken, J., and Matthijs, G. (2001) Congenital disorders of glycosylation. *Annu. Rev. Genomics Hum. Genet.* **2**, 129–151.
8. Rudd, P. M., Elliott, T., Cresswell, P., Wilson, I. A., and Dwek, R. A. (2001) Glycosylation and the immune system. *Science* **291**, 2370–2376.
9. Huang, X., Ushijima, K., Komai, K., Takemoto, Y., Motoshima, S., Kamura, T., and Kohno, K. (2004) Co-expression of Y box-binding protein-1 and P-glycoprotein as a prognostic marker for survival in epithelial ovarian cancer. *Gynecol. Oncol.* **93**, 287–291.
10. Ferrara, N., and Kerbel, R. S. (2005) Angiogenesis as a therapeutic target. *Nature* **438**, 967–974.
11. Burton, D. R., and Dwek, R. A. (2006) Immunology. Sugar determines antibody activity. *Science* **313**, 627–628.
12. Geng, M., Zhang, X., Bina, M., and Regnier, F. (2001) Proteomics of glycoproteins based on affinity selection of glycopeptides from tryptic digests. *J. Chromatogr. B Biomed. Sci. Appl.* **752**, 293–306.
13. Kaji, H., Saito, H., Yamauchi, Y., Shinkawa, T., Taoka, M., Hirabayashi, J., Kasai, K., Takahashi, N., and Isobe, T. (2003) Lectin affinity capture, isotope-coded tagging and mass spectrometry to identify N-linked glycoproteins. *Nat. Biotechnol.* **21**, 667–672.
14. Zhang, H., Li, X. J., Martin, D. B., and Aebersold, R. (2003) Identification and quantification of N-linked glycoproteins using hydrazide chemistry, stable isotope labeling and mass spectrometry. *Nat. Biotechnol.* **21**, 660–666.
15. Pan, S., Zhu, D., Quinn, J. F., Peskind, E. R., Montine, T. J., Lin, B., et al. (2007) A combined dataset of human cerebrospinal fluid proteins identified by multi-dimensional chromatography and tandem mass spectrometry. *Proteomics* 7, 469–473.
16. Pan, S., Wang, Y., Quinn, J. F., Peskind, E. R., Waichunas, D., Wimberger, J. T., Jin, J., Li, J. G., Zhu, D., Pan, C., and Zhang, J. (2006) Identification of glycoproteins in human cerebrospinal fluid with a complementary proteomic approach. *J. Proteome. Res.* **5**, 2769–2779.
17. Gatlin, C. L., Eng, J. K., Cross, S. T., Detter, J. C., and Yates, J. R. 3rd. (2000) Automated identification of amino acid sequence variations in proteins by HPLC/microspray tandem mass spectrometry. *Anal. Chem.* **72**, 757–763.
18. Keller, A., Nesvizhskii, A. I., Kolker, E., and Aebersold, R. (2002) Empirical statistical model to estimate the accuracy of peptide identifications made by MS/MS and database search. *Anal. Chem.* **74**, 5383–5392.
19. Nesvizhskii, A. I., Keller, A., Kolker, E., and Aebersold, R. (2003) A statistical model for identifying proteins by tandem mass spectrometry. *Anal. Chem.* **75**, 4646–4658.
20. Zhang, J. (2007) Proteomics of human cerebrospinal fluid – the good, the bad, and the ugly. *Proteomics Clin. Appl.* **1**, 805–819.
21. Blennow, K., Fredman, P., Wallin, A., Gottfries, C. G., Langstrom, G., and Svennerholm, L. (1993) Protein analyses in cerebrospinal fluid. I. Influence of concentration gradients for proteins on cerebrospinal fluid/serum albumin ratio. *Eur. Neurol.* **33**, 126–128.
22. Blennow, K., Fredman, P., Wallin, A., Gottfries, C. G., Skoog, I., Wikkelso, C., et al. (1993) Protein analysis in cerebrospinal fluid. III. Relation to blood–cerebrospinal fluid barrier function for formulas for quantitative determination of intrathecal IgG production. *Eur. Neurol.* **33**, 134–142.
23. Blennow, K., Fredman, P., Wallin, A., Gottfries, C. G., Karlsson, I., Langstrom, G., Skoog, I., Svennerholm, L., and Wikkelsö, C. (1993) Protein analysis in cerebrospinal fluid. II. Reference values derived from healthy individuals 18–88 years of age. *Eur. Neurol.* **33**, 129–133.
24. Ding, W., Hill, J. J., and Kelly, J. (2007) Selective enrichment of glycopeptides from glycoprotein digests using ion-pairing normal-phase liquid chromatography. *Anal. Chem.* **79**, 8891–8899.

Chapter 18

Mass Spectrometric Analysis of Body Fluids for Biomarker Discovery

David M. Good and Joshua J. Coon

Summary

Hinging on the concept that extracellular proteins and polypeptides will provide information on the physiological state of specific organs, or even entire organisms, proteomic analysis of biological fluids for biomarker discovery has seen rapid expansion in recent years. Although multiple studies have had success using mass spectrometric analytical techniques for determination of proteins within a sample, inspection of naturally occurring species has been difficult, with most analyses using bottom-up methodology. We have applied a new fragmentation method, electron transfer dissociation (ETD), to this problem. We have previously illustrated the benefits to spectral quality and total identifications when using a combination of the complementary fragmentation techniques, ETD, and collision-activated dissociation, for analysis of naturally occurring proteins and polypeptides within biological fluids.

Key words: Biomarker, Body fluid, Mass spectrometry, Electron transfer dissociation

1. Introduction

Owing to several factors, including ease of accessibility (compared with tissue biopsies) and costs of obtainment (both fiscal and patient health), inspection of biological fluids has become the focus for many modern clinical proteomics studies, including those in neuroscience. As such, multiple studies have recently showcased the usefulness of investigating varying body fluids for acquiring information that is diagnostic and/or prognostic of disease, as well as providing the possibility for monitoring treatment response *(1–7)*. Mass spectrometry (MS) has become a central tool for these analyses, and is typically combined with a front-end separations technique, such as capillary electrophoresis (CE),

Andrew K. Ottens and Kevin K.W. Wang (eds.), *Neuroproteomics,* Methods in Molecular Biology, vol. 566
DOI 10.1007/978-1-59745-562-6_18, © Humana Press, a part of Springer Science+Business Media, LLC 2009

liquid chromatography (LC), or two-dimensional gel electrophoresis (2D-E). Though use of a mass spectrometer downstream for detection of molecules separated in a first step is highly sensitive, peptide sequence identifications are usually achieved via employing a second stage of MS (tandem MS) *(8)*. The conventional method used to fragment precursor ions in MS/MS is collision-activated dissociation (CAD). This technique has proven very useful for small tryptic peptide sequencing, but has been less effective for the characterization of naturally occurring peptides (i.e., peptides/proteins which contain internal basic residues, many posttranslational modifications, or are longer than ~15 residues) *(9)*. Although it is not necessary to determine the sequence of a biomarker candidate to create statistically significant, clinically relevant tests *(4, 10, 11)*, one is unable to glean further information on the actual physiological state of the patient without it.

Of recent interest has been the investigation of the role posttranslational modifications (PTMs) being played in disease activity *(12–15)*. Therefore, much emphasis is currently being placed on gaining choice sequence information that allows for the confident characterization of sequences containing PTMs. Although CAD has long provided a solid platform to gain biological information, its limitations in PTM analysis (including neutral losses that prevent localization of the modification) have been outlined in detail *(16–19)*. Also, the prerequisite use of a bottom-up approach *(20)* limits the usefulness of the derived information, as the in vitro enzymatic digestion of the sample increases complexity and therefore lowers the probability of drawing meaningful conclusions on in vivo conditions. In comparison, electron transfer dissociation (ETD) *(21–26)* is a fragmentation method that offers a more robust method to characterize PTMs, including sequence assignment and localization of modifications, which allows interrogation of large peptides or even whole proteins *(27)*. In addition, ETD provides sample information complementary to that generated by CAD *(23)*.

Here we outline our technique for the analysis of biological fluids, specifically human urine. We provide detailed instructions for sample preparation, as well as for using the complementary fragmentation methods of ETD and CAD in an efficient manner. Finally, we discuss data analysis, and provide a template for creating data outputs that are compliant with recently published suggestions for the handling of clinical proteomics data *(28)*.

2. Materials

2.1. Sample Prep

1. HPLC-grade water; Barnstead NANOpure Diamond™ (Barnstead International, Dubuque, IA).
2. Centrisart ultracentrifugation filter devices (20 kDa MWCO; Sartorius, Goettingen, Germany).
3. PD-10 desalting column (Amersham Bioscience, Uppsala, Sweden).

2.2. Sample Introduction

1. Bomb assembly (*see* **Fig. 1**).
2. Nonreactive gas source for use with bomb assembly; standard purity Helium works fine for this application.
3. Short-form style glass vials (2-mL), 0.5 Dr (VWR, West Chester, PA).
4. Capillary material both 365 × 75 μm^2 and 365 × 50 μm^2 (outer diameter × inner diameter) (Polymicro Technologies, Phoenix, AR).
5. C_8/C_{18} 5 µm reversed phase material (Alltech Associates Inc., Deerfield, IL).
6. Irregular (5–20 µm) C_{18} particles (YMC, Milford, MA).

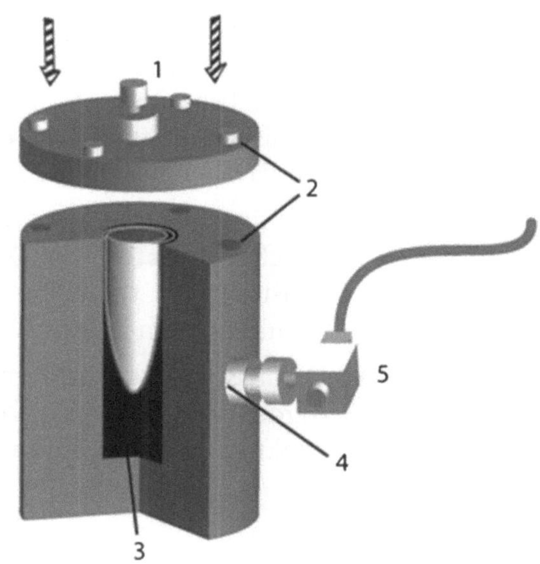

Fig. 1. The bomb assembly consists of a solid stainless steel cylinder separated into a top and bottom section. A hole to fit an NPT 20 – 1/8 fitting is drilled into the top section (1). Four ¼″ – 20 screw holes are bored through both sections (2). A larger hole is drilled into the bottom section, and centered with the NPT fitting, to contain the sample vial (3). A Swagelock NPT-thread male connector is used to connect to a Swagelock SS-41GSX2 valve (4), which is employed to give manual control over placing the bomb under pressure (5).

7. Teflon; 0.012 in. ID × 0.059 in. OD (Zeus Industrial Products, Orangeburg, SC).
8. Laser puller (Sutter Instrument Co., Novarto, CA; model P-2000).
9. LiChrosorb (EMD Chemicals Inc., Gibbstown, NJ).
10. Microflame Gas Torch with Micronox™ and Butane cylinders (Microflame, Inc., Plymouth, MN; model 4400).
11. Disposable calibrated pipets for measuring volume (Drummond Scientific Company, Broomall, PA).
12. Agilent 1100 High Performance Liquid Chromatograph Pump (Agilent Technologies, Santa Clara, CA).
13. 0.25 mm natural peek tubing (Upchurch Scientific, Oak Harbor, WA).
14. Peek Y/T-split (Upchurch Scientific, Oak Harbor, WA).
15. HPLC grade acetonitrile.
16. High purity formic and acetic acid.

2.3. Mass Spectrometry

1. ETD-enabled quadrupole linear ion trap or Orbitrap instruments (Thermo Scientific, San Jose, CA).
2. Fluoranthene (Sigma-Aldrich, St. Louis, MO).

2.4. Data Analysis

1. Bioworks Browser (Thermo Scientific).
2. In-house program for the separation of ETD and CAD spectra. Because of differing search parameters, one must separate ETD and CAD tandem mass spectra prior to analysis. This program should be quite simple when one is using a set data-dependent acquisition method, where every nth MS/MS scan is ETD and every $(n + 1)$th scan is CAD, or vice versa.
3. The final program necessary to carry out data analysis is one which will "batch" the DTA's into single.txt files. Because of constraints of the database search algorithm, these batches should contain at most 2,000 scans per.txt file.
4. Database search algorithm; Open Mass Spectrometry Search Algorithm (OMSSA), free download at http://pubchem.ncbi.nlm.nih.gov/omssa/ *(29)*.

3. Methods

3.1. Sample Preparation

The following sample preparation deals only with urine as the body fluid of interest. Urine has become the body fluid of choice for our clinical proteomic analyses because of multiple factors, including its ease of obtainment, as it is usually obtained by non-invasive procedures, and its excretion is a normal and necessary biological function. Urine can also be obtained in very large volumes (dL to L amounts), and can be collected over a period of time (even offering the ability for collection of multiple samples within 24 h), allowing the possibility of using this fluid to monitor patient therapeutic response over very short time windows *(20, 30, 32–33)*.

1. Use only spontaneously voided (i.e., do not collect urine over a 24-h time period), mid-stream urine samples (*see* **Note 1**).
2. Samples should be stored at –80°C until analysis (*see* **Note 2**).
3. Owing to the sensitivity of mass spectrometers to salts, polymeric compounds, and other contaminants such as detergents, samples must be prepared accordingly to limit the inclusion of such compounds with the sample.
4. For proteomic analysis, thaw a 0.7 mL aliquot of urine immediately before use and dilute with 0.7 mL of 2 M urea, 10 mM NH_4OH containing 0.02% SDS.
5. To remove higher molecular mass polypeptides and proteins (e.g., albumin and immunoglobulin G), filter the sample using Centrisart ultracentrifugation filter devices at 3,000 × *g* until obtaining 1.1 mL of filtrate.
6. Apply this filtrate onto a PD-10 desalting column and equilibrate with 0.01% NH_4OH in HPLC-grade water to decrease matrix effects by removing urea, electrolytes, and salts, and also to enrich polypeptides.
7. Lyophilize and store at ≤4°C until use.
8. Before use, resuspend in 50–100 µL HPLC-grade water.

3.2. Sample Separation/Introduction

1. The first step in the introduction of the sample into the mass spectrometer is the manufacture of reversed-phase microcapillary columns onto which sample can be loaded, and also over which sample can be separated via liquid chromatography (LC) *(30)*. Two distinct types of microcapillary column should be constructed: a precolumn for the loading of sample and an analytical column over which the sample can be separated. The method described here was first reported by Marto et al. *(30)*.
2. To create the precolumn, cut off ~10–12 cm of 365 × 75 µm^2 fused silica (*see* **Note 3**). Using a flame, burn ~2 mm of one end of the silica. The silica should appear charred and black, but there should not be any deformation of the structure of the capillary. If the glass within the capillary is melting upon applying the heat, then use a cooler flame.

3. Wet a Kimwipe with methanol, and firmly wipe off the charred material from the capillary. You should now see the exposed glass within the capillary, as only the polyimide coating should have been burned off.
4. Place the exposed glass end of this capillary into a vial containing Lichrosorb material. After packing the capillary, make sure to wipe off excess Lichrosorb material from around the opening of the capillary using a Kimwipe.
5. With the polyimide coating removed from around the glass, and excess Lichrosorb removed from the outside of the capillary, it should be easy to tell how far up the capillary the material has been packed. Make sure the capillary has been packed with material up to the end of the exposed area (~2 mm).
6. Flame the exposed and packed end of the capillary with high heat (using the Micronox oxidizer and the hottest part of the flame) until the LiChrosorb within the capillary no longer appears to be a powder, but instead looks like a single, clear plug of glass (*see* **Note 4**).
7. After obtaining a stable frit, the precolumn is ready to pack with the separation material of choice. Suspend the desired reversed phase material (C_8/C_{18}) in a 70/30 acetonitrile/isopropyl alcohol slurry within a 2 mL vial (to ensure a fit within the bomb). There is no need to weigh out the material being used, simply ensure that the slurry is an occluded, milky white color (*see* **Note 5**).
8. To pack the precolumn, place the micro-capillary within the bomb apparatus with the unaltered end placed within the packing slurry. Gently increase pressure on the microcapillary until packing material begins to flow up to the frit of the column. A maximum of ~600 PSI should be used (*see* **Note 6**). Also, check that no packing material is leaving the column, which would mean the frit is faulty. It is recommended to take a sample of the eluate of the packing process and inspect this under a light microscope, looking for the presence of packing material. Pack the precolumn with ~5–7 cm of material.
9. To create an analytical column, cut a ~12 cm piece of 365 × 50 μm^2 fused silica. Using a low-heat flame, char a ~1.5 cm area of the silica beginning ~3–4 cm from one end of the silica.
10. Wet a Kimwipe with methanol, and firmly wipe off the charred material from the capillary. You should now see the exposed glass within the capillary, as only the polyimide coating should have been burned off.

11. A bottleneck now needs to be placed within the micro-capillary, to hold in the future packing material. The bottleneck is made using a laser puller, with the exposed glass section of the fused silica being placed in line with the laser source, but offset so that the bottleneck will be pulled very close to the beginning of the exposed area on the longer side, thus leaving space for the addition of the tip later. The bottleneck should be pulled to a point where there is a visible difference between the diameter of the section of glass shaped by the laser puller and the initial diameter of the glass. However, the column still needs to be stable enough to be handled regularly (*see* **Note 7**).
12. To pack the analytical column, irregularly sized reversed phase particles should be suspended in a slurry similar to that used for packing the precolumn.
13. Again using the bomb, pack the analytical column with 1–2 mm of irregular particles followed by ~7 cm of regularly-sized 5 μm reversed-phase particles. It is critical to ensure that no packing material is getting through the bottleneck, but it is just as important to note the packing speed. If the column takes longer than an hour to pack, then its performance may not be on par with those columns which both retain packing material and pack at a more rapid rate.
14. Once the desired amount of reversed-phase material has been packed within the column, the slurry of packing material should be removed from the bomb and pressure should be applied to the column once more, to dry out the column. Application of pressure for only a few minutes should ensure a column that is relatively dry. The outside of the column (especially the exposed glass) should also be inspected to make sure there is no residual liquid – if so, then this should be removed before the next step.
15. An integrated ESI tip is then added to the column, again using the laser puller. This time, however, the laser puller should be set so that the column is heated and pulled to a point where the column actually separates at the point of heating. It is important to keep the distance between the bottleneck and the tip small, thus reducing the dead volume within the column; however, one should take care not to try and pull the tip too close to the bottleneck, as this will result in a deformed tip, and sometimes in the combination of the two features, thus resulting in a tip which will perform poorly.
16. Load sample buffer onto the precolumn by placing it into the bomb apparatus and inserting the precolumn within the sample buffer, with the fritted end exposed to the air.

Pressure is applied, and the column is filled with the sample buffer. It is important to first fill the column with buffer to ensure minimal loss of sample in the column when loading.

17. After a droplet is seen coming from the end of the precolumn, the pressure can be turned off, the precolumned wiped dry, and the sample buffer removed and replaced with the sample.
18. Again apply pressure, and measure the eluent using a calibrated pipet.
19. The precolumn is connected to the analytical column using a butt-joint with a 0.012 in. (i.d.) Teflon sleeve (*see* **Note 8** and **Fig. 2**). It is important to add this sleeve from the back-end of the precolumn, as the exposed glass is very fragile. Make sure the sleeve is long enough to hold the two columns together when pressure from an LC analysis is applied (at least 3 cm of Teflon tubing should typically be employed for this purpose). Also, if the packing material within the precolumn is observed "sliding" within the column, then one may have to connect the two columns under pressure. To do so, one can either place the precolumn in the bomb with simple buffer, or can put the precolumn onto the tee of the LC and apply a small amount of pressure.
20. To achieve nanoflow when using a microflow Agilent 1100 HPLC, the flow is split by a tee to ~100 nL/min, with flow being measured over time using a calibrated pipet as used when measuring sample volume.
21. Spray voltages should typically fall within the range of 1.0–2.0 kV.

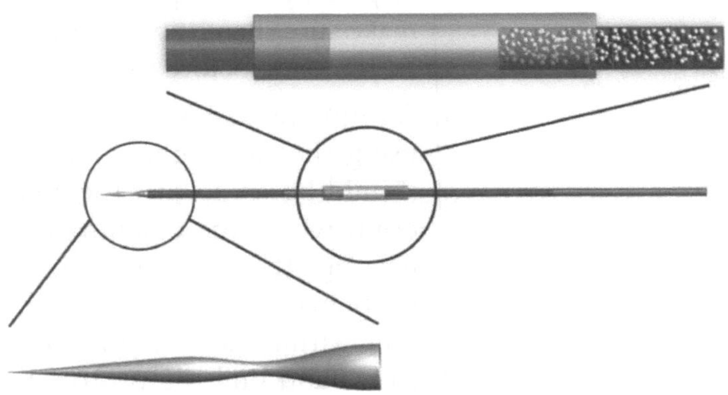

Fig. 2. Blow-up of the total capillary column unit, with combined precolumn (at *right*) and analytical column (at *left*). Note how the analytical column tip consists of both a bottle neck and tip, with the packing material first being sequestered in the column by the presence of the bottle neck. The pre and analytical columns are joined end-to-end within the protective Teflon sleeve. In this case, the packing material does not travel past the glass frit of the precolumn, and the Teflon not only protects the glass frit, but serves to hold the two columns in contact with one another.

3.3. Ion Trap Mass Spectrometry

All methods outlined here have been perfected on an ETD-enabled Thermo Scientific LTQ-XL.

1. Use a data-dependent acquisition method to interrogate the five most intense precursor ions, as identified from each full MS scan, by both ion trap CAD and ETD (two separate sequential events). For example:

Scan event	Data collection	Precursor	Activation
1	MS	N/A	N/A
2	MS^2	1st most intense	ETD
3	MS^2	1st most intense	CAD
4	MS^2	2nd most intense	ETD
5	MS^2	2nd most intense	CAD
6	MS^2	3rd most intense	ETD
7	MS^2	3rd most intense	CAD
8	MS^2	4th most intense	ETD
9	MS^2	4th most intense	CAD
10	MS^2	5th most intense	ETD
11	MS^2	5th most intense	CAD

2. Scan from150–2,000 *m/z* for MS analysis (one can also choose to only inspect the range of 300–2,000, when looking to analyze only larger species, thus decreasing total scan time).
3. For CAD scans, set the AGC target to inject 40,000 peptide cations. Dissociation should be achieved using a q-value of 0.25 with a normalized collision energy of 35% for 30 ms (single scan).
4. For ETD scans, set the precursor cation AGC target at 80,000 ions and the anion AGC value to 100,000 anions. Fix the duration of the ion/ion reaction at 80 ms (*see* **Notes 9** and **10**).
5. The first step to performing ETcaD is to set up an instrumental method to allow for this unique form of activation.
6. The instrumental method can be set up in a similar manner to the one outlined previously, with a third MS^2 event being added to the evaluation of each precursor mass. For example, when analyzing the three most intense precursor ions from the full scan:

Scan event	Data collection	Precursor	Activation
1	MS	N/A	N/A
2	MS^2	1st most intense	CAD
3	MS^2	1st most intense	ETD
4	MS^2	1st most intense	ETcaD
5	MS^2	2nd most intense	CAD
6	MS^2	2nd most intense	ETD
7	MS^2	2nd most intense	ETcaD
8	MS^2	3rd most intense	CAD
9	MS^2	3rd most intense	ETD
10	MS^2	3rd most intense	ETcaD

7. This third MS^2 event should target the ET product species ($[M + 2H]^{+\bullet}$) from the preceding ETD scan.

3.4. Data Analysis

All analysis and data processing should be performed in accordance with the recently published MIAPE guidelines *(28)* and the suggested guidelines for clinical proteome analysis *(31)*.

1. Use Bioworks Browser to generate DTAs. Ensure that the molecular weight range is 0–10,000 and the threshold is set to 100, absolute or 2%, Relative. Group scan and minimum group count should both be set to 0, while minimum ion count should be set to 1. The precursor ion tolerance is generally set to 0.05 atomic mass units.
2. Using the in-house developed program, separate CAD and ETD DTAs into separate folders.
3. Separate these divided DTAs into two more folders, one each for Orbitrap and ion trap DTAs, using the second program developed.
4. Combine DTAs within each folder into batches of 2,000 files (maximum) (*see* **Note 11**).
5. For ion trap-derived CAD and ETD data, searches with a precursor ion mass window of 1.2 Da and product ion mass window of 0.5 Da should be employed.
6. For data obtained using the Orbitrap mass analyzer, searches of both CAD and ETD data should use a precursor ion mass window of 1.2 Da and product ion mass window of 0.5 Da.

4. Notes

1. This is due to the possibility of bacterial growth within the urine over this period. Also, only mid-stream urine should be collected, due to possibilities of contaminants within the first urine exiting the urethra. This urine which is first expelled serves to "flush" the system of these contaminants, which could include things such as bacterial growth in the urethra, unexpelled ejaculate, and other glandular secretions. Finally, when not performing studies that require frequent testing within a 24-h time period, one should sample only the second urine of the morning, as the first urine of the morning is composed of that which has been stored overnight in the bladder, thus offering the possibility of bacterial overgrowth. One should not sample later in the day, as the effects of cardiac output, nutrition, etc. may cause noticeable differences in the composition of the sample *(32)*.
2. In addition to the ease of obtainment and subsequent analysis, there is a need for the biological fluid to be sufficiently stable for its future use in a clinical context. The sample must be sufficiently stable during the procedure, including sample collection, sample transportation from site of collection to site of analysis, sample storage, etc. Furthermore, the reproducible sampling of the body fluid at the bedside is also an issue. In contrast, the proteomes of both CSF and urine have been shown to be quite stable, and thus, give rise to much more reproducible datasets.
3. Because of expense, one may be tempted to use less capillary material, as the final column will only contain ~5–7 cm of packing material. However, we strongly suggest using this longer length, as problems with back-pressures resulting from the joining of the pre and analytical columns may result in the loss of packing material, and subsequently sample, if adequate space is not provided behind the material in the capillary.
4. We recommend the use of a simple light microscope to inspect the result of this polymerization. Make certain that there still exists an opening from the microcapillary after polymerization. Test for this by placing the capillary into buffer solution within the bomb apparatus. Flow should be observed from the tip of the microcapillary upon applying a small amount of pressure (~200 PSI); however, observation of a spray or of an initial "cloud" of material from the capillary usually implies the instability of the frit. In this case, cut off the end of the microcapillary, making sure to remove all parts that were initially affected by the preparation, and then place the microcapillary into an empty bomb. Applying pressure to

the microcapillary will serve to “dry it out,” and will allow for reuse of the microcapillary.

5. While packing generally occurs faster upon higher concentrations of packing material, too high of a concentration may actually decrease packing speed; therefore, we advise that none of the material should precipitate when the solution is left undisturbed for a period of ~5 min. Also, new slurry should be made frequently (at a minimum every week, with fresh slurry every few days being preferred).
6. If pressures above 600 PSI are required, it is advised to start over, as these columns typically do not perform as well as those that pack at lower pressures; ~250 PSI has been found to be an optimal pressure for packing columns.
7. To perform a quick check of the stability of this column, gently flick the top (short end) of the capillary. If the top and bottom of the column do not move as one, then there is a good chance that the bottleneck is too tiny and will either break easily upon regular use, or will be too tight to easily let flow through the opening (thus greatly hindering speed of packing and also possibly affecting LC analyses later).
8. It is sometimes easier to add the Teflon sleeve to the precolumn first, before loading of the sample. In this case, one should fill the void volume within the sleeve with buffer before measuring the volume of sample loaded onto the column.
9. Before doing a run, one should inspect the quality of the signal coming from the fluoranthene anion. To do this, change to negative ion mode while injecting anions from the rear of the instrument. You should now observe the fluoranthene anion peak at 202.16. The injection time should be steady at ~0.1 ms and the signal intensity of the anion should be ~10^6. Also, this peak should be the most intense peak within the spectrum, and there should be no other peaks in the immediate *m/z* range. When observing other, high-intensity peaks, this most likely means that the CI source in the rear of the instrument has become contaminated and should either be thoroughly cleaned or replaced. If no peak is observed for fluoranthene, make sure there is still fluoranthene within the sample vial inside of the GC oven. If there is ample fluoranthene within this vial, inspect the filament within the CI source. Filaments have a finite lifetime and will have to be replaced/repaired depending on use. To check a filament, look for the filament inside of its holder. If the filament is not visible, then it means it has been deflected, and the beam of electrons coming from it is, therefore, not making it through the slot to actually ionize the fluoranthene.
10. As reported in two recent articles by our group *(23, 33)*, ETD has some practical limitations for what species it is

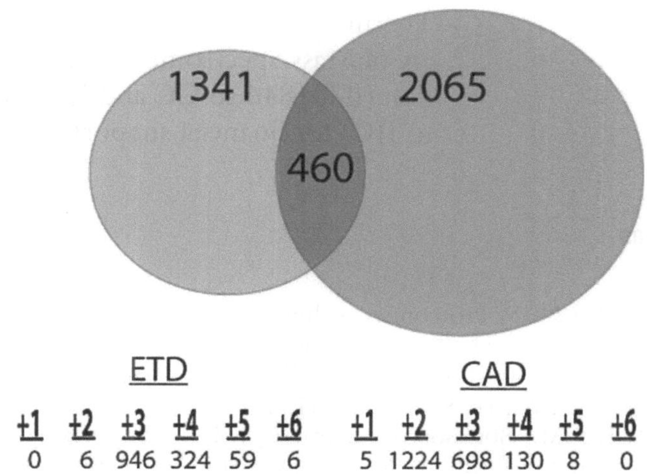

Fig. 3. Total peptide identifications from combined tryptic and lys-c digests of yeast and Arabidopsis, sorted by precursor charge state. These data were collected in a data-dependent top five manner, with alternating ETD and CAD. There were 3,866 total combined peptide identifications for the two fragmentation methods, with an overlap in identifications of 460 peptides, (~12%).

most efficient at fragmenting. In short, as the precursor mass increases, without a related increase in the species' charge, ETD efficiency falls. This is most likely due to noncovalent interactions within the peptide allowing for formation of secondary structure; thus, limiting the ability to separate cleaved fragments from one another upon scanning them out from the ion trap. To deal with this limitation, our group followed up on a technique used with ECD, which is to "tickle" the species with a small amount of vibrational energy (CAD) immediately following fragmentation by ETD. This combined form of fragmentation was termed ETcaD *(33)* (**Fig. 3**).

11. Choice of Database: Perhaps the most important quality of a database is its completeness, i.e., how up-to-date it is. However, we find that it is equally important to have a database that is not highly redundant. High redundancy can lead to problems with data processing later on, after IDs have been assigned. Regardless of the database that is chosen, it is important to stick with the same database for all searches, thus minimizing possibilities for errors associated with the annotation of included proteins.

Acknowledgments

DMG acknowledges support from an NIH predoctoral fellowship, the Biotechnology Training Program (NIH 5 T32 GM08349). We also thank the University of Wisconsin-Madison,

Thermo Scientific, the Beckman Foundation, the American Society of Mass Spectrometry, Eli Lilly, the National Science Foundation (0701846 to GC and JJC), and the NIH (1R01GM080148 to JJC) for financial support of this work.

References

1. Zhang, J. (2007) Proteomics of human cerebrospinal fluid – The good, the bad, and the ugly. *Proteom. Clin. App.* **1**, 805–819.
2. Decramer, S., Wittke, S., Mischak, H., Zurbig, P., Walden, M., Bouissou, F., Bascands, J. L., and Schanstra, J. P. (2006) Predicting the clinical outcome of congenital unilateral ureteropelvic junction obstruction in newborn by urinary proteome analysis. *Nat. Med.* **12**, 398–400.
3. Roessler, M., Rollinger, W., Mantovani-Endl, L., Hagmann, M. L., Palme, S., Berndt, P., Engel, A. M., Pfeffer, M., Karl, J., Bodenmuller, H., Ruschoff, J., Henkel, T., Rohr, G., Rossol, S., Rosch, W., Langen, H., Zolg, W., and Tacke, M. (2006) Identification of PSME3 as a novel serum tumor marker for colorectal cancer by combining two-dimensional polyacrylamide gel electrophoresis with a strictly mass spectrometry-based approach for data analysis. *Mol. Cell. Proteomics* **5**, 2092–2101.
4. Theodorescu, D., Wittke, S., Ross, M. M., Walden, M., Conaway, M., Just, I., Mischak, H., and Frierson, H. F. (2006) Discovery and validation of new protein biomarkers for 4 urothelial cancer: a prospective analysis. *Lancet Oncology* **7**, 230–240.
5. Weissinger, E. M., and Mischak, H. (2007) Application of proteomics to posttransplantational follow-up. *Methods Mol. Med.* **134**, 217–28.
6 Fliser, D., Novak, J., Thongboonkerd, V., Argiles, A., Jankowski, V., Girolami, M. A., Jankowski, J., and Mischak, H. (2007) Advances in urinary proteome analysis and biomarker discovery. *J. Am. Soc. Nephrology* **18**, 1057–1071.
7. Thongboonkerd, V. (2007) Recent progress in urinary proteomics. *Proteomics Clin. App.* **1**, 780–791.
8. Hunt, D. F., Yates, J. R., Shabanowitz, J., Winston, S., and Hauer, C. R. (1986) Protein sequencing by tandem mass-spectrometry. *Proc. Nat. Acad. Sci. USA* **83**, 6233–6237.
9. Huang, Y. Y., Triscari, J. M., Tseng, G. C., Pasa-Tolic, L., Lipton, M. S., Smith, R. D., and Wysocki, V. H. (2005) Statistical characterization of the charge state and residue dependence of low-energy CID peptide dissociation patterns. *Anal. Chem.* **77**, 5800–5813.
10. Zurbig, P., Renfrow, M. B., Schiffer, E., Novak, J., Walden, M., Wittke, S., Just, I., Pelzing, M., Neususs, C., Theodorescu, D., Root, K. E., Ross, M. M., and Mischak, H. (2006) Biomarker discovery by CE-MS enables sequence analysis via MS/MS with platform-independent separation. *Electrophoresis* **27**, 2111–2125.
11. Fliser, D., Wittke, S., and Mischak, H. (2005) Capillary electrophoresis coupled to mass spectrometry for clinical diagnostic purposes. *Electrophoresis* **26**, 2708–2716.
12. Lu, Y., Bottari, P., Aebersold, R., Turecek, F., and Gelb, M. H. (2007) Absolute quantification of specific proteins in complex mixtures using visible isotope-coded affinity tags. *Methods Mol. Biol.* **359**, 159–76.
13. Tomana, M., Novak, J., Julian, B. A., Matousovic, K., Konecny, K., and Mestecky, J. (1999) Circulating immune complexes in IgA nephropathy consist of IgA1 with galactose-deficient hinge region and antiglycan antibodies. *J. Clin. Inv.* **104**, 73–81.
14. Renfrow, M. B., Cooper, H. J., Tomana, M., Kulhavy, R., Hiki, Y., Toma, K., Emmett, M. R., Mestecky, J., Marshall, A. G., and Novak, J. (2005) Determination of aberrant O-glycosylation in the IgA1 hinge region by electron capture dissociation Fourier transform-ion cyclotron resonance mass spectrometry. *J. Biol. Chem.* **280**, 19136–19145.
15. Novak, J., Tomana, M., Shah, G. R., Brown, R., and Mestecky, J. (2005) Heterogeneity of IgG glycosylation in adult periodontal disease. *J. Dental Res.* **84**, 897–901.
16. Shi, S. D. H., Hemling, M. E., Carr, S. A., Horn, D. M., Lindh, I., and McLafferty, F. W. (2001) Phosphopeptide/phosphoprotein mapping by electron capture dissociation mass spectrometry. *Anal. Chem.* **73**, 19–22.
17. Schroeder, M. J., Shabanowitz, J., Schwartz, J. C., Hunt, D. F., and Coon, J. J. (2004) A neutral loss activation method for improved phosphopeptide sequence analysis by quadrupole ion trap mass spectrometry. *Anal. Chem.* **76**, 3590–3598.

18.. DeGnore, J. P., and Qin, J. (1998) Fragmentation of phosphopeptides in an ion trap mass spectrometer. *J. Am. Soc. Mass Spectrom.* **9**, 1175–1188.

19. Wells, L., Vosseller, K., Cole, R. N., Cronshaw, J. M., Matunis, M. J., and Hart, G. W. (2002) Mapping sites of O-GlcNAc modification using affinity tags for serine and threonine post-translational modifications. *Mol. Cell. Proteomics* **1**, 791–804.

20. Delahunty, C., and Yates, J. R. (2005) Protein identification using 2D-LC-MS/MS. *Methods* **35**, 248–255.

21. Coon, J. J., Syka, J. E. P., Schwartz, J. C., Shabanowitz, J., and Hunt, D. F. (2004) Anion dependence in the partitioning between proton and electron transfer in ion/ion reactions. *Int. J. Mass Spectrom.* **236**, 33–42.

22 Coon, J. J., Syka, J. E. P., Shabanowitz, J., and Hunt, D. F. (2005) Tandem mass spectrometry for peptide and protein sequence analysis. *Biotechniques* **38**, 519–523.

23. Coon, J. J., Ueberheide, B., Syka, J. E. P., Dryhurst, D. D., Ausio, J., Shabanowitz, J., and Hunt, D. F. (2005) Protein identification using sequential ion/ion reactions and tandem mass spectrometry. *Proc. Nat. Acad. Sci. USA* **102**, 9463–9468.

24. Good, D. M., Wirtala, M., McAlister, G. C., and Coon, J. J. (2007) Performance characteristics of electron transfer dissociation mass spectrometry. *Mol. Cell. Proteomics* **6**, 1942–1951.

25. Pitteri, S. J., Chrisman, P. A., Hogan, J. M., and McLuckey, S. A. (2005) Electron transfer ion/ion reactions in a three-dimensional quadrupole ion trap: Reactions of doubly and triply protonated peptides with SO_2 center dot. *Anal. Chem.* **77**, 1831–1839.

26. Syka, J. E. P., Coon, J. J., Schroeder, M. J., Shabanowitz, J., and Hunt, D. F. (2004) Peptide and protein sequence analysis by electron transfer dissociation mass spectrometry. *Proc. Nat. Acad. Sci. USA* **101**, 9528–9533.

27. McAlister, G. C., Phanstiel, D., Good, D. M., Berggren, W. T., and Coon, J. J. (2007) Implementation of electron-transfer dissociation on a hybrid linear ion trap-orbitrap mass spectrometer. *Anal. Chem.* **79**, 3525–3534.

28. Taylor, C. F., Paton, N. W., Lilley, K. S., Binz, P. A., Julian, R. K., Jones, A. R., Zhu, W. M., Apweiler, R., Aebersold, R., Deutsch, E. W., Dunn, M. J., Heck, A. J. R., Leitner, A., Macht, M., Mann, M., Martens, L., Neubert, T. A., Patterson, S. D., Ping, P. P., Seymour, S. L., Souda, P., Tsugita, A., Vandekerckhove, J., Vondriska, T. M., Whitelegge, J. P., Wilkins, M. R., Xenarios, I., Yates, J. R., and Hermjakob, H. (2007) The minimum information about a proteomics experiment (MIAPE). *Nat. Biotech.* **25**, 887–893.

29. Geer, L. Y., Markey, S. P., Kowalak, J. A., Wagner, L., Xu, M., Maynard, D. M., Yang, X. Y., Shi, W. Y., and Bryant, S. H. (2004) Open mass spectrometry search algorithm. *J. Proteome Res.* **3**, 958–964.

30. Martin, S. E., Shabanowitz, J., Hunt, D. F., and Marto, J. A. (2000) Subfemtomole MS and MS/MS peptide sequence analysis using nano-HPLC micro-ESI Fourier transform ion cyclotron resonance mass spectrometry. *Anal. Chem.* **72**, 4266–4274.

31. Mischak, H., Apweiler, R., Banks, R. E., Conaway, M., Coon, J., Dominiczak, A., Ehrich, J. H. H., Fliser, D., Girolami, M., Hermjakob, H., Hochstrasser, D., Jankowski, J., Julian, B. A., Kolch, W., Massy, Z. A., Neusuess, C., Novak, J., Peter, K., Rossing, K., Schanstra, J., Semmes, O. J., Theodorescu, D., Thongboonkerd, V., Weissinger, E. M., Van Eyk, J. E., and Yamamoto, T. (2007) Clinical proteomics: A need to define the field and to begin to set adequate standards. *Proteomics Clin. App.* **1**, 148–156.

32. Theodorescu, D., Fliser, D., Wittke, S., Mischak, H., Krebs, R., Walden, M., Ross, M., Eltze, E., Bettendorf, O., Wulfing, C., and Semjonow, A. (2005) Pilot study of capillary electrophoresis coupled to mass spectrometry as a tool to define potential prostate cancer biomarkers in urine. *Electrophoresis* **26**, 2797–2808.

33. Swaney, D. L., McAlister, G. C., Wirtala, M., Schwartz, J. C., Syka, J. E. P., and Coon, J. J. (2007) Supplemental activation method for high-efficiency electron-transfer dissociation of doubly protonated peptide precursors. *Anal. Chem.* **79**, 477–485.

Chapter 19

Traumatic Brain Injury Biomarkers: From Pipeline to Diagnostic Assay Development

Monika W. Oli, Ronald L. Hayes, Gillian Robinson, and Kevin K.W. Wang

Summary

In recent years, the term proteomics is often mentioned together with biomarker discovery, as proteomic studies have the capability of identifying unique and unobvious protein biomarkers from tissues or biofluids derived from animal models or human clinical samples inflicted with various diseases. Proteomics has yielded hundreds of potential biomarker candidates. However, biomarker discovery is only the beginning of a long road for generating a validated, clinically relevant, and FDA-approved biomarker assay. Many technical, financial, legal, and regulatory hurdles have to be overcome before the components can be commercially produced *(1, 2)*. This chapter outlines in a condensed version the steps to successfully develop clinically acceptable biomarkers, given the marker of choice withstands the rigor of developmental challenges along the road.

Key words: Biomarkers, Brain injury, Assay development, Clinical trials, ELISA

1. Introduction

1.1. Biomarker Discovery

The application of Traumatic Brain Injury (TBI) biomarkers in the clinical arena would be invaluable as a tool to aid in the diagnosis of various forms of TBI. Early appearance of a biomarker in mild and moderate TBI could be used, for example, by a coach on the soccer field to guide rational decisions on the continuation or termination of a player after injury. Clinically validated TBI biomarkers will revolutionize the evaluation and treatment of TBI patients and may even provide prediction of outcome of the future impairments of the patients. As troponin has significantly changed the evaluation of cardiac insults *(3)* so will TBI biomarkers revolutionize brain injury diagnostics.

Andrew K. Ottens and Kevin K.W. Wang (eds.), *Neuroproteomics,* Methods in Molecular Biology, vol. 566
DOI 10.1007/978-1-59745-562-6_19, © Humana Press, a part of Springer Science+Business Media, LLC 2009

Our laboratory has been involved in the initial identification of a prototypic TBI biomarker – αII-spectrin breakdown products (SBDPs) produced by calpain and/or caspase-3 proteolysis during the acute and delayed neuronal injury using CSF biofluid from a rat model of TBI *(4, 5)*. We have also confirmed the biomarker release in CSF from human TBI clinical studies *(6)*. However, it was also becoming clear to us that there is room to identify additional unobvious TBI protein biomarkers. We thus embarked on a pursuit with multiproteomic platform [both mass spectrometry CAX-PAGE-RPLC-MS/MS *(7)* and antibody array-based proteomics *(8)*] to identify novel candidate biomarkers.

Applying a proteomic platform to the discovery of biomarkers in close proximity to the damaged tissue (i.e., the brain) has a lot of advantages over searching for biomarkers in the blood. The main advantages are (1) higher concentration of the actual biomarker proteins of choice and (2) less contamination of other, often high abundance, proteins. Preclinical models of traumatic brain injury lend themselves to thorough analysis of brain tissue and allow for collection of CSF and blood to initiate the biomarker validation process.

Figure 1 outlines the path from generation of the TBI biomarkers in the brain tissue to detection of the biomarkers in CSF and blood. During brain injury, neural proteins are released into the extracellular environment and hence the CSF; these biomarkers occur in high concentration in the CSF *(9)*. Eventually the proteins reach the blood stream either via compromised blood–brain barrier (BBB) or filtration of CSF. Clearance and

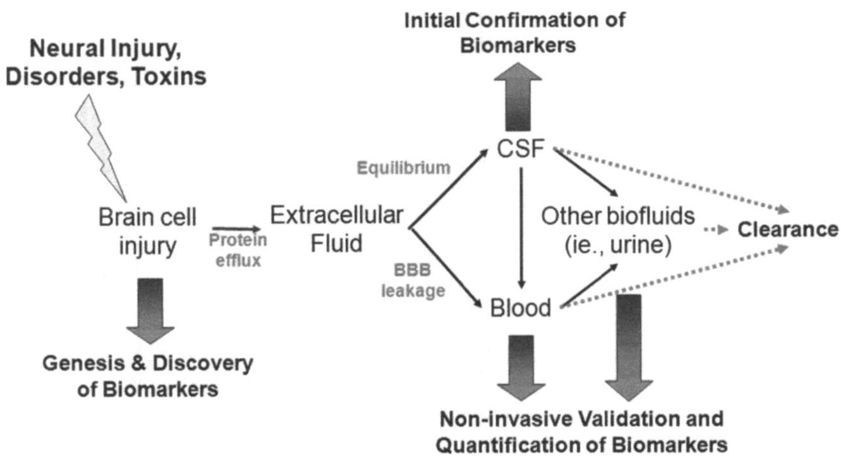

Fig. 1. From generation to detection of TBI biomarkers. Neural proteins are released during traumatic injury of the brain (genesis and discovery). Through exchange with the extracellular fluid in contact, these biomarkers occur in high concentration in the CSF (initial confirmation of biomarkers). Eventually the proteins reach the blood stream either via compromised BBB or filtration of CSF (noninvasive validation and quantification of biomarkers). Clearance of the biomarkers contributes to the final concentration and half-life in the blood (clearance).

half-life of the biomarkers contribute to the final concentration that is present and can be measured in the blood. As the CSF volume of an adult human is about 30–40-fold less than the blood volume (CSF 125–150 ml, blood 4.5–5 l), the brain biomarker concentration is significantly higher in the CSF samples, and therefore they constitute a valuable sample for initial biomarker confirmation and assay development.

1.2. Biomarker Selection Process

Biomarker discovery will always yield a large pipeline of potential biomarkers. As for most investigators and companies, time and resources are the limiting factor. The challenge is to shortlist potential markers to a manageable number for further assay development. The selection process is dependent on many factors including the strength of the biomarker data, robustness as well as the biomarker attribute as a candidate diagnostic marker. There are now emerging also pathway and protein-interaction map software that can help facilitate the selection process. In fact, we adopted a "systems biology" approach that takes many of these factors into consideration with the goal of identifying a small panel of "first-generation" candidate biomarkers that represent nonredundant pathways of hot spots relevant to the pathobiology of traumatic brain injury *(10)* (**Fig. 2**).

The choice of which marker to consider as a lead development product will be ultimately determined by the concentration of the marker in blood and the specificity of the marker to

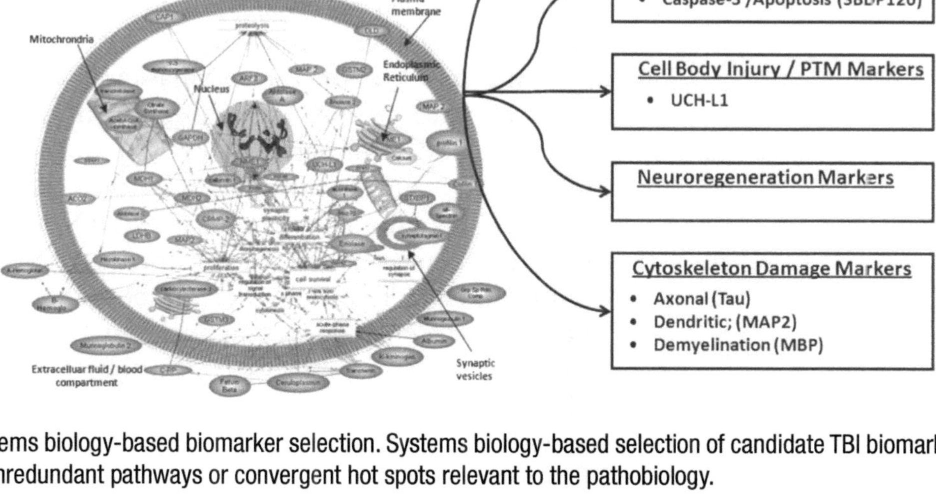

Fig. 2. Systems biology-based biomarker selection. Systems biology-based selection of candidate TBI biomarkers representing nonredundant pathways or convergent hot spots relevant to the pathobiology.

disease state. If the biomarker concentration is so low that serum detection is practically impossible, it will be difficult to justify the development of that marker in the current light of the technology. Second, if the marker is present in normal subjects at a significant level, or if other disease conditions also release the marker, the clinical value of the biomarker is diminished. This is the case for many common biomarkers, for example, S100b is known to appear in the blood after muscular damage and not just after brain injury *(11)*. To complicate the situation even further, in the commercial environment patent (IP) position, licensing opportunities and marketing rights may be the overriding factors that will dictate which markers will eventually reach the market. And as **Fig. 2** shows, the biomarker selection process is only the beginning of a long road of biomarker discovery to clinical application.

1.3. Translation of Preclinical Models to Clinical Applications

The primary challenge facing translation of biomarkers to practical clinical diagnostics resides in the approaches to confirm the clinical utility of putative biomarkers of TBI identified and validated with preclinical animal models. Ideally, the confirmation platform should integrate preclinical validation with clinical assessments of the potential applicability of the candidate biomarkers. A direct comparison of biomarker occurrence between preclinical models and biomarker data from human clinical studies allows investigators to gain considerable insight into the validity (or challenges to the validity) of the employed preclinical animal models. For example, we have shown (unpublished data) similar profiles of αII-spectrin degradation by calpain in both controlled cortical impact (CCI) and severe human TBI. Thus, CCI reproduces important features of calpain-mediated pathology seen in human TBI.

Although there is generally a large similarity between animal models and human applications, some assays will fail as the antibodies may not recognize the biomarkers of different species, and therefore it may be necessary to develop separate assays for animal models and subsequent clinical trials.

Turnover and half-life of the biomarkers, which is an important clinical feature, may be very different in a rodent stationed in a cage and a human TBI patient in the hospital, who is being treated additionally with various medications. Interfering substances (see later) that can significantly influence the assay performance are also known to be different in animals and humans. For example, a laboratory rodent is not usually exposed to environmental antigens that may elicit nonspecific interferences (i.e., HAMA antibodies). Thus, the blood-based matrix of the animal model may be less problematic to work with but may lead to false assumptions about the assay qualification process.

2. Methods

2.1. Functional Research Assay Development

The biomarker selection is obviously highly dependent on a sensitive and reliable readout, most commonly an enzyme-linked immunosorbant assay-based test, ELISA. Thus, the effort spent on developing the biomarker assay is fundamental to the potential success of the biomarker in the respective diagnostic field. The quality of the biomarker assay may ultimately determine the success of the biomarker in the market. Without the assay, even the best biomarker will never make it to a clinical application and the choice of marker is irrelevant if the assay is not performing as desired.

To date, the most common validation technology for neuroproteomics data to biomarker assay relies heavily on antibody-based assays. Since antibodies are by definition epitope-specific binding molecules, they are the operand of choice for assay development. Because of the maturity of immunological methods, antibody-based diagnostic assays have been refined almost to an art; ELISAs are the most commonly used form. Biomarker assay development is based on the formation of an antigen sandwich including two specific antibodies against two epitopes of the same molecule. These two-sided ELISA (or sandwich ELISA; swELISA) require the immobilization of the capture antibody, exposure to protein antigen and detection of complex by a detection antibody. For biomarker-based diagnostic, SWELISA is indeed the method of choice, since it provides antigen enrichment that significantly improves signal fidelity.

In some cases it is possible that biomarker assays kits are already commercially available (search at www.biocompare.com). This can be an advantage for initial screening but caution should be taken as the epitopes of the antibodies are usually not known, and thus it is difficult to assess specificity of the particular assay. For most known proteins antibodies are available, and thus commercial tools can sometimes be used for initial assay development. However, most antibodies are not developed for ELISA applications that require highly purified antibodies (without added BSA) and high-affinity antibodies. Thus, the commercial choices are usually limited and the production of specific antibodies is recommended.

Very commonly brain-specific proteins are present in different isoforms in other tissues. This needs to be carefully examined when choosing antibodies that ought to be specifically detecting brain-specific isoforms. This requirement often necessitates production of specific antibodies to certain epitopes that allow unambiguous detection of the brain isoform, enhancing the specificity of the assay in the long run.

The work flow for developing biomarker SWELISA assays is as follows (**Fig. 3**). Once a putative biomarker is identified

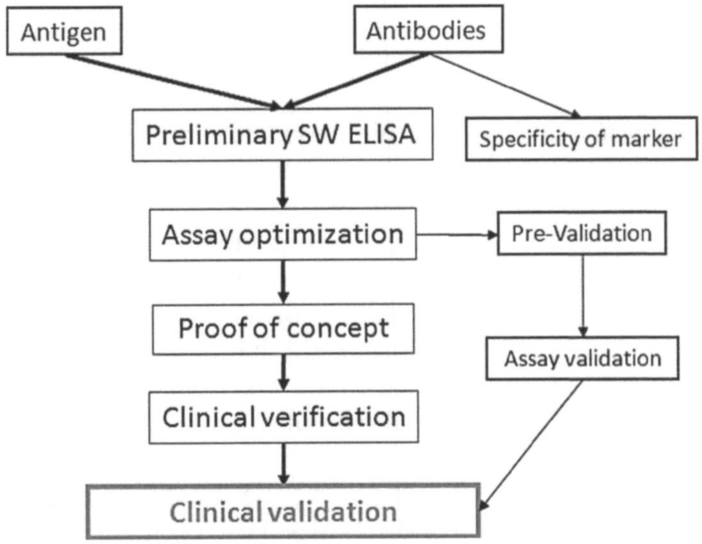

Fig. 3. Biomarker assay development and optimization. Development of biomarker assays is a long process from the selection of antibody tools to a validated clinical platform.

and confirmed, one must obtain a matched pair of compatible high-affinity antibodies. Antibodies from commercial sources can be explored first; however, custom-made antibodies are often needed. Recombinant biomarker protein or a tissue purified antigen must be available to develop the quantitative standard curve of the assay. This combination will allow development of a basic functional SWELISA that will quantitatively detect the biomarker protein of choice.

2.2. Assay Optimization

Assay optimization is probably the most important aspect of the preclinical work regarding the biomarker development (**Fig. 3**). This process can be time consuming and labor intensive, but the optimization can constitute the "make it or break it" decision. A highly sensitive assay with little or no nonspecific background and no matrix effect will always enhance the sensitivity and specificity of the assay as false positives can largely be eliminated. Various iterations of antibody choice, antibody concentrations, buffers used, blocking agents, and sample diluents have to be evaluated. Furthermore, nonspecific binding, largely of matrix proteins, has to be eliminated. This can often be accomplished by dilution of the clinical samples (if the assay is sensitive enough to allow for 5–10-fold dilutions!) and by blocking to reduce HAXA effects (human anti-animal antibody, where X can be mouse, rabbit, rat, etc.) and RF (rheumatoid factor) effects ***(12–14)***. Other parameters such as temperature, incubation time, choice of plastic, etc. can also make a difference in the assay.

It is essential that even at the early stage of assay development normal controls should be included to determine matrix effect of the assay. Furthermore, if recombinant protein is used for the assay development native protein has to be compared in parallel to ensure that the same epitopes are available for antibody binding in the native and the recombinant protein.

2.3. Assay Platform

Another major challenge of clinical utility can be the assay platform as enhanced requirements for detection limits challenge the sensitivity of a regular ELISA-based platform. It has been shown that the best brain injury marker might be very low in abundance (unpublished data). There are several ways to further improve the SWELISA sensitivity, even after extensive assay optimization. Sensitivity can be improved with an enzymatic signal-enhancement method known as tyramide signal amplification (TSA). It is also known that chemiluminescent detection methods can enhance the ELISA signal in addition to improving the linear range of the assay. It is apparent that technological advances in ELISA sensitivity improvement will greatly benefit this field.

A less well-known method of signal amplification is immuno-PCR, whereby the detection antibody is labeled with a DNA molecule and subsequent to the normal ELISA incubation steps a real-time PCR amplification is performed *(15, 16)*. It is claimed that this method can enhance the SWELSIA signal by 100–1,000-fold, which allows detection of ultralow biomarker levels of novel markers and also at early time after generation and appearance of the markers. **Figure 4** compares the different

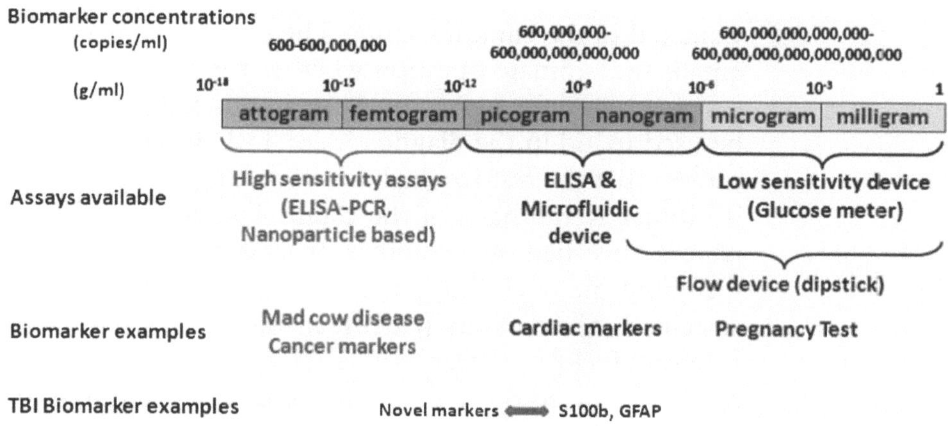

Fig. 4. Biomarker detection on different assay platforms. The most common and least sensitive assay platforms (*from right to left*) allow detection of very high concentration of biomarker analyte, like the detection of FSH in a pregnancy test, usually performed on a dipstick device. Cardiac makers are in the mid range of biomarker concentrations and can readily be detected on ELISA and ELSIA-like platforms (clinical analyzers). Commonly known TBI biomarkers are detected typically in the low to mid ng/ml range on an ELISA platform, but often the specification of the ELISA does not allow for accurate detection of the markers. Less abundant markers, like many cancer markers, can usually not be detected on regular ELSIA platforms and the need for high-sensitive detection method is evident in the biomarker field.

biomarker assays as it relates to analyte concentration. The choice of platform determines sensitivity, time of analysis, and cost – all important parameters that govern the biomarker progression to its clinical utility.

2.4. Assay Validation

An important distinction has to be drawn between biomarker validation and qualification, where validation is the process of assessing the assay or measurement performance characteristics, and qualification is evidently a process of linking a biomarker with biology and clinical endpoints.

Bioanalytical method validation includes all of the procedures that demonstrate that a particular method used for quantitative measurement of analytes in a given biological matrix, such as blood, plasma, serum, or urine, is reliable and reproducible for the intended use. The fundamental parameters for this validation include (1) accuracy, (2) precision, (3) selectivity, (4) sensitivity, (5) reproducibility, and (6) stability. FDA guidelines for assay validation are available at http://www.fda.gov/cder/Guidance/1320fnl.pdf, http://www.fda.gov/cder/guidance/4252fnl.htm.

Assay validation is a rather mundane process of performing the assays in a predetermined manner and calculating certain validation parameters. Even after thorough validation it is understood that assays measuring the same biomarker, produced by different companies will have a variation of assay and clinical parameters.

2.5. Clinical Biomarker Validation

As described earlier, assay validation is an independent process that governs the accuracy and reproducibility of the biomarker value determination. Beyond assay validation the biomarker has to be validated in the clinical setting. The clinical biomarker validation entails the ultimate question whether the data on the biomarker support its use for a given purpose; in this case, can the biomarker be used to aid in the diagnostics of TBI. It has to be statistically proven that the test results have some clinical significance, which is ultimately the basis of the claim. The clinical utility determinant is governed by a complex process of statistical data analysis including many clinical parameters about the demography and sustained injury of the patient, medical history and medication received, inclusion/exclusion criteria, etc. These data can only be reliably obtained through a tightly controlled and well-monitored clinical trial, but they are pertinent to establishing clinical value of the markers. Clinical samples should be analyzed along side with a large matched population of normal samples as well as control samples (i.e., polytrauma patients). Multivariate analysis should be applied to quantify the effect of various parameters on the biomarker data outcome, which should be conducted in collaboration with clinicians, scientists, and statisticians.

To determine diagnostic parameters such as sensitivity and specificity cutoff values, a receiver operating characteristic (ROC) analysis is often developed. The ROC curve is a graphical plot of the sensitivity versus (one – specificity), allowing the determination of optimal cutoff values with their associated % specificity and % sensitivity. ROC analysis is related in a direct and natural way to cost/benefit analysis of diagnostic decision making. An ideal biomarker ELISA assay should strive to achieve high (>90%) selectivity (i.e., the ability to detect the presence of injury) and high (>90%) specificity (i.e., the ability to detect the absence of injury).

Diagnostic criteria employed for the biomarker characterization are *(1)* prognostic stratification (Can the biomarker distinguish severe TBI from mild and moderate TBI?), *(2)* outcome evaluation (Can the initial biomarker level predict the long-term outcome of a patient?), *(3)* predict drug exposure (Is the biomarker indicative of drug treatment efficacy?), and *(4)* diagnose or contribute to diagnosis of pathology (Can the biomarker guide therapy? Can the biomarker detect secondary insults?).

The ultimate goal of a valid biomarker for clinical use: *(1)* it is measured in an analytical test system with well-established performance characteristics (analytical validation), and *(2)* there is an established scientific framework or body of evidence that elucidates the physiologic, pharmacologic, and clinical significance of the test results (clinical validation). Once this is established the collection of assay and clinical data has to be prepared for an FDA submission to get the final approval.

2.6. FDA Approval of a Diagnostic Device

Before the biomarker assay – as good as it may be – can be used in the clinical setting, it has to be scrutinized by the FDA. "Medical devices," which include biomarkers on any platform, are classified by the FDA into three classes: Class I ("General Controls"), Class II ("Special Controls"), and Class III ["Premarket Approval" (PMA)]. Device risks and regulatory control increase from Class I to Class III. Device classification regulations define the regulatory requirements for general device types. Most Class I devices are exempt from Premarket Notification 510(k); most Class II devices require Premarket Notification 510(k), and most Class III devices require Premarket Approval.

Because of the lack of predicate devices, most biomarker submissions have been classified as Class III devices requiring a PMA before the product can be used in the clinical setting. To collect safety and effectiveness data required to support a PMA application submission to FDA, an investigational device exemption (IDE) allows the device to be used in a clinical study to be conducted to support a PMA. All clinical evaluations of investigational devices must have an approved IDE before the study is initiated (http://www.fda.gov/cdrh/devadvice/ide/index.shtml).

Unfortunately, very few biomarkers have crossed the FDA threshold and the lack of adequate clinical validation and the deficiency in assay sensitivity do not allow data to be submitted successfully to the FDA *(17)*. Hundreds of proposed and potential biomarkers can be found in the literature, with little clinical success. This should prompt all biomarker investigators to apply utmost rigor in assay development and clinical validation to ensure success in the future.

References

1. Rifai N, Gillette MA, Carr SA (2006) Protein biomarker discovery and validation: the long and uncertain path to clinical utility. *Nat Biotechnol* **24**, 971–983.
2. Vitzthum F, Behrens F, Anderson NL, Shaw JH (2005) Proteomics: from basic research to diagnostic application. A review of requirements and needs. *J Proteome Res* **4**, 1086–1097.
3. Goldmann BU, Christenson RH, Hamm CW, Meinertz T, Ohman EM (2001) Implications of troponin testing in clinical medicine. *Curr Control Trials Cardiovasc Med* 2, 75–84.
4. Pike BR, Flint J, Johnson E, Glenn CC, Dutta S, Wang KKW Hayes RL (2001) Accumulation of calpain-cleaved non-erythroid αII-spectrin in cerebrospinal fluid after traumatic brain injury in rats. *J Neurochem* **78**, 1297–1306.
5. Ringger NC, O'Steen BE, Brabham JG, Silver X, Pineda J, Wang KKW Hayes RL (2005) A novel marker for traumatic brain injury: CSF αII-spectrin breakdown product levels. *J Neurotrauma* **21**, 1443–1456.
6. Pineda J, Liu MC, Aikman J, Akle V, Lewis S, Wang KKW, Robertson C, Hayes RL (2007) Clinical significance of αII-spectrin breakdown products in CSF after severe traumatic brain injury in human. *J Neurotrauma* **24**, 354–366.
7. Kobeissy FH, Ottens AK, Zhang ZQ, Dave JR, Tortella FC, Hayes RL Wang KKW (2006) Differential proteomic analysis of traumatic brain injury biomarker study using CAX-PAGE/RPLC-MSMS method. *Mol Cell Proteomics* **5**, 1887–1898.
8. Liu MC, Akle V, Zheng WR, Dave JR, Tortella FC, Hayes RL, Wang KKW (2006) Comparing calpain- and caspase-3-degradation patterns in traumatic brain injury by differential proteome analysis. *Biochem J* **394**, 715–725.
9. Romeo MJ, Espina V, Lowenthal M, Espina BH, Petricoin EFI, Liotta LA (2005) CSF proteome: a protein repository for potential biomarker identification. *Expert Rev Proteomics* 2, 57–70.
10. Kobeissy FH, Larner SF, Sadasivan S, Zhiqun Zheng Z, Liu MC, Oli MW, Robinson G, Hayes RL, Wang KKW (2008) Neuroproteomic and systems biology-based discovery of protein biomarkers for traumatic brain injury and clinical validation (review). *Proteomics Clin Appl* **2**, 1467–1483.
11. Anderson RE, Hansson LO, Nilsson O, Dijlai-Merzoug R, Settergren G (2001) High serum S100B levels for trauma patients without head injuries. *Neurosurgery* **48**, 1255–1258.
12. Grebenchtchikov N, Sweep CG, Geurts-Moespot A, Piffanelli A, Foekens JA, Benraad TJ (2002) An ELISA avoiding interference by heterophilic antibodies in the measurement of components of the plasminogen activation system in blood. *J Immunol Methods* **268**, 219–231.
13. Kricka LJ (1999) Human anti-animal antibody interferences in immunological assays. *Clin Chem* **45**, 942–956.
14. Kricka LJ, Schmerfeld-Pruss D, Senior M, Goodman DB, Kaladas P (1990) Interference by human anti-mouse antibody in two-site immunoassays. *Clin Chem* **36**, 892–894.
15. Niemeyer CM, Adler M, Wacker R (2005) Immuno-PCR: high sensitivity detection of proteins by nucleic acid amplification. *Trends Biotechnol* **23**, 208–216.
16. Niemeyer CM, Adler M, Wacker R (2007) Detecting antigens by quantitative immuno-PCR. *Nat Protocol* **2**, 1918–1930.
17. Katz R (2004) Biomarkers and surrogate markers: an FDA perspective. *NeuroRx* **1**, 189–195.

Chapter 20

Translation of Neurological Biomarkers to Clinically Relevant Platforms

Ronald L. Hayes, Gillian Robinson, Uwe Muller, and Kevin K.W. Wang

Summary

Like proteomics more generally, neuroproteomics has recently been linked to the discovery of biochemical markers of central nervous system (CNS) injury and disease. Although neuroproteomics has enjoyed considerable success in discovery of candidate biomarkers, there are a number of challenges facing investigators interested in developing clinically useful platforms to assess biomarkers for damage to the CNS. These challenges include intrinsic physiological complications such as the blood–brain barrier. Effective translation of biomarkers to clinical practice also requires development of entirely novel pathways and product development strategies. Drawing from lessons learned from applications of biomarkers to traumatic brain injury, this study outlines major elements of such a pathway. As with other indications, biomarkers can have three major areas of application: (1) drug development; (2) diagnosis and prognosis; (3) patient management. Translation of CNS biomarkers to practical clinical platforms raises a number of integrated elements. Biomarker discovery and initial selection needs to be integrated at the earliest stages with components that will allow systematic prioritization and triage of biomarker candidates. A number of important criteria need to be considered in selecting clinical biomarker candidates. Development of proof of concept assays and their optimization and validation represent an often overlooked feature of biomarker translational research. Initial assay optimization should confirm that assays can detect biomarkers in relevant clinical samples. Since access to human clinical samples is critical to identification of biomarkers relevant to injury and disease as well as for assay development, design of human clinical validation studies is an important component of translational biomarker research platforms. Although these clinical studies share much in common with clinical trials for assessment of drug therapeutic efficacy, there are a number of considerations unique to these efforts. Finally, platform selection and potential assay commercialization need to be considered. Decisions regarding whether or not to seek FDA approval also significantly influence translational research structures.

Key words: Biomarkers, Brain injury, Translational research, Assay development, Clinical trials

Andrew K. Ottens and Kevin K.W. Wang (eds.), *Neuroproteomics,* Methods in Molecular Biology, vol. 566
DOI 10.1007/978-1-59745-562-6_20, © Humana Press, a part of Springer Science+Business Media, LLC 2009

1. Introduction

Recently, neuroproteomics, like proteomics in general, has been closely linked to the discovery of biochemical markers of central nervous system (CNS) injury and disease. Biomarkers have historically attracted the interest of investigators of acute brain injury such as traumatic brain injury (TBI), spinal cord injury, and cerebral ischemia. Investigators of neurodegenerative diseases have focused on the potential of biomarkers for Alzheimer's disease. Moreover, scientists have only recently systematically focused on the potential of neuroproteomics platforms to discover novel biomarkers of acute brain injury. Much of this research has focused on the potential of these platforms for TBI and included collective efforts of investigators at the University of Florida and Banyan Biomarkers, Inc. ***(1–8)***. However, as this chapter outlines, there are a considerable number of challenges that are faced by investigators who are interested in developing clinically useful platforms to assess biomarkers of damage to the CNS. Aside from intrinsic physiological complications such as the blood–brain barrier, the novelty of the approach requires the development of entirely novel pathways and product development strategies to facilitate efficient translation of biomarkers discovery by neuroproteomics approaches into useful clinical tools for research and patient management. The purpose of this chapter is to outline the major elements of such a pathway.

As indicated earlier, the majority of research to date employing neuroproteomics platforms for biomarker discovery has focused on TBI. Thus, this research area can serve as a useful model for development of similar platforms for other CNS indications. The elements of the platform outlined here will be generally applicable.

Biomarkers are generally defined as measurable internal indicators of changes in organisms at the molecular or cellular level and provide information about injury mechanisms. Biomarkers have already demonstrated proven clinical utility in acute care environments. For example, triponin, often in combination with other biomarkers, is routinely used to facilitate accurate diagnosis of cardiac ischemia and myocardial infarction in patients presenting with chest pain. Incorporation of biomarkers for management of cardiac ischemia has led to considerable refinement of clinical pathways and enhanced efficiency for management of these patients.

Growing recognition of the importance of biomarkers led to the foundation of the Biomarkers Consortium, launched in October 2006 as a public–private partnership including the National Institutes of Health (NIH), the Food & Drug Administration (FDA) as a part of the FDA's critical path initiative, the Centers for Medicare and Medicaid Services as well as industry representatives and nonprofit organizations and advocacy groups. In 2007, Andrew C. Von Eschenbach, Commissioner of Food & Drugs, highlighted the role of biomarkers as one technology the FDA feels "most likely to modernize and transform the development and use of medicines." However, in spite of this broad-based support, there are currently no FDA approved biomarkers for TBI or other acute brain injuries. Finally, a recent NIH workshop on improving diagnosis of TBI for targeting therapies highlighted the need for biomarkers *(9)*.

In the US, TBI accounts for 1.3% of all emergency department visits *(10)*. The Centers for Disease Control and Prevention report that approximately 5.3 million Americans live with the effects of TBI, more than Alzheimer's disease. About half of the estimated 1.9 million Americans who experience TBIs each year incur at least some short-term disability. About 52,000 people die as a result of their injuries and more than 90,000 people sustain severe brain injuries leading to debilitating loss of function. Males are 1.6 times more likely than females to suffer TBI until the age of 65 years, when the female rate exceeds the male. The highest overall incidence rate of TBI occurs in children less than 5 years of age, closely followed by seniors more than 85-years old. Falls represent the most common mechanism of TBI injury, followed by motor vehicle-related trauma *(10)*.

The direct medical costs for treatment of TBI in the US have been estimated to be more than $4 billion annually *(11)*. If costs of lost productivity are added to this figure, the overall estimated cost is closer to $56.3 billion. In addition, mild TBI is significantly under diagnosed, so the likely social burden is even greater *(12)*.

As with other indications, biomarkers can have three major areas of application: (1) drug development, (2) diagnosis and prognosis, (3) patient management.

Drug development has provided the most powerful incentive for establishment of neuroproteomics platforms. "Theranostics" is a recently invoked term that describes the parallel use of a diagnostic test such as biomarkers with therapy development for a human disease to facilitate drug development and clinical trials. This approach has been motivated by the dwindling pharma pipeline and recognition that therapeutic development traditionally has a very high triage rate, with more than 90% of drugs in clinical trials failing. Ideally, linkage of biomarkers in drug development could allow identification of novel therapeutic targets as well as provide the ability to assess therapeutic efficacy noninvasively both in preclinical and clinical studies.

Neurotoxicity represents another important application of biomarkers in drug development. The recent experience of pharma in "Phase IV" studies has highlighted the risk of unexpected toxicity. Although biomarkers of paddock toxicity and toxic affects to other peripheral organs are well developed, there is currently a paucity of sensitive and specific markers for CNS toxicity. Development of such markers is viewed as critical for facilitating novel drugs not only to treat CNS injury, but also for other indications where toxicity represents special challenges including chemotherapeutic agents for cancer.

Biomarkers have compelling potential utility in the design of clinical trials, and this utility is especially apparent in clinical trials of TBI. To date, there are no proven effective therapies for TBI. As recognized by participants in the recent workshop on TBI classification sponsored by the NIH *(13)*, limitations in current diagnostic techniques, including employment of the Glasgow Coma Scale (GCS), have complicated the design and conduct of trials in TBI and potentially contributed to failures in advancing therapies to clinical practice. Biomarkers could importantly supplement the GCS by providing objective biochemical measures of injury magnitude. In addition, biomarkers could provide objective assessments of the effects of secondary insults such as hypotension-induced ischemia on the evolving course of brain injury during the first critical days following hospital admission. As is the case with other disease processes, biomarkers can provide critical insights into the pathophysiological mechanisms of TBI and provide assessments of therapeutic efficacy of specific targets. For example, assessments of proteolytic activity associated with necrotic or apoptotic cell death following severe TBI in humans *(14)* could provide critical surrogate biochemical measures of therapeutic agents targeting those cell death mechanisms. More quantitative assessments of injury processes could ultimately provide earlier, more accurate predictions of outcome, an important component of refined approaches to statistical analysis of TBI trials including the concept of a sliding dichotomy *(15)*. Finally, failure to standardize management across different centers participating in clinical trials could contribute to differences in outcome ultimately confounding therapeutic effects assessed in clinical trials. Biomarkers could provide more objective assessments of management variations between centers and potentially increase clinical trial power.

Diagnosis and prognosis represent other important potential applications of biomarkers. Like chest pain as an indicator of myocardial infarction, mild TBI presents ambiguously and requires further diagnostic tests. Triponin for chest pain, biochemical markers of mild and moderate TBI could assist in accurate diagnosis of CNS injury that is currently not possible with functional neurological assessments or imaging. Similarly, while age and certain characteristics of injury such as the presence or absence of mass lesions can assist in predicting patient outcome, there is broad recognition that biochemical assessments of injury importantly supplement outcome prediction. As indicated earlier, these refinements not only have practical clinical utility but can also assist in design of clinical trials.

It is important to understand how patients are responding to therapeutic interventions and/or management. This consideration is especially critical in TBI since the occurrence of secondary insults can importantly determine the course of patient recovery and ultimate outcome. For example, patients experiencing severe TBI are especially vulnerable to secondary insults imposed by potentially ischemic reductions in circulation related to hypotension and/or increased intracranial pressure. These ischemic events can further exacerbate injury to the brain. At present, patients are monitored exclusively on the basis of maintenance of physiological parameters. Outside of relatively imprecise imaging alternatives such as CT, which are often impractical in ICU environments, there are no reliable methods of assessing patient's responses to management strategies. Thus, biomarkers providing useful information on the progress of brain injury and recovery would dramatically improve clinical decision making in ICU environments.

2. Methods

Other chapters in this book highlight the dramatic improvements made in development and refinement of neuroproteomics platforms to discover biochemical markers of brain injury and disease. These platforms have proven immensely efficient in identifying large numbers of potential candidate biomarkers. However, as is the case with discovery of novel therapeutic molecules, it is critical to develop a rational pathway to select the most promising markers for further development and ultimate clinical application. Outlined below (**Scheme 1**) are the major elements to such a pathway.

2.1. Biomarker Discovery and Initial Selection

Other chapters in this book will address components of neuroproteomic biomarker discovery platforms. However, these platforms need to be integrated at the earliest stages with components that will allow systematic prioritization and triage of biomarker candidates. The major elements of such an approach should include incorporation of preclinical models of injury closely linked to rigorously developed algorithms that consider the relevance of the biomarker to the pathology under consideration as well as a potential for successful assay development. For example, our laboratory has been careful to conduct a number of preclinical assessments of the potential utility of biomarkers in preclinical models of TBI and cerebral ischemia *(16–22)*. These studies have

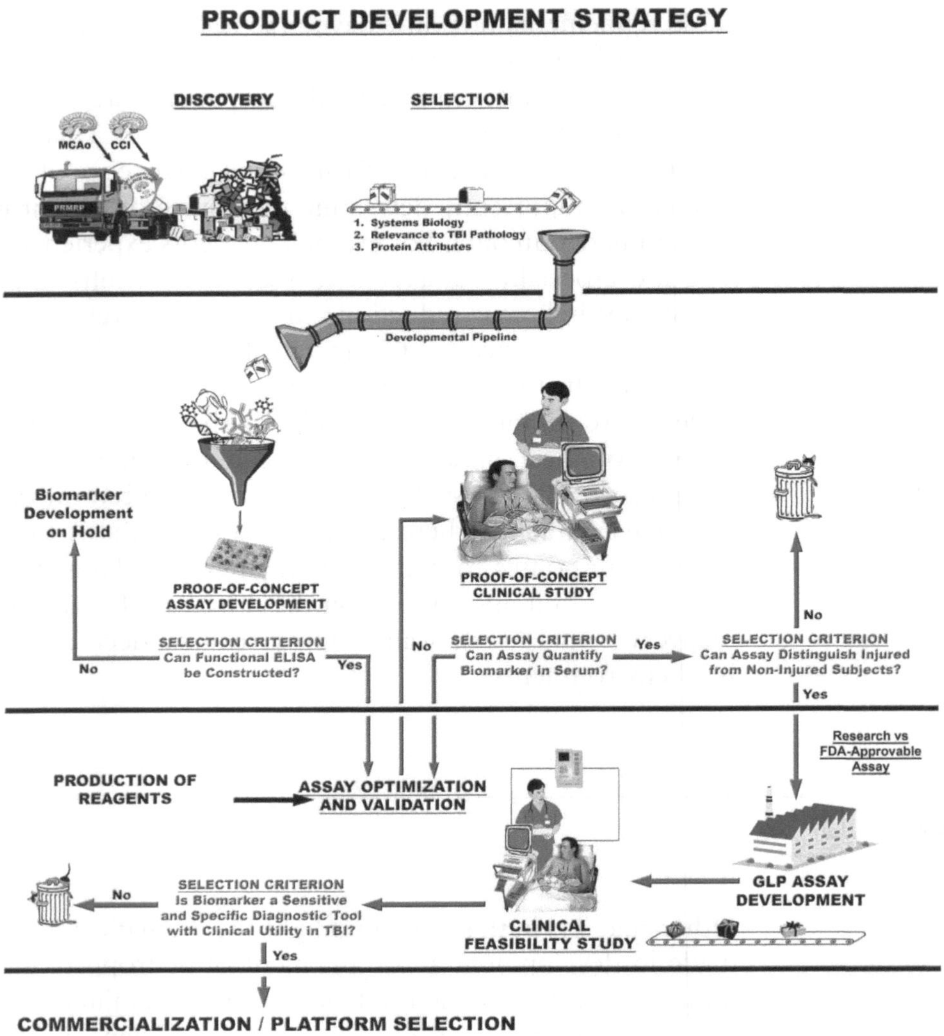

Scheme 1. Pathway for development of clinical biomarkers.

focused on the clinical proof-of-concept studies to confirm that specific biomarkers are, in fact, present in injured brain tissue and potentially diffuse at least into cerebrospinal fluid. As outlined in the studies by Ringger et al. *(18)*, preclinical studies can allow rigorous comparisons of the relationships between biomarker levels, injury magnitude, and assessments of functional outcome not easily done in clinical studies.

An additional component of prioritization and selection of biomarkers is the integration of systems biology analyses of combined proteomic studies to identify candidate brain injury biomarkers *(9)*. This approach allows a rigorous and relatively nonassumptive analysis of the potential relevance of changes in proteins to critical aspects of cell function. Ultimately, these insights need to be linked to an understanding of the pathology of the injury or disease under consideration. For example, in acute brain injury, a variety of biomarkers would be considered relevant if they had relevance to critical components of the pathobiology of TBI including: (1) biochemical markers of proteolytic degradation and cell death; (2) markers of cytoskeletal damage; (3) markers of neuroinflammation; (4) markers of synaptic dysfunction; (5) markers of damage to cell bodies or post translational modifications; (6) markers of neuroregeneration. These considerations should be refined and updated by periodic literature reviews and discussion among research teams.

Other criteria should recognize the ultimate need to develop practical antibody-based assay approaches. Thus, investigators should consider reagent availability both of antigens as well as polyclonal and monoclonal antibodies. Investigators should also consider protein attributes that could affect biomarker utility. These attributes include (1) a comprehensive protein analysis including molecular weight and shape; (2) subcellular localization; (3) complexity of isoforms; (4) cross-species similarities; (5) potential posttranslational modification; (6) brain specificity (mRNA/protein expression profiles); (7) protein abundance in brain.

2.2. Proof of Concept Assay Development, Optimization, and Validation

A number of principles should guide the assay development and validation process. First, assay development and validation should be "fit for purpose," that is tailored to meet an intended purpose. For example, consideration should be given not only to the clinical indication but also to sensitivity and specificity ranges required and the platform on which the assay will ultimately reside. Of course, the assay should be reliable intended application including careful assessments of sensitivity, specificity, and coefficient of variance. There should be a continuous reassessment of data and a continuous optimization based on results of reassessments and refined insights into the needs and uses of new assays. First, investigators must establish a basic assay. This requires a determination of the desired working range of the assay including lower limits of detection. Decisions must be made on assay format (G2-sited ELISA on microplate), assay label (e.g., chemoluminescense), and detection platform. Appropriate reagents need to be procured including antibodies, amplification kit, and purified target protein. At that point, the basic assay can be assembled and tested. Any current assays rely on ELISA technology with HRP detection and signal amplification including tyramide amplification technology.

Subsequently, first phase optimization of assay conditions must be completed. The purpose of first phase optimization is to confirm that the assay can detect the biomarker in relevant clinical samples (e.g., serum). These assessments include testing optimum concentration of all reagents, incubation times, and washing steps. It is also critical to establish dose–response curves and determine dynamic ranges, and upper and lower limits of detection (LOD). Accuracy and precision (LOQ), selectivity, and matrix effects (i.e., minimum dilution) must also be assessed. Precision and reliability of assays are further examined by studies of different day interdate results when compared with same day results. The development of appropriate precision profiles is expressed as coefficients of variance and dose–response curves. Assay results should be compared with other protein detection techniques including immuno-PCR and similar DNA-based enzymatic amplification techniques. ELISA-type assays with nanoparticle detection plus DNA-based amplification should also be examined. In addition to standard ELISA formats and ELISA formats with time-resolve fluorescent detection, it is important to recognize that the first phase of optimization may not always be successful and the assay may not be able to reliably detect biomarkers within clinically relevant ranges and in clinically relevant analytes (e.g., serum, urine, saliva). Thus, a second phase of assay optimization may be required to improve sensitivity and precision and to achieve specifications for limits of quantification (LOQ). This effort may require substantial reworking of the assay. In general, ELISA performance depends on the quality of key reagents. Options for assay improvements critically depend on consideration of reagent quality and performance, e.g., critical to determine the Kd of antibodies and procure better antibodies if possible. HRP-conjugate specific activity and biotinylation efficiency and effect on AB activity are important to assess. It is possible to modify the capture surface of platforms (e.g., magnetic beads) as well the label and detection platforms (e.g., gold nanoparticles and scatter detection).

Finally, it is important to emphasize that assay optimization and validation is typically an iterative process involving systematic examinations of assay performance in preclinical and clinical studies relevant to the ultimate applications of biomarker analyses. This often overlooked feature of assay development requires close integration between assay development teams and clinical and preclinical investigators providing samples for assay validation.

2.3. Design of Human Clinical Validation Studies

Access to human clinical samples is critical to identification of biomarkers relevant to injury and disease as well as for assay development. Establishment of clinical programs to collect these samples requires a major commitment of resources and expertise, as well as consideration of the ultimate application envisioned for the biomarkers. Although these clinical studies share much in common with clinical trials for assessment of drug therapeutic efficacy, there are a number of considerations unique to these efforts. First, careful consideration should be given to developing standard operating procedures for sample collection, shipment, and management. Some clinical studies could rapidly generate several thousand samples. Samples are often collected in challenging clinical environments including emergency rooms and intensive care units. Quality control of all aspects of sample handling is critical. Ideally, the clinical database supporting assessments of clinical parameters should be integrated into the database for sample tracking and management, as well as ultimate assessments of biomarker values. Assessments of biomarkers in blood should probably provide for both serum and plasma samples, since biomarker values can be influenced by those analyte media. Samples are frozen, shipped samples should include records of temperatures of samples during shipment by automated temperature recordings included in the sample shipment.

The design of the clinical study should minimally include the opportunity to assess the biokinetics of samples in relevant analyte compartments. For example, where CSF is available, it is ideal to secure samples with sufficient frequency in both cerebrospinal fluid and blood to make comparisons about the movement of markers across compartments. Generally, these assessments must be made at least every 6 h. Ultimately, assessment of the biokinetics of markers will be critical to interpreting results in human patients. When conducting the clinical database, it is important to insure that relevant clinical dimensions are included. This will require close coordination with clinical experts possibly in a number of subspecialties. For example, in TBI studies, clinical input is necessary from emergency medicine physicians, neurointensivists, neurosurgeons, and physicians in physical medicine and rehabilitation. Finally, it is important to consider whether assays will ultimately be developed for research only or will be subjected to scrutiny for FDA approval. Research only assays are not subjected to the regulatory oversight and rigorous validation required by the FDA. Moreover, the clinical validation of the biomarkers does not require FDA-compliant clinical databases.

2.4. Platform Selection and Assay Commercialization

Considering commercialization and platform selection, a platform should be chosen that best fits selected applications or market goals. Ultimately, the assay and the platform will have to be validated conjointly, both analytically and clinically (**Scheme 2**). Thus, it is important to recognize that assay development and validation processes will need to be repeated on the selected platform. If FDA approval is envisioned, this process must include development of standard operating procedures, quality control, and manufacturing specifications.

Investigators should consider what platforms from which their assays will ultimately reside. Research only assays enjoy considerable latitude in platforms and are not subjected to the constraints typically associated with clinical laboratories. Three categories of assays are widely available for biomarker analyses. First, large footprint and high-throughput assays are commonly employed in clinical laboratories in hospitals and large research

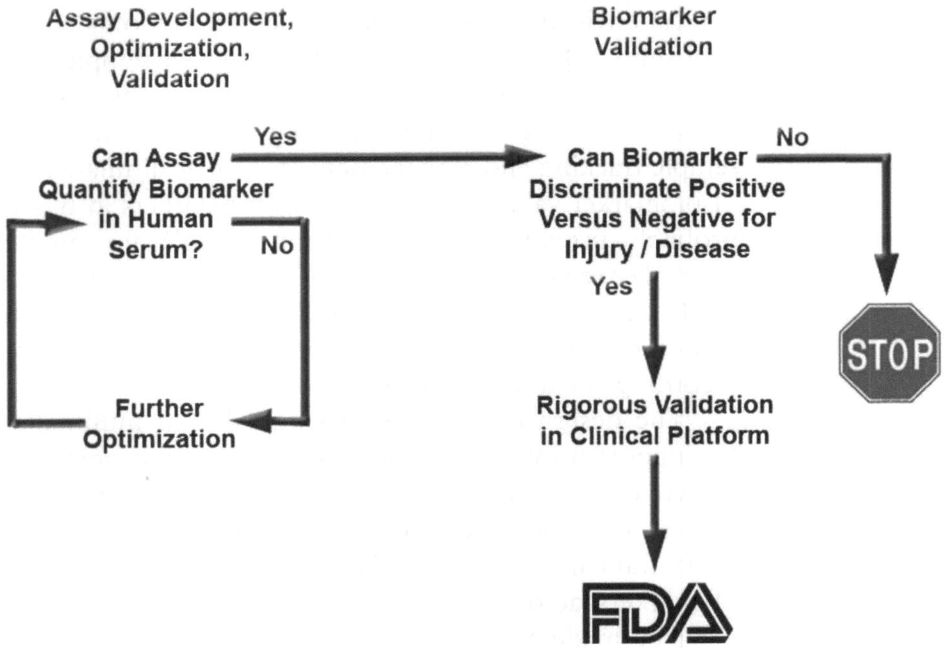

Scheme 2. Clinical validation process.

groups including big pharma. Second, there are smaller assay systems for point-of-care analyses in emergency rooms and other critical care environments. These systems are typically characterized by rapid assay turn around. Finally, a newer generation of assay approaches employing, for example, nanotechnology and microfluidics make assays possible on hand-held devices. The potential need for multiplexing to assay multiple biomarkers concurrently must also be considered.

References

1. Haskins, W.E., Kobeissy, F.H., Wolper, R. A., Ottens, A.K., Kitlen, J. W., McClung, S. H., O'Steen, B.E., Chow, M.M., Pineda, J.A., Denslow, N.D., Hayes, R.L., and Wang, K.K. W. (2005) Rapid discovery of putative protein biomarkers of traumatic brain injury by SDS-PAGE-capillary liquid chromatography-tandem mass spectrometry. *J. Neurotrauma* **22**, 629–644.
2. Ottens, A.K., Kobeissy, F.H., Wolper, R.A., Haskins, W.E., Hayes, R.L., Denslow, N.D., and Wang, K.K. W. (2005) A multidimensional differential proteomic platform using dual phase ion exchange chromatography polyacrylamide gel electrophoresis/reversed phase liquid chromatography tandem mass spectrometry (CAX-PAGE). *Anal. Chem.* 77, 4836–4845.
3. Wang, K.K.W., Ottens, A.K., Liu, M.C., Lewis, S.B., Meegan, C., Oli, M. W., Tortella, F.C., and Hayes, R.L. (2005) Proteomic identification of biomarkers of traumatic brain injury. *Exp. Rev. Proteomics.* **2**, 603–614.
4. Ottens, A.K., Kobeissy, F.H., Haskins, W.E., Golden, E.C., Zhang, Z., Chen, S.S., Hayes, R.L., Wang, K.K.W., and Denslow, N.D. (2006) Neuroproteomics in Neurotrauma. *Mass Spectrom. Rev.* **25**, 380–408.
5. Kobeissy, F.H., Ottens, A.K., Zhang, Z., Liu, M.C., Denslow, N.D., Dave, J. R., Tortella, F. C., Hayes, R.L., and Wang, K.K.W. (2006) Novel differential neuroproteomics analysis of traumatic brain injury in rats. *Mol. Cell. Proteomics* **5**, 1887–1898.
6. Ottens, A.K., Golden, E.C., Bustamante, L., Hayes, R.L., Denslow, N.D., and Wang, K.K.W.

(2008) Proteolysis of multiple myelin basic protein isoforms after Neurotrauma: Characterization by mass spectrometry. *J. Neurochem.* **104**, 1404–1414.

7. Ottens, A.K., Kobeissy, F.H., Fuller, B.F., Liu, M.C., Oli, M.W., Hayes, R.L., and Wang, K. K. (2007) Novel neuroproteomic approaches to studying traumatic brain injury. *Prog. Brain Res.* **161**, 401–418.
8. Papa, L., Robinson, G., Oli, M., Pineda, J., Demery, J., Brophy, G., Robicsek, S.A., Gabrielli, A., Robertson, C.S., Wang, K.K., and Hayes, R.L. Use of biomarkers for diagnosis and management of traumatic brain injury patients. *Exp. Opin. Med. Diag.* **2**, 937–945.
9. Kobeissy, F.H., Sadasivan, S., Oli, M., Robinson, G., Larner, S., Zhang, Z., Hayes, R., and Wang, K.K. (2008)Neuroproteomics and systems biology-based discovery of protein biomarkers for TBI and clinical validation. Clin. App. **2**, (10–11) 1467–1483.
10. Schonberger, S.J., Edgar, P.F., Kydd, R., Faull, R.L., and Cooper, G.J. (2001) Proteomic analysis of the brain in Alzheimer's disease: Molecular phenotype of a complex disease process. *Proteomics* **1**, 1519–1528.
11. Yealy, D.M., and Hogan, D.E. (1991) Imaging after head trauma. Who needs what? *Emerg. Med. Clin. N. Am.* **9**, 707–717.
12. Vollmer, D.G., and Dacey, R.G., Jr. (1991) The management of mild and moderate head injuries. *Neurosurg. Clin. N. Am.* **2**, 437–455.
13. Saatman, K.E., Duhaime, A.C., Bullock, R., Maas, A.I., Valadka, A., Manley, G.T., and Workshop Scientific Team and Advisory Panel Members. (2008) Classification of traumatic brain injury for targeted therapies. *J. Neurotrauma* **25**, 719–738.
14. Pineda, J.A., Lewis, S.B., Valadka, S.B., Papa, L., Hannay, H. J., Heaton, S., Demery, J. A., Liu, M.C., Aikman, J.M., Akle, V., Brophy, G.M., Tepas, J.J., III, Wang, K.K.W., Robertson, C.S., and Hayes, R.L. (2007) Clinical significance of αII-spectrin breakdown products in CSF after severe TBI. *J. Neurotrauma* **24**, 354–366.
15. Murray, G.D., Barer, D., Choi, S., Fernandex, H., Gregson, B., Lees, K.R., Maas, A.I., Marmarou, A., Mendelow, A.D., Steyerberg, E., Taylor, G.S., Teasdale, G.M., and Weir, C. J. (2005) Design and analysis of Phase III trials with ordered outcome scales: The concept of the sliding dichotomy. *J. Neurotrauma* **22**, 511–517.
16. Pike, B.R., Flint, J., Dutta, S., Johnson, E., Wang K.K.W., and Hayes, R.L. (2001) Accumulation of non-erythroid αII-spectrin and calpain-cleaved αII-spectrin breakdown products in cerebrospinal fluid after TBI in rats. *J. Neurochem.* **78**, 1297–1306.
17. Pike, B.R., Flint, J., Dave, J.R., Lu, X.C., Wang, K.K., Tortella, F.C., and Hayes R.L. (2003) Accumulation of calpain and caspase-3 proteolytic fragments of brain-derived αIIspectrin in CSF after middle cerebral artery occlusion in rats. *J. Cereb. Blood Flow Metab.* **24**, 98–106.
18. Ringger, N.C., O'Steen, B.E., Brabham, J. G., Siler, X., Pineda, J., Wang, K.K.W., and Hayes, R.L. (2004) A novel marker for traumatic brain injury: CSF αII-spectrin breakdown product levels. *J. Neurotrauma*, **21**, 1443–1456.
19. Warren, M.W., Kobeissy, F.H., Liu, M.C., Hayes, R.L., Gold, M.S., and Wang, K.K. W. (2005) Concurrent calpain and caspase-3 mediated proteolysis of αII-spectrin and tau in rat brain after methamphetamine exposure: A similar profile to TBI. *Life Sci.* **78**, 301–309.
20. Liu, M.C., Akle, V., Zheng, W., Kitlen, J., O'Steen, B., Larner, S.F., Dave, J.R., Tortella, F.C., Hayes, R.L., and Wang, K.K.W. (2006) Extensive degradation of myelin basic protein isoforms by calpain following TBI. *J. Neurochem.* **98**, 700–712.
21. Warren, M.W., Kobeissy, F.H., Liu, M.C., Hayes, R.L., Gold, M.S., and Wang, K.K. W. (2006) Ecstasy toxicity: A comparison to methamphetamine and TBI. *J. Addict. Dis.* **25**, 115–123.
22. Warren, M.W., Zheng, W., Kobeissy, F.H., Liu, M.C., Hayes, R.L., Gold, M.S., Larner, S.F., and Wang K.K.W. (2006) Calpain and caspase mediated αII-spectrin and tau proteolysis in rat cerebrocortical neuronal cultures after ecstasy or methamphetamine exposure. *Int. J. Neuropsychopharm.* **10**, 479–489.

Index

P

R

S

T

U

MIX
Papier aus verantwortungsvollen Quellen
Paper from responsible sources
FSC® C105338

If you have any concerns about our products,
you can contact us on
ProductSafety@springernature.com

In case Publisher is established outside the EU,
the EU authorized representative is:
Springer Nature Customer Service Center GmbH
Europaplatz 3, 69115 Heidelberg, Germany

Printed by Libri Plureos GmbH
in Hamburg, Germany